Preface

In the year 2020, more than 70 countries have civil space agencies and about one third of them also have military interests in space. Hundreds of corporations around the globe offer space-related products and services. All these organizations—civil, commercial, and military—are exploiting or hope to exploit the many advantages of space assets.

We've titled this book *Space Domain Awareness* (SDA) to distinguish it from space situational awareness (SSA), which, according to the US Space Foundation, "refers to the ability to view, understand and predict the physical location of natural and man-made objects in or around Earth, with the objective of avoiding collisions." A related term, and common in the literature, is space traffic management (STM), which we define as "the set of technical and regulatory provisions and processes used to oversee, coordinate, regulate, and promote safe and responsible space activities."

Our working definition of space domain awareness focuses on SDA practitioners—organizations having the responsibility to maintain space domain awareness. Space domain awareness is the actionable knowledge required to predict, avoid, deter, operate through, recover from, and attribute cause to losing or degrading space capabilities and services.

In essence, SDA practitioners establish a baseline of objects in Earth orbit, note changes to the baseline, understand what changed and why, assess possible outcomes due to the change, and determine a course of action to deal with the change, if necessary. The space objects in the baseline may range in size from entire space stations to golf-ball sized (or smaller) objects, including functional spacecraft, non-functional spacecraft, and to some extent natural and man-made debris that could threaten functional spacecraft. We use the term resident space object (RSO) to denote objects of interest to SDA practitioners.

The goal of this book is to provide an overview of space domain awareness and establish basic principles, philosophies, and approaches that are useful to conduct SDA operations. These operations include, but are not limited to, detecting, identifying, tracking, and characterizing objects and their behaviors, detecting changes associated with selected RSOs, and potentially mitigating threats. Approaches to mitigation are covered in the second volume of this text (TBD).

Space operations face many threats, ranging from on-board anomalies, space weather events, potential collisions, inadvertent re-entry, radio frequency interference, and actions taken by well-meaning operators or nefarious actors.

Personal Perspective

For me, SDA is a dream come true. After 40 years of working in the space community, we were able to persuade the "best-of-the-best" thought leaders in space operations to contribute their wealth of experience and lessons learned to this project. This is good for me, but REALLY good for you, because you'll benefit from the 200-plus years (that's over two centuries!) of collective experience in operating space systems.

Space Domain Awareness is the 30th book in the US Air Force Academy's Space Technology Series. The leadership of this project has been provided by the Department of Astronautics at the Academy continuously for 30-plus years! I wish to express my deepest gratitude to Brigadier Generals Bob Giffen and Mike DeLorenzo and Marty France for their vision and persistence in leading this community effort. Lieutenant Colonel Mark Charlton and later, Colonel Luke Sauter, supplied the hands-on leadership to make SDA a reality. Thank you!

In the Space Technology Series, most of the 30 titles support SDA: *Understanding Space; Space Mission Analysis and Design; Human Spaceflight; Space Propulsion Analysis and Design; Space Launch Systems Design and Operations; Space Mission Operations; Space Systems Engineering; Applied Project Management for Space Systems,* and others provide necessary background information that supports the development of SDA practitioners.

The Space Technology Series receives its sponsorship and funding from the country's space community—and has for over 30 years. More than 23 national organizations have championed the effort since it began in 1988. Leadership, funding, and support from the Air Force Space and Missile Center, multiple program offices and AF wings, the National Reconnaissance Office, NASA Headquarters (including multiple mission directorates), NASA/Goddard Space Flight Center, the Federal Aviation Administration, the Naval Research Laboratory, and a host of others, have made this work possible. Thank you for your continued support.

Here I must recognize the forbearance and suffering of the numerous contributing authors, artists, and editors of this text. For five years they have outlined, drafted, reorganized, reviewed, and re-crafted their material to suit the editors' demands. This book would not be possible without their expertise, persistence, and patience. As editors, our goal has been to create a useful, professional book of which the authors can be proud.

The true heroes of this effort are Anita Shute, Marilyn McQuade, and Mary Tostanoski. As a team, they edited every word and did everything necessary to create a camera-ready copy that truly represents the quality of the author and editor team. Anita Shute personally entered and formatted every page of the book, as she has done for 27 others. Thank you all for your dedication and excellence!

For me, the joy of this endeavor was greatly enhanced by my two exceptional co-editors: Pamela Magee and Moriba Jah. They dedicated themselves to working with their teams to write this book and actually made the five years stimulating and enjoyable! Many of the authors have expressed their gratitude for the leadership, insights, and wisdom provided by the editorial team.

Finally, I want to express my heartfelt appreciation to my family and friends for their unswerving support and encouragement, while suffering my dream with me!

The authors and editors of *Space Domain Awareness* hope that our colleagues, both young and old, will find this book useful in preparing to meet the challenges ahead in managing space operations for national defense, civil application, and commercial enterprise.

Dr. Wiley J. Larson
Managing Editor, USAFA Space Technology Series
June, 2020
719-582-0111
wileyjlarson@mac.com

Table of Contents

Chapter 1

Space Domain Awareness (SDA)

Dr. James Shell, *Novarum Technology LLC*
Dr. Wiley J. Larson, *United States Air Force Academy*
Dr. Moriba Jah, *University of Texas*

Sputnik, the first human-made or artificial satellite to maintain orbit, was launched on October 4, 1957 as part of the Cold War between the Union of Soviet Socialist Republics (USSR) and the United States. This launch fueled the space race that's continued ever since. At this writing, more than 60 years after Sputnik, tens of thousands of artificial satellites are in Earth orbit with thousands of small- and medium-sized spacecraft proposed for launch.

Early on, a few countries developed and operated spacecraft—motivated mainly by military objectives. Later, civil organizations such as the National Aeronautics and Space Administration and the European Space Agency, focused on research and science. Today, commercial organizations keep adding to the number of spacecraft in Earth orbit in order to satisfy business objectives. We recommend you explore the objects in orbit at http://apps.agi.com/SatelliteViewer/. This catalog lists and shows some 23,000 objects; it also lists but usually doesn't show many other objects smaller than a golf ball.

Today, more than 70 countries have civil space agencies and about one-third also have military interests in space. Hundreds of corporations around the globe offer space-related products and services. All these organizations are exploiting or hoping to exploit the many advantages of space assets. Figure 1-1 illustrates typical uses of these assets.

We've titled this book Space Domain Awareness (SDA) to distinguish it from Space Situational Awareness (SSA), which, according to the US Space Foundation, "refers to the ability to view, understand and predict the physical location of natural and manmade

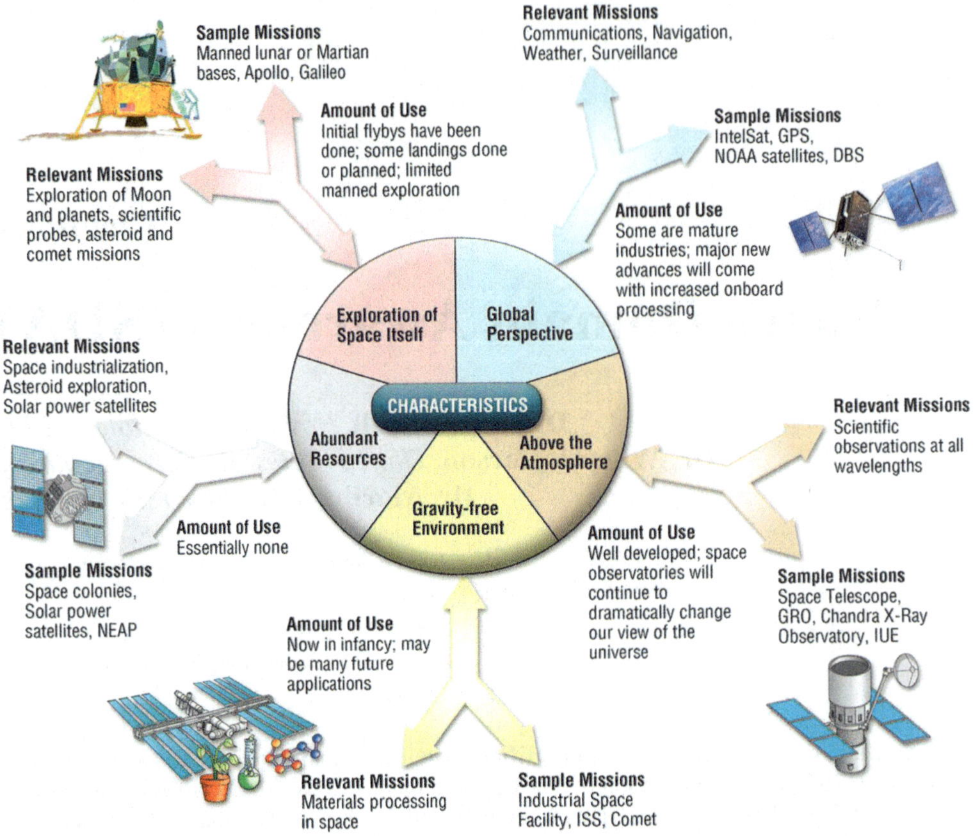

FIG. 1-1. Characteristics of Space Exploited by Various Space Missions. Note the wide variety of missions for the key characteristics of space and that many are only beginning to be used.

objects in orbit around the Earth, with the objective of avoiding collisions." The European Space Agency has a slightly broader description of SSA:

> SSA aims, ultimately, to enable Europe to autonomously detect, predict and assess the risk to life and property due to man-made space debris objects, reentries, in-orbit explosions, in-orbit collisions, disruption of missions and satellite-based service capabilities, potential impacts of Near-Earth Objects (NEOs), and the effects of space weather phenomena on space- and ground-based infrastructure.

Another term related to SSA and common in the literature is Space Traffic Management (STM), which we can define as "The set of technical and regulatory provisions and processes used to oversee, coordinate, regulate, and promote safe and responsible space activities."

The working definition of Space Domain Awareness (SDA) for this book focuses on the SDA practitioner:

Space Domain Awareness (SDA) is the actionable knowledge required to predict, avoid, deter, operate through, recover from, and attribute cause to losing or degrading space capabilities and services.

A major purpose for SDA is to provide decision makers with a quantifiable and timely body of evidence of behaviors attributable to specific space threats and hazards. SDA encompasses all activities of information tasking—collecting, fusing, exploiting, quantifying and extracting—that end in identifying and predicting credible threats and hazards.

Understanding the synergy between the space environment, the interaction of this environment with objects (astrodynamics), the space environment's effects on objects (operational and not), and the available sensors and sources of information, is critical to meaningful SDA.

Included in SDA is collecting raw observables, identifying physical states and parameters (such as orbit, attitude, size, shape), determining functional characteristics (such as active vs. passive, thrust capacity, payloads), inferring mission objectives (such as communications, weather), identifying behaviors, and predicting credible threats and hazards.

Intuitively, SDA is a natural "big data" problem, drawing from a surfeit of metadata and data sources. It has three main parts:

- How we can address these needs rigorously using first principles
- What methods, techniques, and technologies we must leverage from other fields or target for development
- What sensors, phenomenology, sensor tasking, or other data we need to support the SDA mission

Space Domain Awareness and Response (SDAR), for this book, involves detecting objects in space, characterizing them, understanding their current and possible behaviors, identifying their risks, prioritizing these risks, determining potential actions or mitigations for high risks, deciding how to proceed, and completing the mission.

Of course, space domain awareness includes not only appropriate objects in space but also the infrastructure and systems that enable space activities, as shown in Figure 1-2.

All space missions consist of a set of *elements* or *components* as shown in Figure 1-2 [Larson and Wertz, 1999], which planners arrange into a *space mission architecture*. Organizations and programs define their mission elements differently, but most elements are normally present in any space mission. They must survive and use the space and atmospheric environments, shown in Figure 1-3 [Sellers, et al., 2015].

All objects in orbit about Earth must deal with the space environment's elements: gravity, electromagnetic radiation, atmosphere at lower altitudes, vacuum, charged particles, micrometeoroids, and natural and human-made debris. [Sellers, 2015] [Knipp, 2014]

A mission's *subject* is the thing that interacts with or is sensed by the space payload:

- For weather missions: moisture content, atmospheric temperature, or pressure
- For Earth-sensing missions: types of vegetation, water, or geological formations
- For space defense missions: a rocket or intercontinental ballistic missile

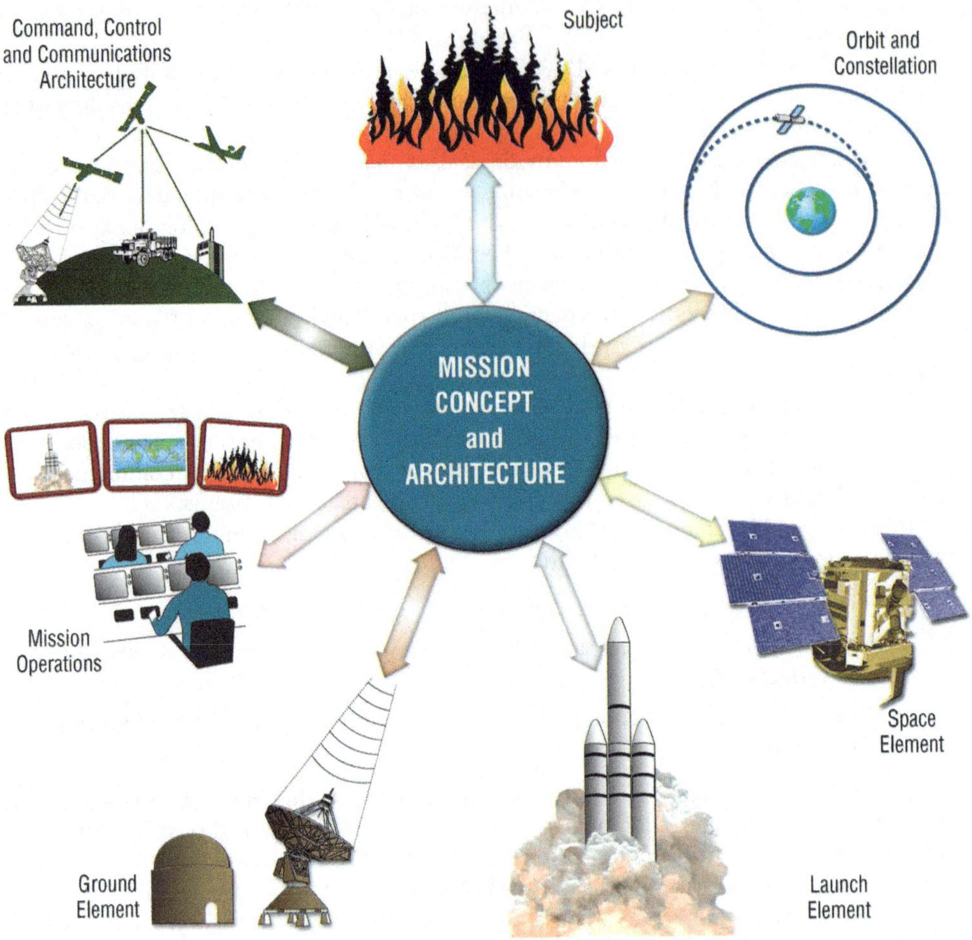

FIG. 1-2. Elements of a Typical Space Mission Architecture. We often focus on the spacecraft when talking about space missions, but the six added elements in this diagram represent the infrastructure and enabling systems needed to do mission operations.

- For commercial communications missions: an antenna and receiver

We must decide what part of the electromagnetic spectrum is appropriate for sensing the subject, thus determining the type of sensor on the spacecraft, as well as its payload mass, size, and power. In many missions, we may trade off the subject. For example, if we're trying to track a missile during powered flight, the subject could be the rocket body or exhaust plume, or both. For communications and navigation missions, the subject is a set of equipment on Earth or on another spacecraft, including communication terminals,

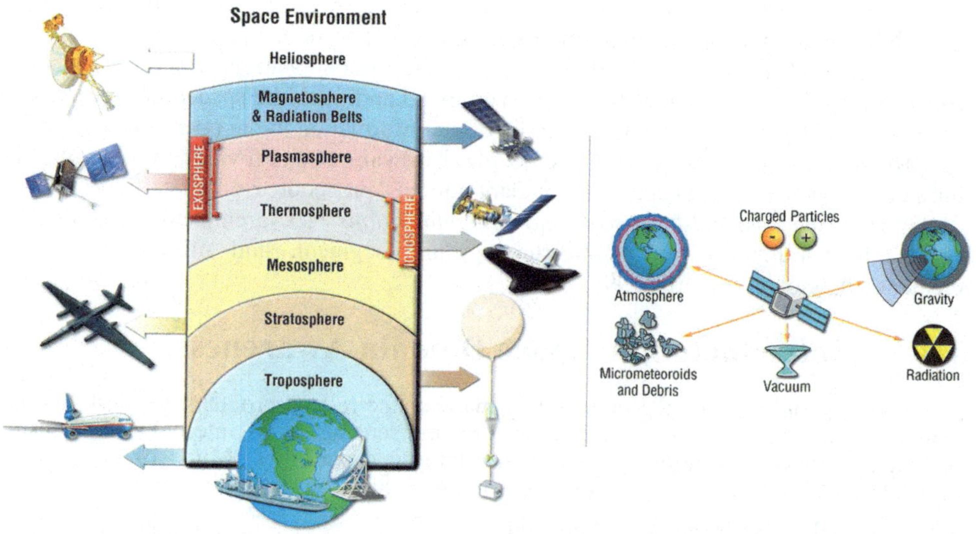

FIG. 1-3. Space Domain Awareness Includes all Atmospheric Layers and the Space Environment. The relative locations of Earth's atmospheric and magnetic regimes, as well as the heliosphere, are shown here along with notional objects. Image is not to scale. The ever-present, ever-changing elements of the space environment are depicted at right.

televisions, receiving equipment for GPS navigation, or other user-furnished equipment. This equipment's parameters characterize the subject for these types of missions.

The *payload* (payload and spacecraft bus make up the space element) consists of the hardware and software that sense or interact with the subject. Typically, we trade off and combine several sensors and experiments to form the payload, which largely determines the mission's cost, complexity, and effectiveness.

The subsystems of the *spacecraft bus* support the payload by providing orbit and attitude maintenance, power, command, telemetry and data handling, structure and rigidity, and temperature control. The payload and spacecraft bus together are called the *spacecraft*, *space element*, or *launch vehicle payload.*

The *launch system* includes the launch facility, launch vehicle and any upper stage required to place the spacecraft in orbit, as well as interfaces, payload fairing, and ground-support equipment and facilities. The selected launch system constrains the spacecraft's size, shape, and mass.

The *orbit* is the spacecraft's path or trajectory. It usually has separate initial parking orbit, transfer orbit, final mission orbit and, sometimes, an end-of-life or disposal orbit. The mission orbit significantly influences every element of the mission and offers many options for trades in mission architecture.

The *communications architecture* arranges components that satisfy the mission's communication, command, and control requirements. It depends strongly on the amount and timing requirements of data to be transferred, as well as the number, location, availability, and communicating ability of the space and ground assets.

The *ground system* consists of fixed and mobile ground stations around the globe—connected by data links. They allow us to command and track the spacecraft, receive and process telemetry and mission data, and distribute information to the operators and users.

Mission operations consists of the people, hardware, and software that carry out the mission, its operations concept, and attendant policies, procedures, and data flows.

Finally, the *command, control, and communications (C3)* architecture contains the spacecraft, communications architecture, ground segment, and mission operations elements. [Larson and Wertz, 1999]

1.1 Introduction to Space Domain Awareness

Today, people around the globe enjoy many space-related products and services that enable global navigation, weather forecasts, remote sensing, radio, internet and television and many others, in addition to civil and defense capabilities. Much of the world's population now depends on space products and services in seven main categories:

- Travel, lifestyle, and entertainment
- Energy, resources, and environmental monitoring
- Governance, education, and infrastructure
- Retail, finance, and corporate services
- Transportation, logistics, and manufacturing
- Homeland security, defense, and intelligence
- Science, biotechnology, and healthcare

In 2017, global space-related revenues exceeded $340 billion, which we can divide roughly into commercial space products and services (~40%); commercial infrastructure and support industries (~38%); US government space budgets, including DOD, NASA and others (~13%); and non-US government space budgets (~9%). [US Space Foundation Report, 2016]

In 2017, more than 50 countries from all continents operated more than 1,260 spacecraft in Earth orbits. Six decades of spaceflight have left in orbit an estimated 750,000 objects larger than 1 cm, and more than 100 million objects smaller than 1 cm. [International Academy of Astronautics, Space Traffic Management Study, 2017]

The authors identify five key trends that will affect the numbers and mix of orbiting spacecraft as well as how they are managed:

- Continued and even accelerated global climate change, expected to lead to significant costs needed to respond, adapt, and cope with increasing and intensifying natural disasters. These actions will cause more migration and conflicts.
- Population growth to an estimated nine billion people on Earth in the next decades, which is likely to lead to problems in food supply, energy supply, housing supply and urban development.
- Increased need to create, distribute, use, integrate, and manipulate information to support economic, political, and cultural activity.

- Military conflicts, which are likely to grow in number and intensity. More states operate satellites for military purposes or as part of their military infrastructure, including telecommunications, navigation, reconnaissance and early warning.
- Rise of new powers with resulting shifts in the geopolitical balance. These power shifts may influence both the well-known actors and patterns of space activities and the traditional approaches to policy and (international) law.

Now that we understand the big-picture perspective of Space Domain Awareness and a view of our future in space, let's see how the space community got to this point.

1.1.1 Brief History

The escalating Cold War during the 1950s set the stage for Space Domain Awareness (SDA). The threat of intercontinental ballistic missiles (ICBMs) delivering nuclear payloads prompted technology development to enable detection and attack warning for such threats. Indeed, the whole doctrine of "Mutually Assured Destruction" rested on the foundation of knowing soon enough that an attack had started, so the attacked country could launch a retaliatory strike. World War II-era air defense radar systems weren't up to tracking ballistic missiles, given the required target ranges and sensitivity. So development of radar systems that could handle ballistic missile warning was vigorous in the mid-1950s. [Stone and Banner, 2000]

The earlier than anticipated launch of Sputnik in October 1957 caused a flurry of activity to enable space-object tracking. Three different satellite-tracking technologies had been progressing, and each was rapidly forced into service. The Smithsonian Astrophysical Observatory enabled optical tracking of objects using an international network of Baker-Nunn film-based cameras. Radar systems were mainly developing to track ballistic missiles, but the Millstone Hill Radar operated by MIT Lincoln Laboratory saw early satellite-tracking service. The Naval Research Laboratory's "Minitrack" system measured satellites using passive radio frequencies and a cooperative transmitter. By 1959, they had mated Minitrack interferometric receiver technology with an active RF source, thus beginning more than 50 years of service for the Navy Space Surveillance System. [Easton and Fleming, 1960] Finally, amateur observers helped track objects using optical systems and ham radios. In fact, after Sputnik launched, amateur radio operators provided some of the best early data because the Minitrack system didn't operate on the right frequencies for Sputnik. Teams of amateur optical observers organized under the Smithsonian Astrophysical Observatory's "Moonwatch" program [Hayes, 1962] also made important early observations.

Timely detection of ballistic missiles with low false alarms requires filtering out known Earth-orbiting objects, which inextricably linked the missions of space object tracking and Ballistic Missile Early Warning (BMEWs). In fact, BMEWs radar systems make up most of the United States' space surveillance radar network.

A public-domain "space object catalog," as it's often called, has recorded all space objects and their basic orbital parameters since the beginning. Indeed, the tracking of Sputnik and contributions by ham radio operators and the Moonwatch team established this paradigm, with updates through the Smithsonian Astrophysical Observatory's (SAO) news conferences. Starting in the 1960s, NASA's Goddard Space Flight Center published the

"Satellite Situation Report," released every two weeks, which lists all space objects with their basic orbital information and data "computed and compiled by Goddard Space Flight Center, NORAD, and the Smithsonian Astrophysical Observatory." [NASA, 1962] Figure 1-4 shows part of one of these compilations.

OBJECTS IN ORBIT

OBJECT	CODE NAME	SOURCE	LAUNCH	NODAL PERIOD	INCLI-NATION	APOGEE S. MILES	PERIGEE S. MILES	TRANSMITTING FREQ. (MC/S)
1958 ALPHA 1	EXPLORER I	US	1 FEB 58	106.0	33.19	1084	217	
1958 BETA 1	ROCKET BODY	US	17 MAR 58	138.3	34.26	2684	406	
1958 BETA 2	VANGUARD I	US	17 MAR 58	133.8	34.24	2449	404	108.022
1959 ALPHA 1	VANGUARD II	US	17 FEB 59	125.3	32.87	2079	313	
1959 ALPHA 2	ROCKET BODY	US	17 FEB 59	129.6	32.90	2254	370	
1959 ETA 1	VANGUARD III	US	18 SEP 59	129.7	33.30	2299	332	
1959 MU*	LUNIK I	USSR	2 JAN 59	450D	0.01	1.3177AU	0.9766AU	
1959 NU*	PIONEER IV	US	3 MAR 59	398D	0.127	1.1421AU	0.9871AU	
1959 IOTA 1	EXPLORER VII	US	13 OCT 59	101.1	50.31	669	344	
1959 IOTA 2	ROCKET BODY	US	13 OCT 59	100.9	50.30	656	345	
1960 ALPHA*	PIONEER V	US	11 MAR 60	311.6D	3.35	.9951AU	.8061AU	
1960 BETA 1	ROCKET BODY	US	1 APR 60	99.1	48.41	463	429	
1960 BETA 2	TIROS 1	US	1 APR 60	99.1	48.39	467	429	107.997
1960 BETA 3	NONE	US	1 APR 60	97.8	48.46	445	375	
1960 BETA 4	NONE	US	1 APR 60	99.8	48.15	503	434	
1960 GAMMA 2	TRANSIT 1B	US	13 APR 60	94.7	51.28	401	229	
1960 GAMMA 4	NONE	US	13 APR 60	96.8	51.21	458	299	
1960 EPSILON 1	SPUTNIK IV	USSR	15 MAY 60	91.6	65.02	275	160	
1960 EPSILON 3	NONE	USSR	15 MAY 60	93.0	64.89	352	170	
1960 EPSILON 4	NONE	USSR	15 MAY 60	POSITION	UNCERTAIN			
1960 ZETA 1	MIDAS II	US	24 MAY 60	94.2	33.01	310	299	
1960 ETA 1	TRANSIT 2A	US	22 JUN 60	101.6	66.77	649	389	162;216;
1960 ETA 2	GREB	US	22 JUN 60	101.6	66.77	657	381	
1960 ETA 3	ROCKET BODY	US	22 JUN 60	101.4	66.77	642	383	
1960 IOTA 1	ECHO I	US	12 AUG 60	116.1	47.30	1314	563	
1960 IOTA 2	ROCKET BODY	US	12 AUG 60	118.0	47.22	1049	932	
1960 IOTA 3	METAL OBJECT	US	12 AUG 60	118.2	47.20	1049	941	
1960 IOTA 4	METAL OBJECT	US	12 AUG 60	118.2	47.23	1071	922	
1960 IOTA 5	METAL OBJECT	US	12 AUG 60	118.3	47.20	1060	940	

FIG. 1-4. **Space Domain Awareness Includes All Atmospheric Layers and the Space Environment.** An excerpt of the early satellite catalog as presented in NASA's January 2, 1962 Space Situation Report.

The growing use of space and population of orbiting objects increased international concerns through the mid-1960s. The United Nations (UN) developed multilateral space treaties, two of which are important for space object tracking. The *Convention on International Liability for Damage Caused by Space Objects*, adopted in 1972, established the legal framework for damages from space objects, with liability incurred by the launching state. The *Convention on Registration of Objects Launched into Outer Space*, or simply the *Registration Convention*, adopted in 1976, requires the launching state to report launched objects to the UN. The Registration Convention formalized initiatives from as early as 1961, which encouraged states to inform the UN General Assembly of space launches. [Jakhu, Jasani and McDowell, 2018]

These treaties require nations to agree on the accurate identification of orbiting objects. The United States' public domain space object catalog, which assigns each object an international designation and launching state, is the accepted authority for this common international understanding.

Early object tracking was manageable because satellites were in low-Earth orbit (LEO). But in 1963 NASA's Syncom satellites were the first to launch into geosynchronous orbit (GEO), whose altitude and range to sensors of about 36,000 km ushered in a new era of space surveillance challenges. Trackers typically used cooperative ranging from radiofrequency emissions for these satellites. Remarkably, new techniques enabled Millstone Hill Radar to actively track the 70 cm × 40 cm cylindrical Syncom satellites, but only in a cued manner and with some 45 minutes of incoherent integration. The Baker-Nunn optical telescopes, originally designed to track objects in low-Earth orbits, became increasingly employed to track geosynchronous objects. The Soviets didn't launch a geosynchronous satellite until 1974, ending the assured cooperative tracking of western-only geosynchronous satellites.

An international incident in 1978 brought space object tracking to public attention. [Weiss, 1978] Cosmos 954, one in a series of nuclear-powered satellites launched by the Soviet Union, failed on orbit resulting in an uncontrolled reentry predicted for early 1978. The potential hazards from radioactive materials were threatening, so the National Security Council and US intelligence agencies had to decide whether to publicize the incident and how to approach the Soviets to learn more about the imminent reentry. The US eventually contacted the Soviet Union and then notified allies and other countries. Cosmos 954 reentered over Canada and spread radioactive materials only in a sparsely populated area. As a result, the UN applied its 1972 Space Liability Convention for the first time by requiring the Soviet Union to pay Canada for the cleanup effort. With the Cosmos 954 reentry still in the public memory, the uncontrolled reentry of Skylab in 1979 produced a lot of excitement and public concern.

Also in 1978 NASA scientist Don Kessler published analysis about what is now termed "Kessler's Syndrome": the self-sustained increase in orbital debris because of debris-on-debris collisions. [Kessler and Cour-Palais, 1978] But significant focus on orbital debris didn't occur until the 1980s. By 1984, some 75 objects were known to have fragmented, often from unexpended propellant, and produced many more debris objects. [NASA, 2008] In 1988, the US National Space Policy released under President Ronald Reagan included a statement on limiting orbital debris and direction to work toward applying a space debris policy. But the US government didn't issue standards for mitigating orbital debris until 2001. Meanwhile, the new Inter-Agency Space Debris Coordination Committee (IADC) [IADC, 2016] enabled working relationships among NASA and space agencies in Europe, Russia and Japan. [Weeden, 2016] In 2007, the IADC published their Space Debris Mitigation Guidelines, and the United Nations endorsed a variant of these guidelines later that year. [Liou, 2017] Orbital debris had become the objects most requiring tracking and screening for potential collisions with operational satellites.

Current State of Space Domain Awareness

U.S. military actions in Iraq in the early 1990s set the stage for what would become an important motivator for Space Domain Awareness: the need to protect and defend military satellites. Often deemed the first "modern" war, the Iraq conflict gave satellites a prominent role in supporting military operations. Global Positioning System (GPS) satellites enabled precision strikes. Reconnaissance and communications satellites supported critical operations. In short, US warfighting depended on the enabling space infrastructure. Decades later, antisatellite weapon developments by Russia and China seek to gain an advantage that could cripple US global reach and warfighting. Although the Soviet Union and United States had each demonstrated antisatellite technologies in the 1980s, these new efforts are much more far-reaching.

Protecting satellites from harm has become one of the main reasons to advance space domain awareness. It also motivates much of the government-sponsored investment in upgrading infrastructure for sensing, processing, and supporting military decisions. [National Academy of Engineering, 2016] The commercial space sector also has grown rapidly, and now spends more on space programs than governments do.

Commercial space interests used to focus on geosynchronous communication satellites. But miniaturized satellite components and lower-cost smaller satellites are fueling a commercial space renaissance, supported by the United States' National Space Policy. We now have high-capacity, reliable internet connectivity because of increasing commercial launches, persistent remote sensing, environmental monitoring, and early stages of "mega constellations" of LEO-based communication. This commercial emphasis has resulted in new scrutiny of orbital-debris policies and their effect on corporate profits. Direct satellite damage or destruction from debris collisions will cost a lot of money, as will making satellites meet debris-mitigation requirements and paying higher insurance premiums for orbital-debris risks.

Growing threats to defense satellites and more commercializing of space have prompted interest in transferring "Space Traffic Management" from the US Department of Defense (DoD) to a US civil agency. A civil agency could add collision warnings to the space object catalog and relate better to other countries and corporations. [SAIC, 2016] Also, if DoD doesn't have to maintain the catalog, it could focus better on protecting and defending key satellite systems from attack.

Commercial firms also are offering space domain awareness services that improve collision-avoidance decisions and help space operators limit, identify, and resolve radio frequency interference. [Youtube, 2017] Starting in the 2000s, commercial firms developed inexpensive large-format, solid-state detectors that allowed modest telescope systems to detect many space objects. Similarly, solid-state RF technologies enabled better radar systems at lower cost. Finally, the never-ending increases in computational power, coupled with the commercial sector's fast development cycles, have improved space surveillance without government sensing networks.

Space: "Congested, Contested, Competitive"

The 2011 US National Security Space Strategy [DoD, 2011] described the strategic environment of space as "congested, contested, and competitive" (3Cs)—an accurate, commonly repeated soundbite description. Although the "3Cs" are independent elements, they're also closely linked. Increasing orbital debris makes space more "congested," which challenges the long-term use of space. State-sponsored antisatellite systems make space more "contested." Some of these systems may destroy satellites, generate more debris, and thus add greatly to congestion. Finally, space is more "competitive" because the cost of entering it is falling because of small satellites plus rapid growth of commercial launches, led by SpaceX and others.

Space has become more and more congested, contested and competitive over the last decade and in 2019, the National Defense Authorization Act for 2020 redesignated the Air Force Space Command as the U.S. Space Force and established it as an independent branch of the U.S. Armed Forces. The Space Force is responsible for the following activities, among others:

- Space domain awareness (military, civil and commercial)
- Offensive and defensive space control
- Command and control of space forces and satellite operations
- Missile warning and space support to missile defense operations

There are several fundamental concerns below that contributed to the decision to form the Space Force.

Orbital Debris: A key to Space Domain Awareness

Today, we often hear about "space junk" through mainstream news outlets and publications that warn of the long-term consequences of orbital debris and highlight events of interest with growing frequency. [Shell, up through 2018] The 2013 movie *Gravity*, though it included some excusable technical inaccuracies, gave the public a point of reference to appreciate issues involving orbital debris. Understanding the nature of orbital debris is essential to space domain awareness. Most objects in orbit are debris, and no-one tracks many of the objects that can kill a mission if they strike a satellite.

NASA's Orbital Debris Program Office led the world by studying and highlighting the long-term concerns and consequences of orbital debris. [Kessler and Cour-Palais, 1978] Now, the nongovernmental Inter-Agency Space Debris Coordination Committee (IADC) is an effective international organization that offers recommendations, guidance, and compliance standards for mitigating orbital debris, which in turn form the basis for the United Nations' positions. [UNOOSA, 2018]

Small debris not in an orbital catalog are characterized using statistical sampling from ground-based sensors as well as onsite measurements. Radars sample the number of untracked near-Earth objects, [Hamilton, et al., 2017] whereas telescopes sample deep space objects. [Krisko et al., 2009] Although we may detect these small objects, we don't have enough sensing resources to reliably track and maintain them.

Nations broadly agree orbital debris will continue to grow even with high adherence to mitigation standards. [IADC, 2013] Such projections have spawned an interest in "active debris removal": technologies targeting the removal or accelerated reentry of orbital debris. [McKnight, 2010] These technologies also have an inherent military application, so they're tangled in complicated policy considerations.

Geostationary Slot Allocation: Important to managing SDA

Later in this book we will discuss the function of the International Telecommunications Union (ITU) as it relates to SDA. The ITU manages geostationary slot assignments as well as frequency allocations, both very important to the conduct of SDA. Needless to say, the slot allocation and frequency assignment processes are somewhat arduous and highly political. Some geostationary slots are over subscribed and highly sought, while others have limited use, so there is strong competition for key slots. Space missions and systems require efficient and secure communications so it is very important for military, civil and commercial operators to secure the proper RF frequencies to conduct operations. So, frequency management is key to SDA operations and RF interference poses a significant threat to all operators.

Key events—"wake up calls"

Some recent key events have sounded the alarm on how orbital debris will affect the long-term sustainability of space. On January 11, 2007, China destroyed one of its own retired weather satellites (Feng Yun 1C or FY1C) in LEO with a direct ascent missile antisatellite (ASAT) test. This event resulted in more than 3000 pieces of debris the US Space Surveillance Network could track and maintain, almost doubling the number of debris fragments tracked at that time. Because the satellite was at an ~860 km altitude, the debris unfortunately will last a long time and has required collision avoidance maneuvers by the International Space Station and other satellites. [NASA, 2014]

An equally disturbing event occurred on February 10, 2009, when an operational satellite (Iridium 33) collided with a defunct Russian satellite (Cosmos 2251). This was the first provable instance of an operational satellite being destroyed by a naturally occurring impact with another satellite. The result was more than 2000 pieces of debris that could be tracked and maintained. The prevailing "big sky" theory no longer gave us statistical assurance for orbital safety of flight. The Chinese ASAT test and the Iridium/Cosmos collision nearly doubled the space object population below 1000 km. [NASA, 2013]

1.2 Setting the Stage

Today global civil, commercial and defense organizations are discussing space domain awareness in light of their needs, objectives, capabilities, risks and concerns for the future. Each type of organization has a perspective on the space domain and how its elements affect the organization, its country, and their citizens. Let's explore a few situations.

1.2.1 Civil arena

Imagine that, however unlikely, NASA has detected an asteroid aptly named IceCube as they routinely monitor space. The asteroid appears to have a trajectory that would bring it dangerously close to Earth within about seven years. Many of us have seen documentaries or science fiction movies that introduce a plot involving such an event. What would the citizens and leaders of the United States and other countries around the globe do? How might they respond?

NASA, working with other national and international resources, would start tracking IceCube, refining their approach, and predicting the asteroid's possible trajectories. They would consider the limited estimated information about the asteroid's physical characteristics, as well as how the Sun and other celestial bodies affect its trajectory. Some trajectory estimates predict this relatively small asteroid, about the size of a Volkswagen

Bug, would hit Earth, whereas others suggest it would pass millions of kilometers from Earth. NASA would keep tracking and refining their estimates as the asteroid proceeded toward Earth. In all likelihood, they would determine the asteroid isn't a threat, but here let's imagine it maintains an estimated trajectory toward Earth.

Scientists around the globe would want to know more about the asteroid's size, shape, surface color and internal structure. Some countries would try to use space-related resources to get more detailed information about IceCube. Others might launch new space systems to better image the asteroid's physical characteristics. As the asteroid gets closer, still others might try to rendezvous and land on it to do seismic studies on its internal structure. Many asteroids appear to have a rubble-pile structure; others may be somewhat homogeneous. In the end, scientists might try to deflect IceCube from Earth or destroy it. And the story goes on...

1.2.2 Commercial Arena

Operators of commercial space systems around the globe have tremendous respect for space weather and orbital debris, two important elements of our space domain. Most geostationary communications satellites rely on radio frequency (RF) communications to provide services (communications, television, radio and internet) to customers and users, as well as to send commands and receive telemetry that helps spacecraft operate. Radio frequency communications are their "bread and butter." So intentional or unintentional disruptions cause dire consequences for the spacecraft operator, customers, users, and the spacecraft's overall mission. Brian Weeden of the Secure World Foundation presents this useful example. [Weeden, 2016]

> **Galaxy 15 "Zombiesat" – April to December – 2010**
>
> In April 2010, a malfunction caused Intelsat S.A. to lose contact with Galaxy 15, a geosynchronous satellite that provided communications services over the Americas. The malfunction prevented Intelsat from maneuvering Galaxy 15, and the spacecraft started to drift slowly past other operational satellites in the region. Galaxy 15's receiver and transmitter equipment was still functioning, meaning it could pick up and re-broadcast signals aimed at other satellites as it drifted past, potentially causing serious radio frequency interference. Ultimately, Intelsat was able to regain control of Galaxy 15, but only the close communication and cooperation between satellite operators prevented any significant harmful effects.

As part of the electromagnetic spectrum, radio frequency (RF) communications are essential to satellites. Satellites use radio waves to receive commands from ground controllers (uplink) and relay information about their health and status in return (downlink). Many satellites also use radio waves to do certain functions, including retransmitting television broadcasts or transmitting imagery or scientific data they've collected. Unintentional RF interference can arise from a number of sources. Natural interference can occur from solar storms and other forms of space weather, interaction with the Earth's atmosphere, and sometimes even clouds and rain. Unintentional human-made interference can result from a satellite transmitting too close to another satellite on the same frequency or from ground communications systems operating on the same or similar frequencies.

Intentional RF interference, often referred to as jamming, is a way of temporarily or reversibly disrupting a satellite's normal functioning without destroying it and thus creating long-lived space debris. Intentional interference is also relatively easy to do, often requiring nothing more than an antenna and a transmitter. Applications for jamming range from blocking undesirable radio and television broadcasts from being transmitted into a country, to blocking satellite navigation signals that keep an employer from tracking movements, to degrading an adversary's ability to use precision munitions, among others.

International and national mechanisms exist to regulate RF communications. But they focus more on allocating spectrum and assigning frequencies than on preventing interference. They also lack enforcement powers. As instances of unintentional and intentional RF interference increase because of crowding and congestion on orbit, these regulatory shortcomings strongly challenge sustainable use of space over the long term.

1.2.3 Defense Arena

Spacecraft and their infrastructure (launch vehicles, launch sites, mission operations systems, ground systems, sensor systems) strongly support national and international defense. To illustrate how spacecraft and infrastructure play a role in defense, let's examine a recent space-related scenario. A Southeastern Asia Country (SAC) has been unusually active in the recent past, owing to a hostile military leader determined to secure a noteworthy place in the third world. Rumors suggest SAC has secured a small amount of radioactive material from the black market, but without confirmation from intelligences sources.

Meanwhile, other countries have supplied SAC with a rudimentary ability to launch scientific and military payloads. In the past five years, they've launched twelve missiles with dummy and scientific payloads. The last two, which have had dummy payloads, apparently hit within one kilometer of the intended target.

SAC has built one scientific and one military space launch facility. In the past five years, they've launched six scientific payloads, with one failure that re-entered shortly after launch. They successfully launched three military payloads whose missions haven't been determined because of encoded telemetry. Analysis is ongoing, but our launch detection systems have identified the launch sites and can distinguish which sites were involved. The first space launch was into a sun-synchronous orbit at low altitude (230 km ′ 500 km) from the military launch facility. Even with a lack of insight into the telemetry, the initial assessment suggests the spacecraft is an optical reconnaissance vehicle. All nations want their own; this capability is high on the list of desired military missions. Narrowband radar signature analysis of this mission determined the vehicle's longitudinal axis was oriented to nadir. It also identified solar panels oriented toward the Sun and observed the longitudinal axis was pitched off the ground track that could be associated with high-value ground targets. (This is where I believe we could use SMAD to postulate missions that match the observed characteristics and compute the resolution, power levels, etc. versus the current state of technology. For unknown missions, the observed characteristics could be used to propose a range of missions and assign a probability that could be used along with other observed information to combine in the Bayesian approach in the Hughes study to compute likelihoods.)

SAC's government news agency has announced the current launch as a scientific weather satellite, but further data analysis found the launch came from the military launch site. Initial narrowband and optical signatures show the payload isn't like those from any of the other three military space launches. We've tasked wideband images but will need several hours to obtain and process them.

Over the next 12-24 hours, we process wideband images and discover four objects associated with the launch: the payload, a cylinder (likely the last stage tank), and two smaller objects that appear to be fragments and are tumbling. The wideband images show the payload is horizon stable with solar panels and a rectangular shape with deployed solar panels. Optical signature analysis shows dark blue flat panels, which could be more solar collectors, cover the spacecraft's body. Strangely, the wideband images also reveal antennas on top (anti-nadir) pointing toward the depths of space. We've spotted other antennas on the nadir side of the payload and assumed they're for telemetry or other communications. Computations that estimate the radio frequencies being used pass to the intelligence community so telemetry collection can begin. We've detected no signals. Tasking and reporting data from the sensors takes about six hours, after which results go to the intelligence community for tasking the appropriate collectors.

The new satellite's orbit is sun-synchronous, with its ascending node about two hours before sunset over the northern hemisphere. This makes the orbit good for an imaging mission, for ground scenes, or for a weather mission. The apogee and perigee are 800 km × 500 km, with the argument of perigee in the southern hemisphere near 20 degrees south.

Given the payload's size from the wideband images (2 m × 2 m × 4 m), plus dimensions and altitudes, we can propose a resolution. From the observed size of solar panels and body-mounted cells, we can calculate total power available.

As time passes (say a few days), the drag on the satellite and the frontal area in the direction of travel yield an estimate of the payload's mass. (Previous results have been +/– 10% of the actual mass.)

Eventually, the intelligence community uses dimensions from external observables to identify the signals coming from the payload. They postulate (not yet proved) the antenna structures on top of the vehicle are to relay data in real time.

The story continues from here, but we've offered enough information to begin an exercise in space domain awareness. Let's establish a process that will allow us to think through how we would deal with the SAC situation.

The three scenarios above follow the same basic process, even though the organizations may be concerned about different elements of the space domain. In their own way, each organization MONITORs the space domain's elements of interest, establishes a BASELINE for these elements, DETECTs changes to the baseline, TRACKs new objects or changes to the baseline, PREDICTs variations, ANALYZEs and investigates the situation, DECIDEs how to respond to changes in the baseline, and ITERATEs through this process. This book offers insights and focuses on these activities.

1.3 Context for Space Domain Awareness

By this point we've described the nature of challenges to space domain awareness, mentioned the tens of thousands of natural and human-made objects that surround Earth, reviewed briefly how we got here, and offered some examples of tense situations that may require intervention. We're now going to name and describe key stakeholders and actors engaged in the daily activities of our Distributed Space Situation Operations Center (DSSOC), shown in Figure 1-5. The circle in the center represents the DSSOC System, which includes all hardware, software, processes, procedures and items the team uses to carry out space domain awareness. The stakeholder groups shown around the outside of the circle represent the many systems, organizations, countries and capabilities included in SDA. We'll describe each item, beginning with the Space Object Catalog and continuing clockwise around the circle.

The Space Object Catalog supplies information on Earth-orbiting objects, both natural and artificial, which come from the United States Space Surveillance Network's efforts to detect, track, catalog and identify artificial objects orbiting Earth. Examples include active and inactive satellites, spent rocket bodies, and fragmentation debris. In general, space surveillance enables us to do eight things:

- Predict when and where a decaying space object will re-enter Earth's atmosphere
- Prevent a returning space object, which to radar looks like a missile, from triggering a false alarm in missile-attack warning sensors of the U.S. and other countries
- Chart the present position of space objects and plot their anticipated orbital paths
- Detect new human-made objects in space
- Correctly map objects traveling in Earth's orbit
- Produce a running catalog of human-made space objects
- Determine which country owns a re-entering space object
- Inform NASA if objects may interfere with satellites and the International Space Station's orbits

For space domain awareness, we compare changes to the baseline established by the current Space Object Catalog.

Space Assets include all active and inactive spacecraft (we'll use "spacecraft" for artificial, human-made objects and "satellite" for anything else orbiting Earth—crewed or robotic, launch and stage vehicles, ground stations, and infrastructure). Space assets also include the organizations and nations that own and operate them. The DSSOC team can task selected space assets for capabilities and information.

Space Operators include space mission operations systems and people for civil, commercial, academic, and defense organizations that operate missions for their systems. Figure 1-7 shows what spacecraft operators usually do within the commercial, civil, and defense arenas.

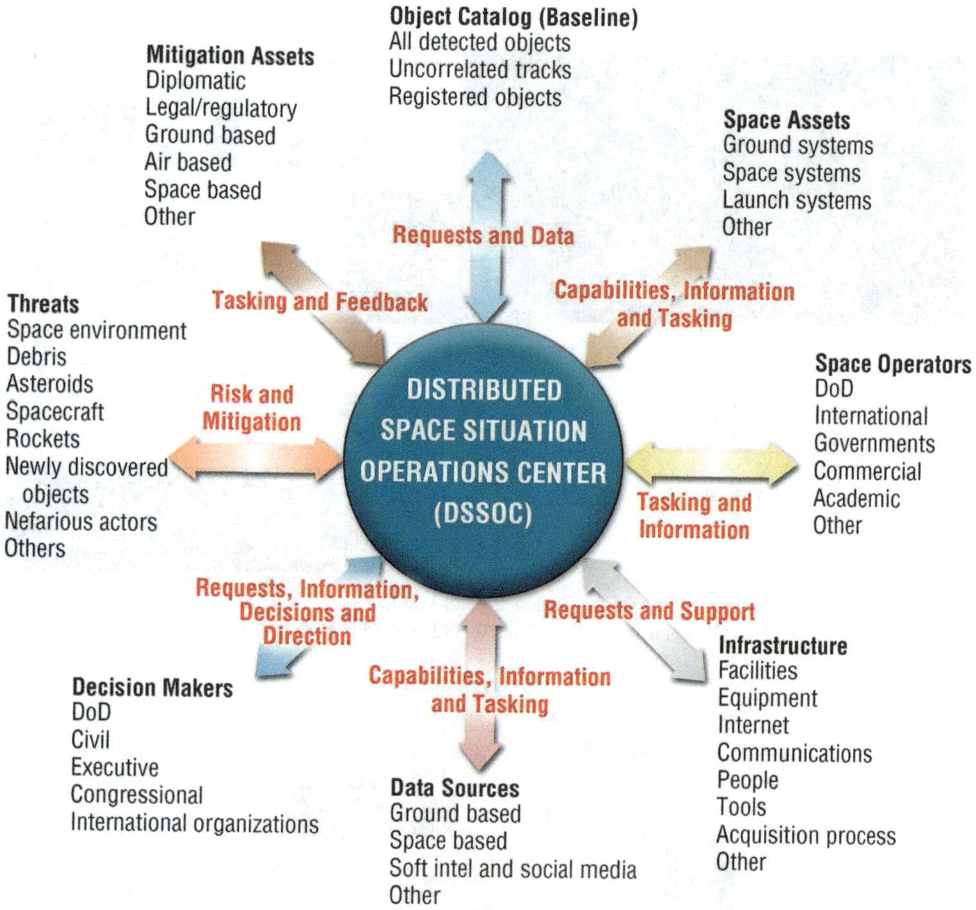

FIG. 1-5. **Active Stakeholders of the Distributed Space Situation Operations Center (DSSOC).** The DSSOC must engage with many of these key stakeholders every day to satisfy the needs of space domain awareness.

Within this stakeholder group we include commercial spacecraft operations and service providers. Larger space-related service providers include Intelsat, SES, Eutelsat, Telesat and Skyperfect JSAT Corp. These organizations and their smaller counterparts offer the world many services:

- Global and regional telecommunications: broadcast and point-to-point
- Secure communications and data for all aspects of society and business, such as radio, television, and other networks—reaching such domains as medicine, banking, normal business operations, and religion
- Spacecraft operations and consulting

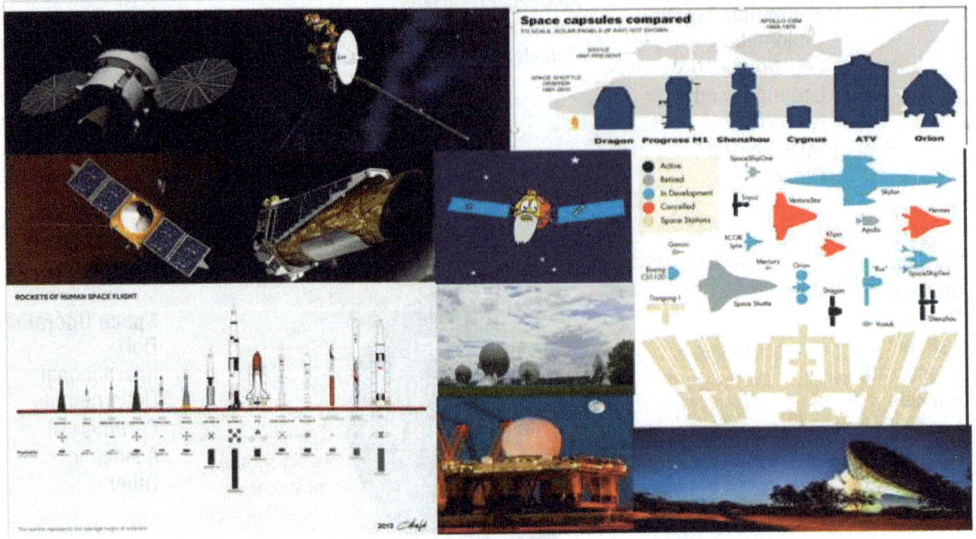

FIG. 1-6. **Space Assets for SDA.** Space assets include active crewed and robotic spacecraft, launch systems, ground stations, tracking stations, and organizations and personnel.

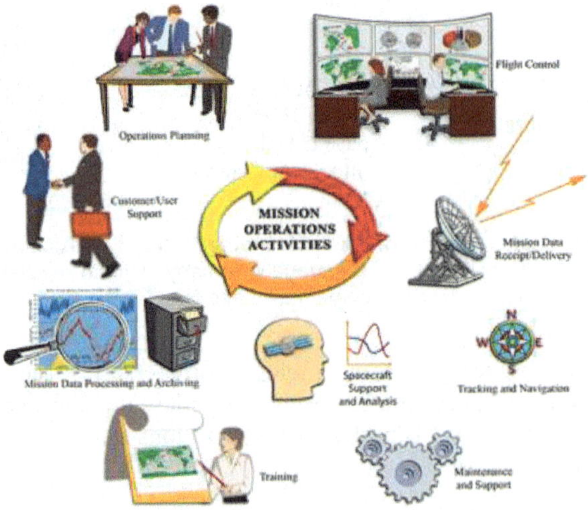

FIG. 1-7. **Space Operator Functions.** What space operators do for active mission spacecraft, crewed and robotic.

- Mobile communications
- And much more

These space service providers are essential to global societies, civil space, commercial space and defense.

Infrastructure for the space domain includes facilities for orbit operations, launch, and development, assembly and test. All communications architectural elements are included along with the internet. Organizations and people who operate and manage the infrastructure as well as the acquisition process, also are part of the infrastructure. Finally, the infrastructure includes the very important element of cyber security.

Data Sources for SDA include all stakeholders, with a focus on ground-based and space-based systems, as well as soft intelligence and social media.

Decision makers may be in the DSSOC or be available to consult elsewhere. The DSSOC will need a structured approach to making decisions at various organizational levels that depend strongly on the type and severity of the threat. Each stakeholder, whether associated with defense, civil, commercial or all of them, will need a well thought out decision process. The key will be to have the right data and information available for the decision makers at the right time. A large part of this book's later chapters is devoted to this issue.

The **threats** to space systems include the space environment, debris of all kinds, asteroids, other spacecraft, rockets, newly discovered objects, nefarious actors on the ground, in the air or in space, and cyber security. These threats to the spacecraft, launch systems, ground systems, or mission operations systems), whether real or perceived, require the Distributed Space Situation Operations Center to take action.

The possible courses of action or mitigations depend on the nature and severity of the threats or changes to the baseline. The space environment supplies an example. In the case of an imminent solar flare, with little or no warning, space operators may choose to place their spacecraft in a power-safe-hold mode while the storm passes. Here, there is no time to discuss possible actions or mitigations—the operators need to act immediately.

Another example is a possible collision between the International Space Station and a robotic Chinese spacecraft, which poses a threat to both vehicles. In this situation, the threat arises early enough for operators to discuss possible actions. Considerations include likelihood of collision; necessary actions for both vehicles; the relationship between owners of the two vehicles, (in this case, NASA and the Chinese Government); credibility of orbit predictions; space environment (density); knowledge of the ballistic coefficients of both vehicles; and others. Figure 1-8 provides a typical threat-response diagram that shows interactions among key stakeholders. Later we discuss the decision process that leads to a course of action. In this case we assume all parties agree to maneuver the Chinese spacecraft to a slightly different orbit to avoid the collision.

The situations above all follow the same basic process, even though the organizations may be concerned about different elements of the space domain. In their own ways, they MONITOR the space domain elements of interest, establish a BASELINE for elements of the space domain, DETECT changes to the baseline, TRACK new objects or changes to the baseline, PREDICT..., ANALYZE and investigate the situation, DECIDE how to respond to changes in the baseline and ITERATE through this process.

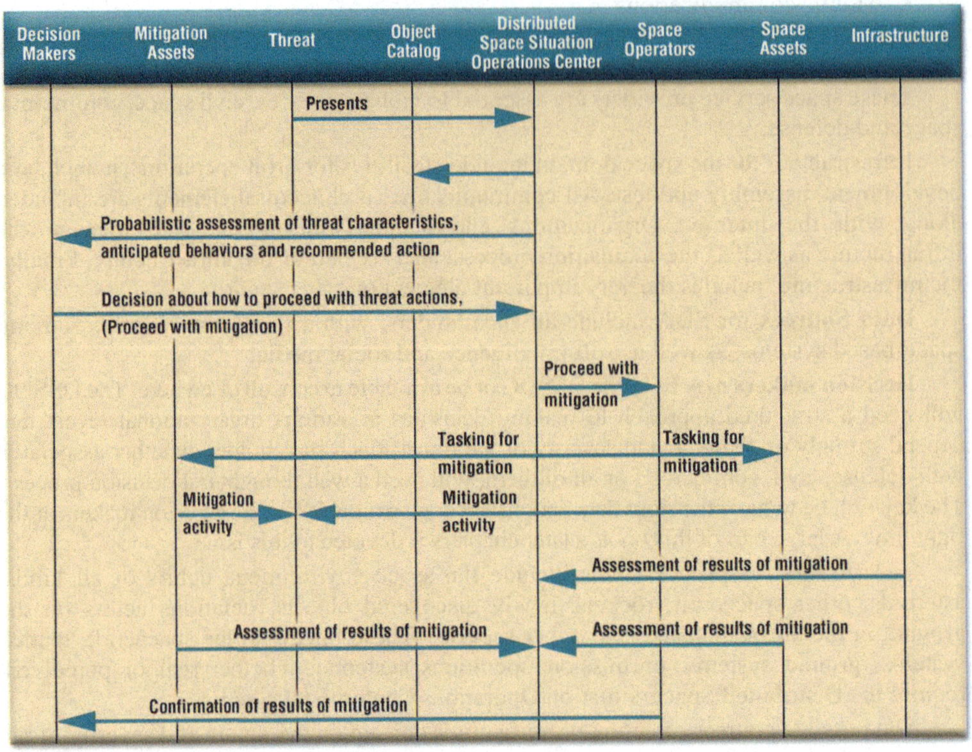

FIG. 1-8. **Threat-Response Interaction Diagram.** This figure shows the hypothetical interaction among key SDA stakeholders and the Distributed Space Situation Operations Center from the time a threat presents itself through mitigation and documentation.

1.4 Functional Needs for Space Domain Awareness

Isolating functional categories of space domain awareness allows methodical analysis and permits a useful study approach. The categories mirror the logical progression of steps in a process. In broad terms, we must detect and track space objects using sensors and characterize them to understand their nature. We also must identify events involving these space objects (often quickly) to enable responses. Events range from routine conjunctions to accidental collisions, to deliberate attacks. Finally, we must integrate and use many data sources efficiently so we can understand events and make timely decisions. This construct is useful for all SDA customers, including government and commercial enterprises. To summarize, we divide SDA into four functions:

- Tracking objects
- Characterizing objects
- Identifying events

- Fusing and using data

These four functions align closely to those for space situational awareness as outlined in the United States' military doctrine. [NASA, 2013] They undergird any investigation into the many elements of SDA. We use the hypothetical Distributed Space Situation Operations Center, Figure 1-9, to show how to carry out such operations.

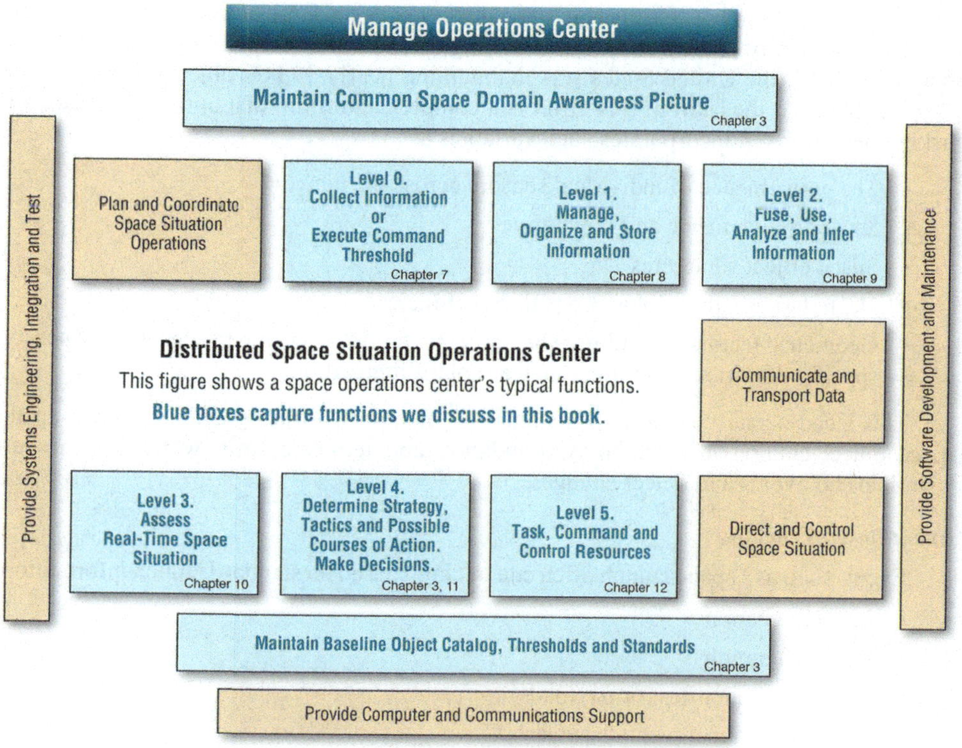

FIG. 1-9. **Operational Space Domain Awareness Functions.** Each box in this figure represents an operational SDA function. Blue boxes and their text show the SDA functions explored in this book.

1.4.1 Tracking Objects

Tracking space objects is the most basic element of SDA, but it's still complex and challenging. Tracking begins when a sensor detects an object and we establish an orbit accurately enough for a sensor to confidently reacquire it in the future. We need repeated tracking to maintain positional accuracy because astrodynamic force models may not account for orbital perturbations that degrade its fidelity.

Steady-state Tracking

Space object tracking maintains the position of orbiting objects to within a required tolerance. Degradation of orbital accuracies affects our ability to predict space object conjunctions, resulting in a higher number of collision warnings to crewed space flight and satellite operators. Maintaining accurate object positions therefore reduces the number of collision false positives (false alarms) and helps us reacquire the track. Radars do most tracking for objects in low-Earth orbit; optical telescopes track most higher orbiting objects.

The complexity of the steady-state tracking task becomes evident when one considers that, as of 2017, the United States was maintaining nearly 19,000 objects in Earth orbit. [NASA, 2018] At the heart of this effort is a complex algorithm that optimizes sensor and tasking operations. It incorporates such variables as

- The performance of individual sensors across the network
- Sensor maintenance and down time
- Space object signatures
- Degradation rates of object position accuracies
- Geometric factors (including when objects should be visible to sensors, and sensor-specific detection conditions such as solar illumination)

This steady-state tasking and maintaining of object orbital positions makes up the space object catalog. It's the SDA foundation and template from which we measure unanticipated events and detect changes.

Event-based Tracking

Events such as a space launch often cue tracking based on standard launch information:

- Launch location
- Launch azimuth
- Number of satellites (or payloads)
- Number of launch vehicle stages
- Intended orbit

This information is usually available to all nations, even for military satellite launches, and is consistent with launch safety requirements and compliance with United Nations treaties. Although countries may not disclose intended orbits for military launches, reasonable estimates are still possible based on the launch vehicle used and the launch azimuth.

Tracking objects from the time of launch is critical for establishing sources and creates the basis for enforcing the UN Space Liability Convention if the object harms another entity's citizens or property. A launch may deploy multiple satellites, so discriminating which satellite is which may take several days—often by working with the satellite's owner-operators. This identification process supports the first ground contacts with the satellite and early commanding.

Unanticipated events, such as the breakup of a satellite or rocket body into multiple pieces, may also prompt tracking. Steady-state tracking usually detects a breakup or debris-producing event by observing several objects when only one was anticipated. This in turn cues other sensors to collect more observations, so we can establish the orbits of as many object fragments as possible. In 2017, commercial optical sensors captured two debris-producing events as they occurred.

Satellite maneuvers also require increased tracking to regain an accurate orbital state. We must identify the maneuver to solve a new orbital state because sensor observations before and during a maneuver result in measurements that don't reflect natural orbital motion. Rapidly identifying maneuvers also enables us to recognize potentially harmful actions in space, such as attacks from other orbiting objects. A maneuvering object, especially if the owner-operator doesn't share data, may warrant increased tasking and more frequent sensor revisits to maintain the object's orbital state.

Unidentified Object Tracking

Given the large amount of orbital debris, as discussed in Chapter 3, we often detect and track objects of unknown origin. But we need multiple tracks with time in between them to establish an initial orbit. This method, called uncorrelated track (UCT) processing, uses sophisticated algorithms. Once we produce an accurate-enough orbit, steady-state tracking can begin. Because these objects are often physically small, with low optical and radar signatures, only a few sensors may detect them. So they often fall out of routine maintenance and may be "lost." Later tracking and rediscovery, sometimes years later, may find the same objects again.

Because space contains so much debris, it's common to track unknown objects near the edge of the network's sensor sensitivity. As we add more sensitive sensors, we may track and maintain more objects. Such is the case with the US Air Force's eagerly anticipated "Space Fence," an S-band phased array radar slated to start operations by 2020 and widely expected to double or triple the space object catalog.

Once we have an accurate orbit for an object, we can propagate its position backward in time to determine its probable origin. The success of this process depends on how well the orbital perturbations are modeled, which may include previously uncorrelated observations that refine the object's orbital history. For some objects and orbits, it may take several years. We must determine an object's origin to disclose it publicly in the United States catalog, but the US now tracks and maintains several thousand other objects.

Characterizing Objects

One track of an object partly characterizes it with such data as the radar return magnitude (radar cross section) or radiometric brightness (visual magnitude). Characterizing encompasses all the information for an object, of which the orbital parameters derived from tracking are a small subset. This information may be "hard data" derived directly from sensors but often includes "soft data" from public media and other sources. Examples of hard data are measurements of electromagnetic energy scattered from or emitted by an object, orbital perturbations, and the coupled inferences between orbital dynamics and object signatures.

Characterizing an object's type is a first-order task. We can categorize space objects as satellites (active or inactive), rocket bodies and upper stages, mission-related debris, or fragmentation debris. Usually, we identify satellites, rocket bodies, or upper stages during launch, and then keep custody of the objects' tracks to maintain their identification.

Measuring the electromagnetic (EM) energy scattered from or emitted by an object is a main way to characterize it. All sensors used for metric or positional tracking supply these signatures. As mentioned, radars usually give us a cross section, and optical telescopes measure radiometric brightness, but we have to calibrate these sensors properly for accurate measurements. Passive radio frequency (RF) receivers monitor RF frequencies in use, as well as frequency changes due to Doppler shifts from orbital motion. These are all examples of non-spatially-resolved or non-imaging radiometric measurements.

We can get a lot of information from these measurements, using significant science to extract it from the EM emissions. This information includes amplitude (signal strength), frequency (or wavelength), and polarization. The short- and long-term changes in these quantities are of interest. For instance, we can learn about an object's rotational period from the short-term (typically minutes) temporal change in the amplitude. Measuring the amplitude of its non-resolved signature enables us to use fewer sensors for steady-state tracking.

Spatially-resolved measurements of an object produce an image that people can interpret directly. Ground-based imaging by optical telescopes and wideband radars may confirm solar array deployment and sizes, as well as an object's rotational state and spin axis. This imaging also helps us accurately identify objects from a new launch. We have to use these sensors carefully because they're often not available and can't produce many measurements. Sometimes, we can use anomalies from non-resolved sources to cue sensors that offer more-resolved images.

An object orbit's perturbations also supply important information. For example, we can use the perturbing forces of solar radiation pressure and atmospheric drag to infer physical properties such as object mass and area cross section. We can then compare estimates of an object's area and size from astrodynamic perturbations to those from non-resolved radiometric or RCS measurements.

Finally, we can use an object's routine patterns to recognize abnormal behavior. These behavior patterns are observable in signature orbits and maneuvers. For instance, one satellite owner-operator may maintain a geosynchronous satellite within a given longitude and inclination tolerance, which we may clearly observe. Another operator may do station keeping maneuvers on a given day of the week. For the signature domain, active geosynchronous satellites present a systematic, repeatable optical signature to ground-based optical telescopes. Deviations in these signatures may indicate an anomaly or other activity, which in turn may cue more sensitive sensors for investigation. The ability to identify maneuvers and signature changes as "normal" or "abnormal" is very valuable.

This discussion leads to the concept of a space object fingerprint that allows the DSSOC operator to identify each object with some certainty. Elements of the space object finger print are shown in Figure 1-10.

Essential elements of information required to identify and characterize space objects include:

- **Type of object**—natural, debris, non-functional spacecraft, functional spacecraft, launch vehicle or stage, other

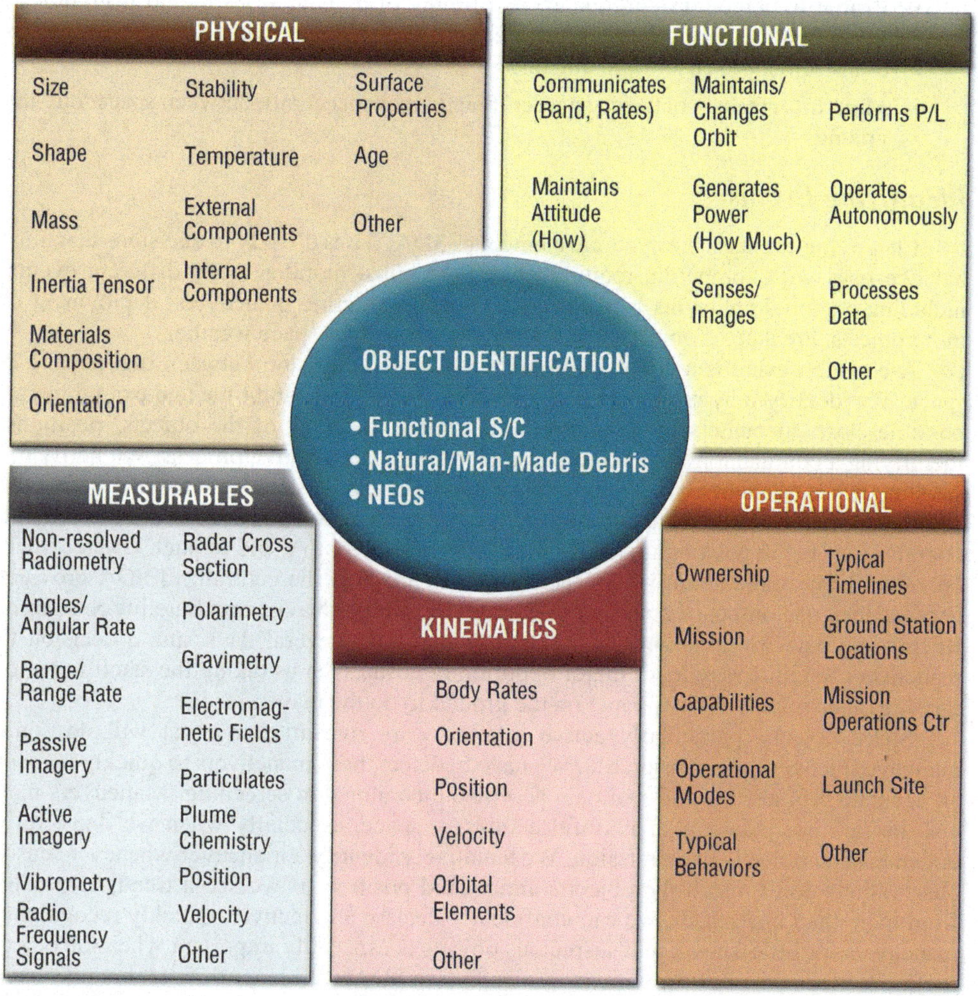

Fig. 1-10. Space Object Fingerprint. We use these types of information about space objects to identify and characterize. Need reference in text.

- **Basic information**—epoch, orbital elements, registration, type, stated mission, owner, status, date of launch, date of first observation, launch site, launch vehicle
- **Estimated or confirmed characteristics**—visual magnitude, radar cross-section, size, mass, dimensions, stability, pointing direction, size of arrays and technology, estimated power produced, nozzles, propulsion type, estimated delta V, instruments (type and size), sensors (type and size), RF frequencies, modulation, attitude control approach and accuracy

- **Probable behaviors**—frequency and timing of transmissions to and from object; frequency and timing of rotations; frequency and timing of maneuvers; view angles of instruments, sensors and antennas
- **Main interfaces**—between spacecraft and other spacecraft, between spacecraft and ground

Identifying Events

Identifying events is the dynamic element of SDA; it ties directly to decisions that must balance risks with competing requirements. SDA must monitor many different events, including potential collisions between space objects, satellite maneuvers, deployment of more objects, breakup of objects, and natural events such as space weather.

A common event is a conjunction, or the potential collision between two objects in space. We identify it by propagating objects forward in space and time and examining all possible close encounters at distances matching the accuracy of the objects' positions. Identifying a conjunction with an active satellite starts an iterative process. We notify the satellite owner-operator about the possible conjunction and task more sensors to refine the object's orbital state and accuracy of our conjunction assessment. In parallel, the satellite operator may begin positioning the satellite for a maneuver to avoid impact. Often owner-operators know more about where their satellites are than the centralized SDA provider because they use on-board positioning through the Global Navigation Satellite System or RF ranging data. Notification and refinement may take several days, and a decision to maneuver a satellite may incur major opportunity costs, such as taking the satellite out of mission, plus more labor by people on the ground to do the maneuver.

Space objects, particularly active satellites, are dynamic, so most will do some maneuvering over their mission life. We have to detect these maneuvers to quickly recover an accurate orbital state and maintain acceptable conjunction screening. Maneuvers may also be the first indicator of harmful actions in space, especially when we don't fully understand an object's main mission. We tentatively identify a maneuver whenever sensor observations don't match the object's anticipated position. If we see a trend, especially from more than one sensor, we can confidently declare a maneuver. Quickly recognizing maneuvers for little-known and suspicious objects is especially important when the object is close to other satellites. As with many elements in SDA, we can work faster but may have a higher false alarm rate, which may not be acceptable given the opportunity cost to operators. An object's maneuverability influences how often we should sense it. If it can accelerate quickly, sensors must revisit it often enough so the first sensing after the maneuver doesn't result in an uncorrelated object. Finally, a "maneuver" may not be intentional. Collisions with untracked debris, [Krag et al., 2017] as well as internally released energy (such as from a pressurized tank), will look like a space object maneuver.*

Space objects may deploy other objects as part of their mission. For instance, the International Space Station has deployed hundreds of objects. In some cases, a satellite not well understood may unexpectedly deploy another object. Such events first look like we've detected multiple objects when we expected only one object from tracking, with the

* Examples of these maneuvers include the Russian Blitz satellite and US GOES XX.

secondary objects recognized from the lower optical or radar signature. Occasionally, catastrophic breakups occur because satellites or rocket bodies contain residual energy. Propellant tanks in rocket bodies or satellite batteries that don't undergo proper passivation (sealing of cells) after disposal are subject to these events. Such events may produce hundreds of objects that have enough signatures for tracking and higher velocities than the main object. Similarly, low-energy events may occur where an object releases pieces with relatively low velocities. For example, as thermal control blankets degrade in space, they release pieces—many with high area to mass ratios. We often call such low-energy events anomalous. They leave the object largely intact and may leave it operational.

Some events, such as radiofrequency interference (RFI), are specific to satellite missions. Providers of commercial communication satellites must identify and resolve RFI to make money, so they've organized to improve how they reduce and resolve RFI events.* [SDA, 2018] RFI identification is of increasing concern to mega constellations in LEO and is a good example of how cooperation between traditional providers of SDA information and owner-operators benefits everyone.

Natural events, such as space weather, can affect SDA. Heightened solar activity produces more high-energy particles, which can cause electronic upsets and satellite anomalies. So recognizing solar events helps resolve these anomalies. Solar events also expand the thermosphere, which increases drag on objects with lower perigee and degrades orbital accuracies. Finally, solar events alter the ionosphere, which degrades radar calibration and the quality of object observations. Identifying space weather events factors into many SDA components.

Fusing and Using Data

Finally, we must have a way to integrate all the hard and soft data sources and derive the information needed to serve SDA customers. Critical decisions hinge on this information, so it's not a trivial task. Challenges include verifying sensor quality, particularly important for metric data used to update orbital states. The ability to integrate data across several data networks while preserving information security also poses unique challenges. Sensor and data providers may limit the scope and redistribution of their data, which may therefore restrict our information products. Thus, we must have a means to leverage the value of non-traditional and soft data sources.

The decisions SDA customers make must drive how we use this data. Algorithms and applications for data sources must be timely and have quantified confidence levels for decision makers. For instance, conjunction notifications must have enough advance warning to enable a maneuver if the operator considers it necessary but be accurate enough so they don't burden operators with false alarms. An extremely time-critical case is an attack by one satellite on another. An abnormal, high-velocity maneuver could be a threat, but we couldn't declare it an attack without refining the maneuvering state and incorporating other data sources.

---

* The genesis of the Space Data Association is largely founded on RFI mitigation. http://www.space-data.org/sda/

1.5 Challenges

Space domain awareness faces many challenges. We can readily recognize most technical challenges, but policy challenges are more complex. Rampant commercial usage and the threat from orbital debris have elevated SDA to an international issue. This elevation increasingly demands a mutual understanding of the space environment overseen by such entities as the United Nations.

Although nation-states have often disagreed about the use of space for defense, the commercial use of space has accelerated at an unprecedented rate in the past decade. These commercial activities also extend to the space infrastructure that nation-states have always controlled. Commercial launch services, space surveillance sensing, and SDA services are in the offing. The "cubesat" revolution has spawned miniaturized space hardware components and has resulted in many new entries and microsats being launched. Should launches of cubesats with no maneuverability be allowed into orbits that don't meet the 25-year reentry rule for orbital debris? Who should oversee and enforce such guidelines if they become internationally accepted? SDA has a central role in these issues.

Meanwhile, state-sponsored capabilities for national defense continue to be employed for vital functions such as detecting missile launches, secured communications, reconnaissance, and global navigation services like GPS. Growing evidence of counter-space programs to negate satellite systems [NAS, 2016] adds to the tension and the resistance to publicly disclosing orbital information on some systems. How could we share more SDA data internationally to lessen concerns about national security?

Despite the United States providing a public domain Space Object Catalog, visibility and insight into space domain activities have been limited. This lack of insight has hindered growth of "norms of behavior" that entities have adopted in other domains such as air and maritime. No accepted international rules define irresponsible behaviors in space, other than generally accepted measures for combatting space debris. How close is too close for one satellite to approach another? Under what conditions are approaches to other satellites acceptable? SDA offers the essential awareness to quantify these behaviors, mature norms, and most important, enforce proper conduct in space. Calls within the United States are to designate a civil agency to oversee "Space Traffic Management" [SAIC, 2016] [IDA, 2016] and divorce the Department of Defense from routinely maintaining the Space Object Catalog.

Space domain awareness also confronts many technical challenges. As previously discussed, large amounts of untracked space debris could collide with and instantly disable an operational satellite. We must field more capable sensors, such as the US Space Fence, to track this small debris and warn of potential collisions. We also must improve orbital accuracies for these objects to avoid unnecessarily identifying improbable conjunction events. These requirements prompt reflection on a key policy issue. Once Space Fence more than doubles the number of tracked objects, how many will be attributed to their original source? Will the UN Liability Convention not apply to such objects? Or might adjudication be necessary to verify independent claims of an object's origin?

More sensitive sensors naturally detect more space objects—mainly small debris from past fragmentation events. Effectively generating orbits for these objects requires advanced uncorrelated track algorithms that produce states accurate enough to enable reacquiring them. Many high area-to-mass ratio or "HAMR" objects in deep space are particularly challenging to maintain. We need to better understand the dynamic forces attributed to solar radiation pressure in order to improve state estimates for these objects. For near-Earth or LEO orbits, we must enhance calibration of dynamic atmospheric effects to refine drag forces on objects.

Finally, we need a way to effectively integrate "non-traditional" data sources, or those outside the bounds of historical sources, which have been subject to quality reviews. The growth of commercial sensors has produced large volumes of data that we may leverage if we can resolve concerns about their quality and possible corruption. Similarly, satellite operators often maintain positional information on satellites with a greater precision than we can get from external sensors. But we need to integrate this data into conjunction screenings in order to decrease demands on sensor tracking.

References

Stone, M. L. and G. P. Banner, "Radars for the Detection and Tracking of Ballistic Missiles, Satellites, and Planets," Lincoln Laboratory Journal, Vol 12, No 2, 2000. https://www.ll.mit.edu/publications/journal/pdf/vol12_no2/12_2detectsatellitiesplanets.pdf

Easton, R. L. and J. J. Fleming, "The Navy Space Surveillance System," Proceedings of the IRE, April 1960, pp 663-669. https://doi.org/10.1109/JRPROC.1960.287444

Hayes, E. N., "The Smithsonian's Satellite-tracking Program: Its History and Organization, Part II," Smithsonian Institution Annual Report for 1963, pp. 331-357. http://siarchives.si.edu/collections/siris_sic_3188

Satellite Situation Report," Vol 2, No 1, January 2, 1962. https://ntrs.nasa.gov/search.jsp?R=19630012120

Jakhu, R., B. Jasani and J. McDowell, "Critical issues related to registration of space objects and transparency of space activities," Acta Astronautica 143 (2018) pp. 406-420 https://doi.org/10.1016/j.actaastro.2017.11.042

Weiss, G. W. "The Life and Death of Cosmos 954," declassified CIA Article, 1978. https://www.cia.gov/library/readingroom/document/cia-rdp85b01152r000200260006-4

Kessler, D. J. and B. G. Cour-Palais, "Collision Frequency of Artificial Satellites: The Creation of a Debris Belt," Journal of Geophysical Research, Vol 83, No A6, June 1, 1978. http://dx.doi.org/10.1029/JA083iA06p02637

"History of On-Orbit Satellite Fragmentations," 14th Edition, NASA Orbital Debris Program Office, June 2008. NASA/TM-2008-214779. https://www.orbitaldebris.jsc.nasa.gov/library/satellitefraghistory/tm-2008-214779.pdf

"Terms of Reference for the Inter-Agency Space Debris Coordination Committee," IADC-93-01, revised September 28, 2016. http://www.iadc-online.org/Documents/IADC_TOR_rev_11.3_public.pdf

Weeden, B. C. "The Evolution of U.S. National Policy for Addressing the Threat of Space Debris," 67th International Astronautical Congress, Guadalajara, Mexico, 26-30 September 2016. IAC-16.A6.8.3. https://swfound.org/media/205624/iac-16a683-weeden_evolution_us_space-debris_policy-paper.pdf

Liou, J.-C. "Orbital Debris Briefing," NASA Orbital Debris Program Office, briefing to Executive Office of the President/Office of Science and Technology Policy (EOP/OSTP); 8 Dec. 2017; Washington, DC; United States https://ntrs.nasa.gov/search.jsp?R=20170011662

"National Security Space Defense and Protection: Public Report," National Academies Press, 2016. ISBN 978-0-309-44748-5, DOI: 10.17226/23594. https://www.nap.edu/catalog/23594/national-security-space-defense-and-protection-public-report

"Orbital Traffic Management Study," Final Report, 21 November 2016. Prepared for NASA by SAIC. http://www.spacepolicyonline.com/pages/images/stories/Orbital%20Traffic%20Mgmt%20report%20from%20SAIC.pdf

Commercial SSA video compilation used for the 2017 AMOS Space Surveillance Technology Conference, https://www.youtube.com/watch?v=JPQmzjJ_oH8

2011 National Security Space Strategy, http://archive.defense.gov/home/features/2011/0111_nsss/

For instance, see articles as curated from one of the author's Twitter accounts, https://twitter.com/shell_jim

Kessler, D. and B. G. Cour-Palais, "Collision Frequency of Artificial Satellites: The Creation of a Debris Belt," Journal of Geophysical Research, Vol 83, No A6, pp 2637-2646, June 1, 1978. https://ntrs.nasa.gov/search.jsp?R=19780057167

United Nations Office for Outer Space Affairs, Space Debris: http://www.unoosa.org/oosa/en/ourwork/topics/space-debris/index.html

Hamilton, J. et al., "Radar Measurement of Small Debris from HUSIR and HAX," NASA JSC-CN-39228, April 2017. https://ntrs.nasa.gov/search.jsp?R=20170004608

Krisko, P. et al., "The Geosynchronous Environment for ORDEM2010," NASA JSC-CN-18954, October 2009, IAC-09-A6.2.02. https://ntrs.nasa.gov/search.jsp?R=20100002022

"Stability of the Future LEO Environment," Inter-Agency Space Debris Coordinating Committee, January 2013. https://www.iadc-online.org/Documents/IADC-2012-08,%20Rev%201,%20Stability%20of%20Future%20LEO%20Environment.pdf

McKnight, D. "Pay Me Now or Pay Me More Later: Start the Development of Active Orbital Debris Removal Now," 2010 AMOS Technical Conference, http://amostech.com/TechnicalPapers/2010/Posters/McKnight.pdf

"Fengyun-1C Debris Cloud Remains Hazardous," NASA Orbital Debris Quarterly News, Vol 18, Issue 1, January 2014. https://www.orbitaldebris.jsc.nasa.gov/quarterly-news/pdfs/odqnv18i1.pdf

"An Update of the FY-1C, Iridium 33, and Cosmos 2251 Fragments," NASA Orbital Debris Quarterly News, Vol 17, Issue 1, January 2013. https://www.orbitaldebris.jsc.nasa.gov/quarterly-news/pdfs/odqnv17i1.pdf

Joint Publication 3-14, Space Operations, 29 May 2013, http://www.jcs.mil/Portals/36/Documents/Doctrine/pubs/jp3_14.pdf

Orbital Debris Quarterly News, Vol 22, Issue #1, February 2018. https://orbitaldebris.jsc.nasa.gov/quarterly-news/pdfs/odqnv22i1.pdf

For example, in 1970 the rocket body used to launch a Nimbus 4 fragmented six months after launch. As of 2008 there were 373 pieces of debris cataloged. However, in 2017 an additional 65 debris objects from this event were added to the US catalog, 47 years after launch!

The United States Space Surveillance Network uses three generic categories: "Payload" for satellites, "Rocket Body" and "Debris."

H. Krag, et al., "A 1 cm space debris impact onto the Sentinel-1A solar array," Acta Astronautica, Volume 137, 2017, Pages 434-443, ISSN 0094-5765, https://doi.org/10.1016/j.actaastro.2017.05.010.

Examples of these maneuvers include the Russian Blitz satellite and US GOES XX.

The genesis of the Space Data Association is largely founded on RFI mitigation. http://www.space-data.org/sda/

National Academies of Sciences, Engineering, and Medicine. 2016. National Security Space Defense and Protection: Public Report. Washington, DC: The National Academies Press. https://doi.org/10.17226/23594.

"Report on Space Traffic Management Assessments, Frameworks and Recommendations," 21 November 2016, prepared by Science Applications International Corporation (SAIC) for NASA. http://www.spacepolicyonline.com/pages/images/stories/Orbital%20Traffic%20Mgmt%20report%20from%20SAIC.pdf

"Evaluating Options for Civil Space Situational Awareness (SSA)," IDA Science & Technology Institute, August 2016, IDA Paper NS P-8038. https://www.ida.org/idamedia/Corporate/Files/Publications/STPIPubs/2016/P-8038.ashx

Knipp, Delores, Understanding Space Weather and the Physics Behind It, 2nd Ed, CEI Publications, 2014.

Larson, Wiley J. and James Wertz, Space Mission Analysis and Design, 3rd Ed., Microcosm Press, Inc, 1999.

Larson, Wiley J., Space Mission Analysis and Design Core, CEI Publications, 2014.

Possible Video clips

https://www.youtube.com/watch?v=M9adk2C2pMc Breakup of spacecraft reentering, ESA

Chapter 2

Strategy and Tactics for Space Domain Awareness (SDA)

Dr. Donald A. Lewis,
The Aerospace Corporation

Comprehensive awareness of the space domain is challenging at best and nearly impossible under many scenarios. Space domain awareness (SDA) requires consistent and capable surveillance and observation, as well as the appropriate analytical tools and tradecraft. The space domain includes not only the region of space typically characterized as outer space (that region 100 or more kilometers beyond Earth's surface), but also interfacing terrestrial elements, such as launch sites, satellite communication (SATCOM) terminals, jammers, and anti-satellite weapons. The increasing complexity of space (increasing numbers of objects in space and their actions as well as the growing number and types of terrestrial interfaces and innovative uses of space for good and bad) is confounding the challenge of achieving useful levels of space domain awareness. It is nevertheless imperative to develop and evolve fundamental strategic principles for achieving the best level of SDA within the limits of available technology and resources.

This chapter offers structured and disciplined approaches for developing SDA strategies and tactics. The primary motivation behind all requirements for SDA is the need to be sensitive to conditions and changes in the space domain that require responses. "Require" means that to be unaware of or to ignore such changes may lead to either inimical circumstances for one's space equities or capabilities, or missed opportunities for gain (economic or otherwise). Ultimately, we must match the applied SDA strategies with the gravity of the options for courses of action or consequences of inaction.

There are many reasons for achieving space domain awareness across the defense and security, civil, and commercial sectors. These sectors have many requirements in common but are often concerned with different aspects of the space domain. The primary common application of SDA is detecting changes in the domain that may have important consequences. These include loss or degradation of space assets or impacts to terrestrial infrastructures from space environmental conditions. Another common and important application of SDA is providing a basis for diagnosis and subsequent prognosis of observed events and actions in the domain, insights that are valuable to decision making. Space domain awareness is also an important contributor to mission planning, assessment of courses of action, and diplomacy.

Much of what SDA provides is a basis for decision making. And this role mandates a disciplined approach to achieving SDA with sufficient confidence to allow appropriate courses of action or other decisions. The space operator, after having gathered, organized, and processed data from a myriad of sources, must assess the prevailing situation and understand the underlying context and behaviors. This enables the operator to make decisions and recommendations and to manage the results. Many of the options for mitigating courses of action, particularly concerning the protection of space assets and related space security measures, have significant implications and potential risks.

The remainder of this chapter is structured as follows: First, we introduce a process framework for SDA that captures the fundamental functional elements contributing to comprehensive space domain awareness (regardless of the specific application). We then discuss specific strategies for SDA for the commercial, civil, and defense / national security sectors.

2.1 SDA Framework Model

Comprehensive awareness of the space domain requires the managed confluence of three concepts: (1) observations, data, and information concerning space, (2) synthesis and analysis of that information, and (3) transformation of the first two into actionable inputs for decision-making or command and control processes. We then construct a process framework from these concepts that the SDA practitioner can use to provide the necessary intellectual and practical discipline across the execution of all three; see Figure 2-2. At its most basic level this framework defines common functions regardless of the end-user application or requirement for SDA. As a generalized model, it is intended to be tailored to the specific requirements within the defense and security, civil, and commercial space sectors.

Domain awareness continuously takes into account the above three basic concepts. Simply acquiring and amassing large amounts of data and information on the space domain does not itself constitute awareness. This takes continual analysis and comparison of these data with baselines, thresholds, and expectations relevant to the user's requirements. Figure 2-1 shows the relationships among these three concepts as they contribute individually and as a whole to space domain awareness.

Each of the three framework concepts entails strategies and tactics that must be aligned with the requirements for specific SDA applications, as we will explore further. There are

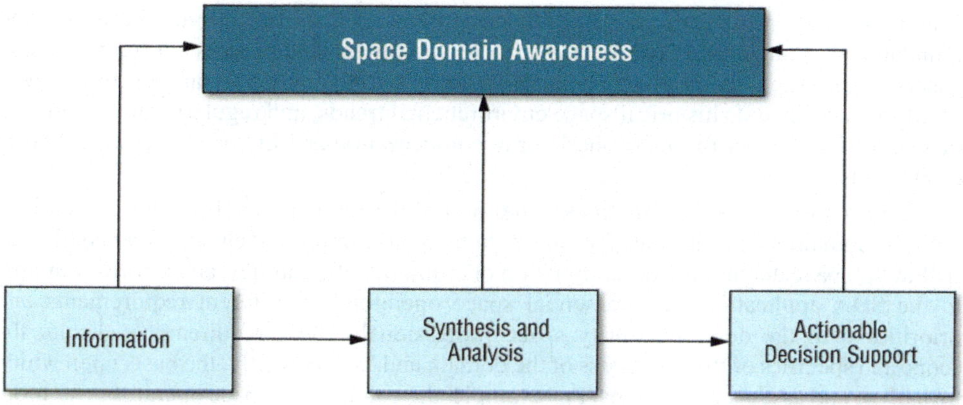

FIG. 2-1. Enabling Concepts for Space Domain Awareness. It's important to tailor these concepts to the user's needs.

no hard delineations among these concept regimes; they contribute to a continuum of increasing awareness leading to support of decisions appropriate to the circumstances.

The process of SDA starts with acquiring and accumulating ***information*** on the domain as the foundation for awareness. When combined, that information comprises past and present characterizations of the space domain. These, along with the prevailing context and standards, help us assess the present status of the domain and become aware of significant changes therein. Examples of such information are the current status of the counter-space forces of our adversaries or detecting a satellite maneuver that may pose a collision risk. To enable comprehensive SDA, the information pool should include:

- Observations and measurements of the space domain
- The prevailing context or big-picture backdrop causing or affecting events in the domain
- Historical baselines of behaviors and conditions within the domain
- Foundational knowledge of the physics, systems, players, etc. relevant to the domain
- Standards and thresholds that articulate requirements and concerns regarding the domain

Examples of current observations collected from various sources include satellite orbital tracks and maneuvers, space environmental conditions, the status of space launch vehicles, health and status of on-orbit satellites, and RF jamming signals. These data provide one of the two here-and-now components of SDA. The other is the prevailing contexts of current activities and observations. They include geopolitical, environmental, economic, and engineering considerations. Combined, these data enable us to be aware of the ***current state of the domain***.

In parallel, the information pool must also contain information on historical events, conditions, and capabilities of space-related systems. These data provide baselines and foundational knowledge against which we evaluate and assess current domain observations—the ***comparative foundational basis*** for the domain. Examples of this type of information include historical space environmental trends, and regularity and frequency of satellite maneuvers for atmospheric drag compensation at LEO or ITU-assigned GEO orbital slots.

User requirements are also an essential part of the foundational basis for comparison with the domain's current state. The ability to recognize important changes and conditions within the space domain is a direct function of requirements, equities, and priorities unique to the SDA application. A commercial space operator has different requirements and priorities than the defense/security space professional. These requirements dictate the scope and specifics of the awareness of the domain and, consequently, the basis upon which to analyze and assess observations. For example, the commercial space operator is far more likely to be concerned about the potential for collision with another space vehicle than about the next launch of a foreign military communication satellite.

Following the acquisition of SDA-relevant information is ***synthesis and analysis*** of that information to yield answers to such questions as who, what, where, when, how, why, and what's next? Diagnosis and prognosis of prevailing space domain situations or conditions are the fundamental analytical functions of the SDA process framework. This step is typically initiated through the detection of significant changes in the domain, followed by a comprehensive understanding of the potential root causes of those changes and the perspective of what will happen next.

The detection and subsequent understanding of changes in the space domain largely rest on a priori criteria for significant deviations that may warrant responses. The first step is detecting changes in the domain that exceed nominal natural variations in key parameters. Exceeding pre-established thresholds of "normal" may trigger diagnostic processes. The objective of these processes is to ascertain with the highest degree of confidence "what just happened" and the root causes involved. The biggest concern stemming from the changes in the space domain is whether mitigating action is necessary to minimize further risks or costs. Diagnostic processes provide an awareness of the "what, where, and why" of the situation. This typically leads to a prognosis on what may happen next, including the when and where. Although prognoses may be integrated with preceding diagnoses, they are usually the analytical basis for determining the nature of any needed response.

The third SDA concept integrates current data and observations with foundational and historical baselines combined with diagnostic and prognostic perspectives. These support planning and carrying out specific responses. Such integration constitutes ***actionable decision support***. An often overlooked necessity of SDA processes is to convey the results of diagnostic and prognostic analysis and supporting evidence to those responsible for making decisions. This communication must also include the veracity of the underlying observations, the confidence in the analytical assessments, and any plausible alternative explanations and prognoses. Being aware inherently includes an appreciation and sense of the uncertainty associated with extrapolating a relatively small number of observations to an accurate characterization of the domain.

The individual and combined output of these three concepts is an SDA picture. This picture should reflect the implications of any courses of action that would perturb the space domain, i.e., actions that would result in further changes in the space domain. Figure 2-2 shows the relationships among the informational inputs and the analytical processes that contribute to an SDA picture.

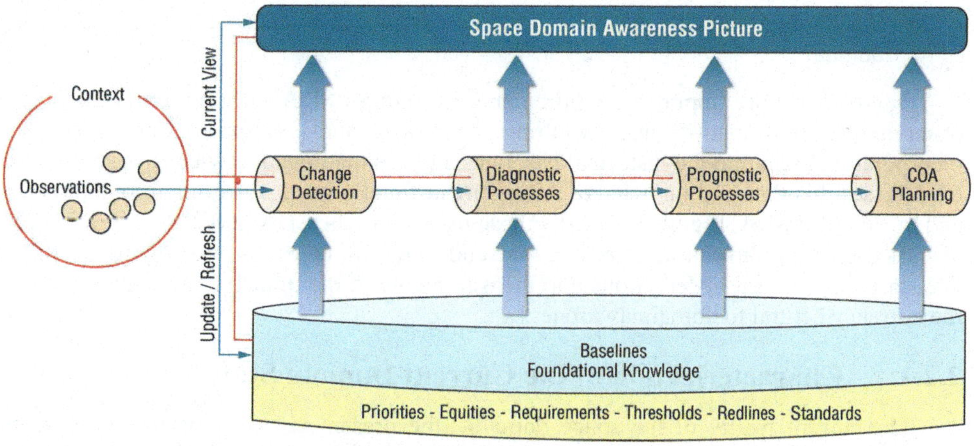

FIG. 2-2. Framework for Achieving Space Domain Awareness. Although depicted individually, the processes must act as a whole for effective SDA. (COA = course of action.)

We now examine in more detail the elements of these three SDA enabling concepts and how they should seamlessly interact to yield a continuous space domain awareness picture.

2.2 SDA Information Strategy

We now address the fundamental basis for awareness, namely the observations, measurements, and other current information about the space domain, along with how to recognize relevant conditions. Relevance has to do with the requirements, equities, and priorities specific to each SDA practitioner or user. Thus, a commercial space operator has different requirements and priorities than the defense space professional. These requirements dictate the scope and specifics of the awareness of the space domain and, consequently, the basis upon which to analyze and interpret observations collected on the domain. SDA information falls principally into categories of information related to (1) current state of the domain and (2) the foundational basis for awareness. The first represents current conditions while the second provides the means to assess these conditions and forecast potential outcomes.

Current Domain State

- Observations and measurements of the space domain

- The prevailing context or big-picture backdrop affecting events in the domain.

Foundational Basis for Awareness

- Historical baselines of behaviors and conditions within the domain
- Foundational knowledge of the physics, systems, players, etc., relevant to the domain
- Standards and thresholds, which articulate our requirements and concerns about the domain

One of the more important requirements for current SDA information is taking the observational information and resulting analytical and assessment outcomes and incorporating them into the historical baselines and foundational knowledge base. Chapter 6 addresses this data and knowledge management function and it is critical to continuous and meaningful SDA. Figure 2-3 shows these inputs into the SDA process framework. We discuss each in turn to characterize its scope and some strategies for maximizing its utility. We start with current observations and measurements of the domain, since these provide the currency central to domain awareness.

2.2.1 Characterization of the Current Domain State

The current reality of the space domain—the present status of orbital objects, the prevailing environmental conditions, the status of counter-space forces, the status of space launch activities for putting new satellites in orbit, etc.—is the current domain state. This state changes continuously, sometimes predictably but often in seemingly random, unforeseen ways. To achieve SDA, we must continuously sample, observe, and measure various portions of the domain in detail. We must also step back and understand the prevailing context in which we collect these observations and measurements.

Observations and Measurements

A key of domain awareness writ large, whether it is the maritime, air, land, cyber, or space domains, is collecting the domains' observables. Observables are data and information concerning events, activities, situations, locations, status, etc., that are sensed (observed) in real or near real-time in an operationally relevant timeframe. Observables are essential to operational SDA functions and products. Events, conditions, and behaviors that are not sensed, detected, or measured cannot, by definition, be observables. This is not to say that the absence of some observables is not valuable information. Observables, if recorded and properly assimilated, can (and should) ultimately become part of historical baselines.

The following are a few examples of observables relevant to SDA that highlight some of the categories applicable to the SDA challenge:

Related to on-orbit space objects

- Physical parameters of their orbital motion
- Changes in an object's radar cross-section
- Health and status of a satellite's sub-systems

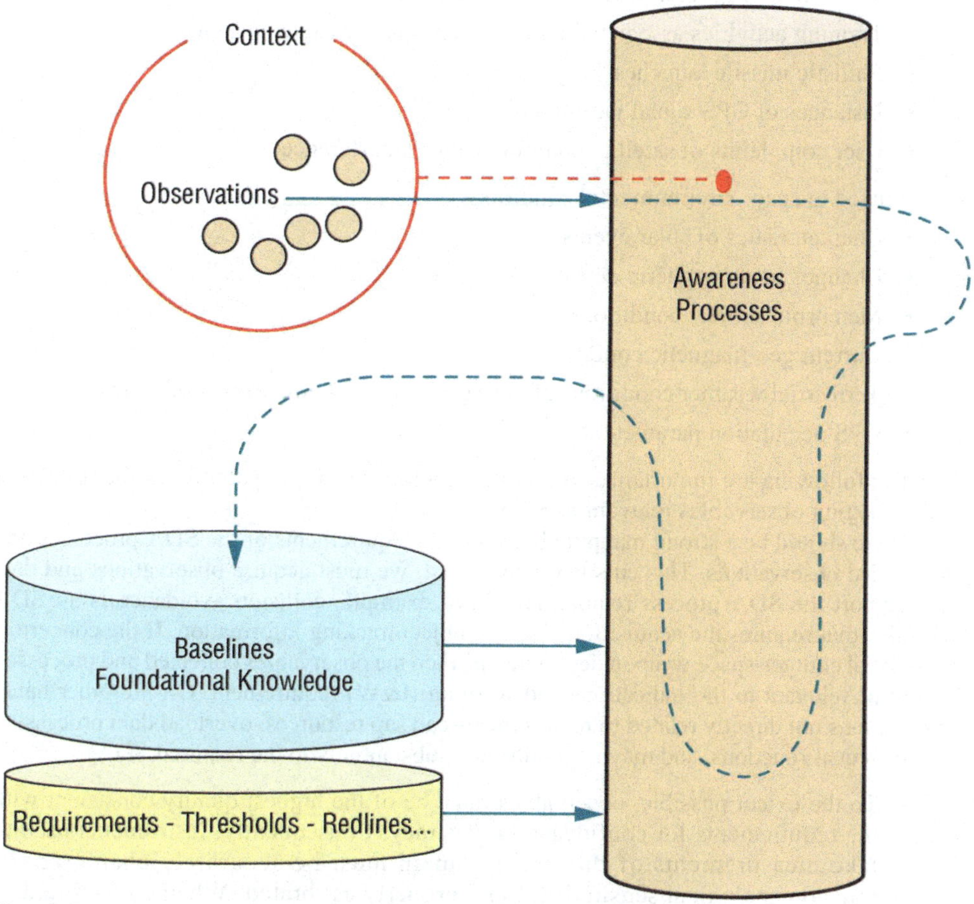

FIG. 2-3. **Observations and Measurements.** We incorporate current observations and measurements with what we already know, to keep our SDA up to date.

- Radio frequencies in use by a specific satellite
- A satellite's orbital maneuvers
- Orbital parameters of objects engaged in rendezvous and proximity operations
- Imagery of a space object from a terrestrial telescope

Related to terrestrial activities associated with the space domain

- Launch manifest at a space launch facility
- Launch timeline and orbital insertion plans for a satellite launch
- Status of ground-based satellite control centers
- Status of ground-based space surveillance sites

- Movement of ground-based counter-space weapon systems
- Training activities associated with counter-space weapon systems
- Ballistic missile launches
- Instances of GPS signal jamming
- User complaints of satellite communication interference

Related to space environmental conditions

- Characteristics of solar events
- Changes in atmospheric density
- Meteoroid shower conditions
- Current geo-magnetic conditions
- Terrestrial weather conditions affecting RF atmospheric transmission losses
- GPS occultation parameters

The following are important considerations when developing strategies for collecting and managing observables relevant to SDA:

There should be a strong mapping between the requirements of the SDA processes and the needed observations. This cuts two ways. First, we must acquire observations and data that support the SDA process requirements. For example, collision avoidance is an SDA objective that requires the acquisition of space object tracking information. If the concern is adversarial counter-space weapon deployments, then the observables collected and processed should be relevant to the indications and warning (I&W) requirement. On the other hand, observations not directly related to requirements can sap resources, overload data processing and analytical functions, and may not contribute substantially to the required SDA.

- To the extent possible, observables must be of the highest quality consistent with the requirements for confidence or accuracy. Consequently, instruments used to take measurements of the space domain must be accurately characterized (phenomenological sensitivities) and properly calibrated. When collecting data from multiple sources, we should take into consideration differences in the sensitivity and accuracy of the instruments used. This may often require the acquisition and communication of not only the primary SDA observations but also meta-data like the technical specifications of the collection instrumentation or processes as well as the conditions and context surrounding the observations.
- Observations must be made and recorded at operationally appropriate intervals. For example, if a parameter or activity in the domain changes hourly, then the sampling of that parameter should ideally be less than an hour to maximize its utility. The ops tempo of future activities is anticipated to be extremely high. This will mandate an ever higher sampling frequency for some operational domain parameters.
- It is often critical that observations made within the space domain be very accurately time-tagged. This permits accurate fusion and correlation with other observations and collection activities. Satellites traverse Earth's surface at velocities up to 7.5 km/s for those in LEO orbits. An error of a fraction of a second

can seriously impair accurate geolocation of ground-based events of SDA significance, such as sources of RF interference.

- The observational data and information amassed to underwrite SDA must be discoverable. Discoverable means that the information can be found with minimal effort in a timely manner (operationally relevant) via prevailing information search methods and technology. For example, information that is not keyworded or indexed in a database is not particularly "discoverable." Observations that are made but are not discoverable by the SDA practitioner in an operationally relevant time period have little utility. This is particularly important for observations that are not mainstream or that occur under high ops temp conditions. Data and information collected but not discoverable do not contribute to awareness; at best, they remain potential background information.
- An open mind about what constitutes a relevant observation for any particular SDA application is important. The domain is complex; sensors and other collection systems making observations only touch upon a small fraction of the observable activities, conditions, and events within the domain. We should look for potentially undiscovered observations that have been made but are not necessarily within the traditional SDA information stream. This may be particularly important for the resolution of ambiguities and enigmas, as we discuss later.
- Today's observations can and should become part of the historical record and provide the basis for baselines. Disciplined management of processes associated with the ultimate disposition of collected observables after their initial use goes a long way to establishing the required baselines and foundational knowledge needed for future SDA applications.

Context

Understanding the context in which events occur in the space domain is essential to the timely and accurate interpretation of direct or indirect observations of those events. This is critical to appropriate decisions regarding potential courses of action. Taken out of context, many observations may be overlooked, ignored, or misinterpreted. Context is the sum of all of the conditions and circumstances that cause, contribute to, or lead to a specific situation. Context is inherently dynamic and is constantly evolving with local and global activities, events, and other changes in the space and other domains. The context in which events and activities occur in the domain incorporates the roles that the physical environment, prevailing geopolitics, civil and commercial objectives, and defense and security activities contribute to observables.

Context helps provide a rationale for observations that might otherwise be ambiguous or enigmatic. Context also helps focus the alignment of observations with present circumstances to increase their relevance. The need for a clear understanding of context puts an additional burden on the SDA assessment processes by requiring information that may initially appear not to be related to the space domain nor easily obtained by traditional SDA observational systems.

To better understand the role that context plays in achieving space domain awareness, we walk through a discussion of five major overlapping contextual regimes that often influence space activities, as shown in Figure 2-4.

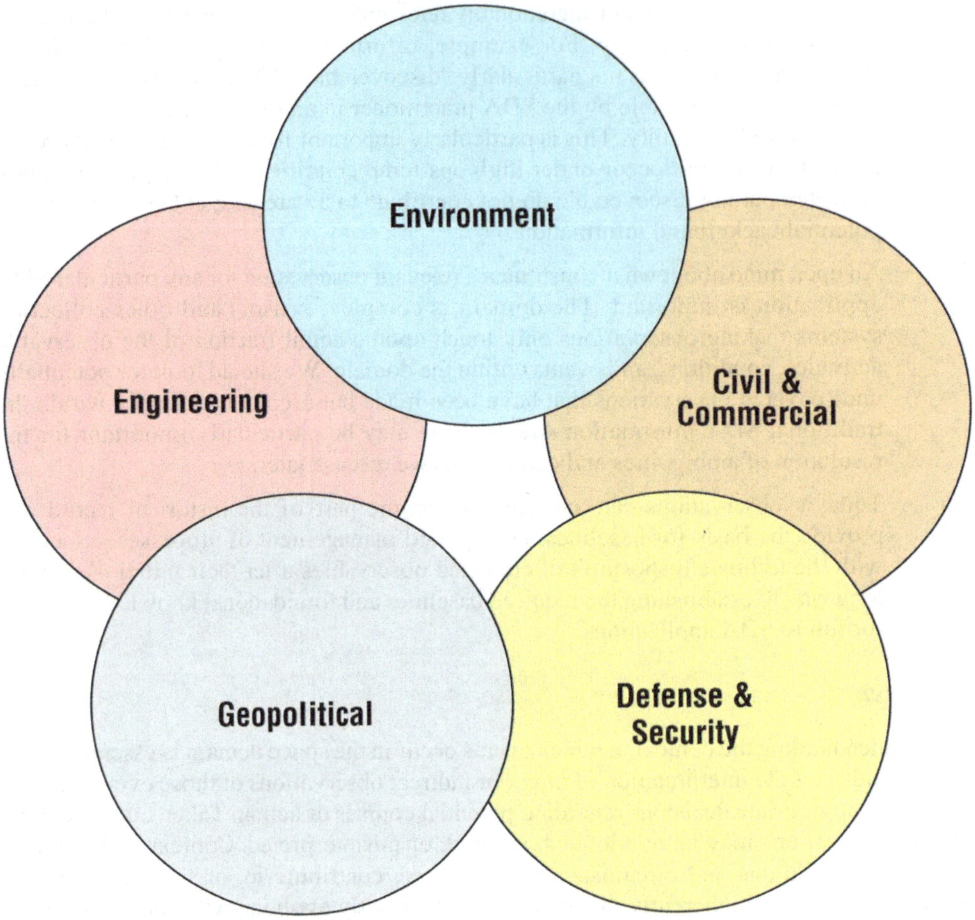

FIG. 2-4. **Elements of Context for SDA.** Understanding these contextual elements for all parties in the space domain is essential for comprehensive SDA.

Space Environment Context

The space environment contributes significantly to the behavior and operation of space assets as well as some critical elements of terrestrial infrastructures. From a contextual perspective, we are concerned with the influence that the space environment has—directly or indirectly—on behaviors, activities, and events we observe within the domain. The space environment context is generally about energy and objects along with their source or cause, some natural and some of human origin. Examples include:

Energy-related (Chap. 3)

- Radiation from space and the sun
- RF emissions from space and terrestrial sources
- Geomagnetic field conditions

Object-related (Chap. 4)

- Orbital objects, such as spacecraft, rocket bodies, debris
- Meteoroids

Engineering Context

Space systems are complex, engineered to operate in the hostile environment of space while being hundreds or thousands of kilometers from their points of control. The high cost of building and launching spacecraft imposes significant requirements for various levels of autonomous operations and other features to ensure mission success under a wide range of conditions. The operational behaviors and associated limitations of spacecraft depend directly on their engineering designs.

Observed anomalous behavior in satellites may be the result of several root causes. Understanding the engineering context of a spacecraft is essential in determining whether the cause was how the spacecraft was engineered or operated, or whether it was external factors.

For the SDA practitioner, access to relevant engineering context can be through comprehensive baselines on satellite engineering designs and past performance metrics. If that is not possible, simply being aware of, for example, the age of a spacecraft and its design lifetime, may help to diagnose some observed behaviors.

Geopolitical Context

Geopolitical context is often associated with big-picture status designations of peacetime, periods of tension, heightened tensions, and ultimately combat. Understanding the geopolitical context helps us judge whether an observed behavior is likely to be hostile or not.

For example, counter-space weapons, if used, are likely to incur high costs both in the actual cost of the deployment and use and the consequential and opportunity costs associated with harming another's satellite. If the applicable geopolitical context is peaceful, then observed behaviors are less likely to be hostile. On the other hand, if tensions are high, the geopolitical context becomes an extremely important optic through which to evaluate observations.

Understanding how the geopolitical context affects adversarial decision making is an important input to SDA diagnostic processes. It may drive adversary decisions regarding the nature of the mitigation used.

Civil and Commercial Context

The context of civil and commercial space operations is significantly different from that of the defense and security sector. At the risk of over generalization, civil and commercial space endeavors are often driven by first-order returns on investment. To underwrite civil space programs means diverting resources from other investments in civil infrastructure and services; minimizing opportunity costs is the driver. Maximizing the value of a service or product for their constituents (such as earth observation, meteorological information, etc.) is their mission objective. For the commercial sector, return on investment is the driving metric, while profits from their services are their business objective.

Understanding the financial and economic constraints and objectives for both of these sectors is an important contextual input to SDA analytical processes. For example, decisions regarding collision avoidance in the commercial sector are different from the security sector. Maneuvering a commercial satellite to avoid collision with another space object incurs an opportunity cost measured in lost revenues during the maneuver and subsequent stabilization period. On the other hand, understanding that a commercial operator may not take part in defensive measures during a conflict to avoid becoming targeted is an important contextual factor.

Civil space programs often have tight budgets and may choose operational behaviors that limit their risk exposure, since replacement satellites may be many years in the making. Civil programs also tend to fewer requirements for resiliency or sustained operations under stressing conditions.

SDA practitioners in the civil or commercial sector should already be well aware of their operational priorities regarding mission success and risk tolerance or avoidance. On the other hand, SDA practitioners involved with the security sector should ensure that they fully understand the context in which civil and commercial space systems operate, especially for those systems that support the military/security sector.

Defense and Security Context

Related to the geopolitical context is the defense and security context. Whereas the geopolitical context provides insights into motivations and intentions, the defense and security context addresses the doctrinal and tactical status of blue, red, and gray actors, including their associated capacity and readiness to engage in or support defense or security activities.

- **Blue**—the actor conducting the SDA processes in the context of natural and manmade threats and risks. (This chapter addresses the perspectives of the blue actor.)
- **Red**—adversarial actors potentially conducting actions inimical to the blue actor.
- **Gray**—neutral parties with space activities and equities that may or may not support in a neutral manner blue and red activities in space.

This context is probably the most important for aiding in assessments of indications and warnings of potentially hostile actions.

There are two areas of defense and security contextual concerns. The first relates to potential hostile threats against space assets and services; this is ***counter-space***. The second focuses on the use of space capabilities and services in support of overall military activities; this is ***force enhancement***. Understanding these contextual elements for all parties in the space domain, whether they be red, blue, or gray, is essential for comprehensive SDA.

2.2.2 Foundational Basis for Awareness

To arrive at comprehensive SDA, we must join or compare the understanding of the current domain state with foundational information that enables us to know the significance of changes or events within the domain. This involves historical baselines that capture norms of behavior, thresholds of concern for changes, and foundational information that all help in diagnostic and prognostic processes.

Baselines

We begin our discussion of baselines with two rather obvious contributions to domain awareness. The first is the nature and character of steady-state conditions. In the context of the space domain, this might mean the number of satellites presently in geosynchronous orbit or the current orbital state vectors of a specific space object. Of course, the concept of steady state is a function of how long things truly remain unchanged, as well as the user's definition of the term. The second concept is change in heretofore steady-state conditions, such as the change in orbital parameters or the launch of a new satellite that increases the space object population. To be able to become "aware" of the latter (i.e., changes), we must be aware of the former. This brings us to the critical need for baselines to enable capturing "normal" or steady-state conditions as well as the basis for detecting changes in the domain.

Baselines are intrinsically historical—they are based on observations and assessments of activities, events, behaviors, and so on that have occurred in the past—sometimes a very long time ago, and sometimes only a few minutes ago. Often baselines are called "historical baselines," implying almost ancient history. But in fact, baselines associated with the space domain may need to be reset or recalibrated on much shorter timescales to accurately reflect new "normal" conditions.

Baselines are sets of data and information that allow us to characterize "normal." Most of the elements of the space domain, as with all operational domains, are constantly evolving. For example, terrestrial and space weather is always changing, spacecraft performance degrades with age, ground station antennas routinely change pointing angles, and some spacecraft adjust their orbits to account for atmospheric drag. This often makes establishing and maintaining meaningful baselines a challenge.

Accurate and useful baselines require disciplined recording of all relevant observables and properly determining the normal deviations due to inherent temporal, operational, and contextual variations. Temporal variations may result from daily or seasonal factors or user behavioral patterns. For example, some baselines may need to include the variations within routine movements in military exercises. Quantitative observables may yield baselines having parameters that characterize normal upper and lower limits and rates of change. One example is orbital parameters (e.g., apogee and perigee) of a low-Earth-orbit satellite.

The fundamental requirement of SDA baselines is that they be current and accurately reflect as closely as possible what "normal" is. Normal is the status quo condition that does not indicate any significant change. To that end, baselines must always be assimilating new observables to refine the normal picture. As an example, for quantitative observables that change slowly relative to events of concern, a running average might be suitable and appropriate. Other baselines may need to be updated with just the most recent observations. However, without frequent updating, the usefulness of baselines may degrade rapidly.

Baselines are essential for high-quality space domain awareness, but they are not always easy or inexpensive to establish or maintain. So what constitutes a relevant SDA baseline? How should they be created and maintained? We explore the answers to these questions below.

The immensity of the space domain precludes anything approaching a comprehensive set of baselines that would be of value to all SDA users. Baselines should be appropriate for their intended use. There should be a clear mapping between a specific requirement for a space domain awareness application and the underpinning baselines. If the SDA requirement by a commercial space operator is to be aware of all spacecraft with orbits that may come close to their assets, then an orbital baseline of satellites posing potential collision risk should be accessible. Conversely, changes in the health and status of foreign military satellites are probably not of much interest to a commercial satellite operator.

Baselines not only allow us to detect potentially significant changes, but they are also essential in diagnoses and operational planning. The following are some concepts and practices that may improve the viability, usefulness, and efficiency of baselines used for SDA applications:

- When deciding what baselines are important, consider potential significant scenarios, even if they have never happened before. Understand potentially significant cause and effect relationships. This includes information on conditions whose change would signal an important future condition.
- Consider multiple sources in compiling baselines. Multiple sources help minimize the risk of ambiguities that occur when insufficient data are available.
- Although enigmatic situations are inherently difficult to predict, baselining operational information that is readily discoverable may help in diagnoses and prognoses.
- The information in baselines, like the other information input into the SDA process, must be discoverable. Baselines that are difficult to access, query, or understand or that contain calibration or measurement errors, will not be particularly useful.
- The SDA practitioner should have experience with and understand the critical attributes of the data and information being baselined to help in its discoverability. (It doesn't help if you can't find the information for comparison quickly.)
- Make efficient the processes by which current observations and measurements become incorporated into the discoverable baseline.
- Understand how to properly age the information in a baseline consistent with the collection frequency and the inherent rate of change in the baseline information. Much historical information contained in baselines is no longer relevant (and could lead to serious oversights or diagnostic errors).

Foundational Knowledge

Foundational knowledge is a body of information having traits similar to contextual background and baselines. Like baselines, foundational knowledge is inherently historical since it represents knowledge accumulated over time. And, like context, foundational knowledge is multi-disciplinary and pulls insights from all that is known to apply to SDA. The SDA practitioner relies on foundational knowledge as the basic intellectual understanding of everything in the space domain.

Four areas of foundational knowledge are especially useful in SDA analytical processes, with many more that usefully contribute.

- Foundational intelligence
- Foundational physics
- Foundational space operations
- Foundational military art and science

Foundational intelligence encompasses the knowledge primarily of competitor or adversary motivations, intentions, capabilities, and activities associated with the space and other relevant domains. We use it to understand and interpret their observed actions and expected responses to events in the domain. The term "foundational intelligence" here has the broadest possible meaning, yet we are not referring to current intelligence. That is part of the current pool of intelligence observations of the domain. Foundational intelligence has much longer time constants compared to the tactical environment. It has the benefit of longer periods for acquiring information and subsequent analysis leading to a base of knowledge whose accuracy is largely unaffected by immediate circumstances. The intelligence community provides foundational intelligence by acquiring information from multiple sources and applying the expertise of subject matter experts to assimilate and transform that information into knowledge. Still, this knowledge has limits in depth, confidence, and even relevance.

Another area of foundational knowledge critical to SDA analytical processes is the body of foundational physics. These are the laws of physics that dictate the motion of orbital bodies around Earth as well as other important behaviors that contribute their own contextual backdrop against which all activities in space are evaluated. The most important of these are Kepler's laws of orbital mechanics, which provide the basis for much of the behavior of orbiting objects. Other relevant laws include those related to thermodynamics, the electromagnetic spectrum, generation and propagation of energy in space, and nuclear energy. Without understanding the limitations that these physical laws levy on space activities, we can't make accurate and appropriate interpretations of space domain observations.

To be maximally "aware," space domain awareness specialists must fully understand the physics behind the activities and situations in the space domain that are most significant to their equities and responsibilities. For example, understanding and appreciating the physical context is critical to the accurate and timely assessment of some observed spacecraft behaviors. A maneuver to compensate for atmospheric drag experienced by a LEO satellite is a routine operation to sustain its mission; it probably does not indicate an orbital change for inimical purposes. The observer must learn how to distinguish between the innocent and provocative, since they may not be drastically different in character.

The third body of foundational knowledge for SDA is the fundamentals of space operations, including experience-based tradecraft. Most space operations are conducted pretty much the same way throughout the domain. Space launch vehicles are prepared, integrated with their payloads, fueled, checked out, etc., in a very similar manner no matter where they are launched from. This is because space systems are expensive and time-consuming to build. Minimizing risk is central to the common practices. The same holds for operating spacecraft on orbit. Satellites are tracked and monitored by their owners. When events or conditions arise that put the spacecraft at increased risk due to internal engineering or external environmental conditions, spacecraft operators respond in much the same way.

Understanding how space operations are commonly conducted is as important to SDA as understanding the differences. Different countries conduct their space operations differently due to differences in availability of space tracking or command and control resources, differences in the spacecraft themselves, differences in risk tolerance, and other factors. What is an anomalous situation for one country's spacecraft operation may be perfectly normal for another's.

The fourth area of foundational knowledge important to the security sector is that of the military art and science, including operational art, of the space domain. Although still in its developmental phase, this body of knowledge is critical to accurately interpret or diagnose observed changes in the space domain relative to defense and security activities. What makes sense from an operational art perspective is critical to correctly interpreting indication and warning signatures (which we discuss later) associated with potential adversarial actions. There is a well-developed body of knowledge concerning preparing for and conducting war. Understanding the logical sequences of essential actions as they apply to the space domain is important for both the diagnostic and prognostic processes we address in Section 2.3.

Standards and Thresholds

We include a short mention of standards and thresholds as an essential component of the informational inputs to the SDA process. Standards and thresholds reflect the requirements of space domain awareness to provide relevant and actionable insights. Whereas baselines are used to detect changes in the domain (significant or insignificant), standards and thresholds represent the a priori decisions by the SDA practitioner on what constitutes a significant change or event that may warrant further attention or action. Standards and thresholds are entirely the responsibility of the end user to define and establish; they do not necessarily have universal merit or value.

Some example of standards and thresholds applicable to SDA include:

- Allowable separation between two space objects
- Thresholds in orbital parameters for reporting maneuvers
- Geolocation accuracy for RF interference
- Confidence requirements

- Pending launch and launch warning timelines
- An event resulting in space debris
- Changes in spacecraft health and status
- GEO COMSAT station-keeping requirements
- RF spectrum usage
- Solar event magnitudes
- Spacecraft anomalies
- Behaviors and activities associated with weapons testing or deployment

To determine if standards have been met or thresholds exceeded, such standards and thresholds must map to the reality of what we can reliably and accurately observe and for which adequate baselines exist. To a great extent, the SDA equity holder is responsible for coordinating between the establishment of standards and thresholds and the availability of relevant and accessible SDA observations and baselines.

Another point worth noting is that understanding other satellite operators' standards and thresholds may help diagnose observation of their behaviors as well as prognosis of their anticipated response to changing conditions or circumstances within the space domain.

2.3 SDA Synthesis and Analysis

The mere collection and compilation of observations, data, and other information on the space domain yields very little in the way of awareness. That information, along with our knowledge of the historical dimensions of the domain and its activities, must be synthesized into useful awareness for which we can justify appropriate courses of action. In this section we explore that portion of the SDA process framework where we synthesize and analyze our information on the domain and come to a level of awareness we did not previously possess.

We synthesize awareness of the space domain from the information available, i.e., the current observations of the domain along with relevant baselines and foundational knowledge and the prevailing global or environmental context. The principal contributor to awareness is simply the ability to know the day-to-day state and prevailing conditions within the domain. Being aware of the state of the domain—including those physical quantities and elements that may change very slowly relative to immediate operational needs—contributes a critical backdrop to all space domain awareness applications.

We also focus our synthesis through the lens of our concerns, equities, thresholds, and other requirements that drive our need for SDA in operationally relevant time contexts. Thus, in addition to being generally aware of the state of the domain, we achieve further and deeper awareness through three fundamental processes:

- Change detection
- Diagnosis
- Prognosis

These three processes are not mutually exclusive, and they are rarely sufficient alone to yield the desired levels of SDA. In some situations, they function synergistically and logically follow one another, and in other cases, there may be highly iterative interactions among them.

2.3.1 Change Detection

The detection of changes in the domain is conceptually straightforward. We compare incoming collected or sensed observables against points of reference, standards, and expectations. If we detect significant differences between them, then potentially important changes in the domain may have occurred that warrant further attention or action. Or the changes could be completely insignificant or a consequence of routine variations or noise in the environment. Although change detection is a simple concept, its execution is generally not so easy or straightforward.

Three things are necessary for simple change detection. The first is observations or data points reflecting the current conditions within the domain. Next comes some basis or reference for detecting a difference from prior conditions. This is some previous reference condition, standard, status, or level within the domain. The third is a threshold factor or measure to define or characterize the significance of any detected change. This measure may be qualitative or quantitative, depending on the nature of the observable and its reference. The objective of SDA change detection is to trigger follow-on, subsequent investigation when one or more thresholds have been exceeded in one or more observables. Setting a priori thresholds too low will result in false alarms and setting them too high may result in missing important changes. The quality and accuracy of the observables directly contribute to the accuracy and sensitivity of the change detection process. However, even observables that are not sufficiently accurate for further diagnostic purposes may be enough to trigger over-threshold warnings. Multiple or redundant observations from different or independent sources increase the confidence that thresholds have been exceeded.

For some space domain awareness requirements, detection of a change in a single parameter may be sufficient; an example is detection of orbital maneuvers. For those cases, a simple set of observables, historical baseline values, and thresholds may be all we require. However, there are SDA requirements that involve detecting changes in and throughout much more complex arrays of inputs. SDA practitioners must adapt their tradecraft to ensure that they can confidently detect significant changes in those situations.

Complex challenges can arise in situations requiring multiple observations of different types of parameters or activities simultaneously to correctly detect important changes. Changes in any one or a few parameters might not be particularly important, but changes in a broad array of parameters may portend a more significant event. The key to complex change detection scenarios is a deep understanding of the underlying relationships among the parameters involved in the subject or target behaviors. This means knowing what to look for and how much joint change among parameters is significant. The more one knows about what can change and under what situations, as well as the ability to unambiguously make sense of any observed changes, the better will be the resulting awareness. This is where foundational knowledge is useful.

The following are some considerations for executing SDA change detection processes:

- The baselines used for change detection must reflect current normal, pre-change conditions. This will help to reduce the effect of natural variations in observables. Continual incorporation of new observations into the baselines is essential.
- We should try to minimize Type I errors—also known as false alarms. Up-to-date baselines and thresholds are essential. The diagnostic challenges discussion in Section 2.3.2.
- The accuracy and confidence in the observations provided as evidence of change must be known and clearly articulated. This also includes the means and circumstances under which the observations took place.
- We should try to minimize Type II errors—missing or not recognizing a real change of significance. Not only are up-to-date baselines and thresholds crucial, but foundational knowledge and adequate contextual inputs are also important. Appropriate SDA end-user engagement with setting threshold requirements is essential.
- Communicating the basis for a detected change to subsequent diagnostic and prognostic functions is important. All observables, thresholds, and other inputs used to validate a significant change should be reported.
- Understand when available observations are not sufficient for a confident detection of change. Know when to request more information to resolve ambiguities and to minimize Type I and Type II errors.

Even the timely and confident detection of significant change within the space domain is not necessarily useful to the command and control processes executing suitable courses of action. The simple detection of change may be sufficient in some cases, but comprehensive SDA usually requires that command and control elements comprehend the implications of changes. Detected changes must be mapped to the potential consequences that require action. This requires an analysis or diagnosis of the detected change to understand the causes involved and their significance. The analysis or diagnosis provides operational decision-makers with an appropriate basis on which to proceed. This is a far greater challenge than the simple detection of change because of the need to determine and implement, for example, confidence measures and short response times.

2.3.2 Diagnosis

Diagnosis is, at its simplest, the identification of the underlying causes and comprehensive characterization of an observed situation or condition. In the context of a change detected in the space domain (for example, a change in a satellite's orbital parameters) the diagnostic process attempts to answer questions such as "What just happened?," "Where did it happen?," "Why and how did it happen?," and "Who was involved or responsible?"

Foundational intelligence, baselines, contexts, and historical cases are essential to accurate diagnostic assessments. Diagnoses are often initiated by an unexpected observation or change that triggers further investigation. We use various diagnostic approaches to assess observed changes in the space domain, each with its own applicability and levels of confidence.

The SDA practitioner must understand the scope of these approaches and their appropriate application. Apart from ensuring that important changes are actually detected, few things are more important than arriving at an accurate diagnostic perspective. Incorrect assessment of the root cause of a situation may, at the very least, result in the ineffective application of mitigating measures. At the other extreme, erroneous attribution of an observed event in space to a potential adversary may result in serious escalation of conflict, with disastrous consequences. The SDA practitioner must be cognizant at all times of the challenges, limitations, and constraints on the accurate diagnosis of observed events in the space domain. This can be particularly important in time-constrained situations where a rapid diagnosis is necessary to support an immediate decision to act. Some of the more important considerations include:

- The quality and quantity limitations of relevant observables made available to the diagnostic process
- The limitations on the ability to acquire independent corroborating observations and data
- The lack of historical norms of behavior relative to space defense (inadequate baselines)
- The limitations of experience and heuristic models to many potential space events of significance
- The often short timelines imposed on the delivery of diagnostic results to decision makers

At the core of diagnostic approaches for SDA is a set of logical reasoning perspectives. Understanding how to apply these perspectives is a key responsibility of the SDA practitioner. This chapter is not a text on the philosophy and science of logic, but the application of the following logical perspectives is fundamental to good SDA diagnostics (and prognostics, as we address later):

- Deductive reasoning: Inferring a true diagnosis by applying one or more premises (truths). For example, from the assertion that Kepler's laws of motion apply to all orbital objects and the observation that a specific satellite is orbiting Earth, we can infer that the satellite's motion will accord with Kepler's laws.
- Inductive reasoning: Inferring a diagnosis based on a set of observations that are consistent with that diagnosis. For example, inferring that a satellite's change in orbit resulted from an intentional drag compensation maneuver based on the observations that (1) the satellite was increasingly subject to drag at its operational altitude, (2) it appeared to be operational after the maneuver, and (3) historically the satellite made similar drag compensating maneuvers. This diagnosis may be highly probable, but it is not certain. Other plausible (although perhaps much less likely) reasons exist for its orbital change; for example, the satellite operators intentionally put it into an orbital regime that would bring it closer to another satellite.

- Abductive reasoning: Inferring increasingly likely hypotheses that fit, or are consistent with, a set of observations refined through subsequent hypothesis testing. Acceptance or rejection of hypotheses refines the set of plausible diagnostic hypotheses. Abductive reasoning differs from inductive reasoning in that it focuses on the explanatory power of hypotheses rather than the identification of any hypothesis that is consistent but not necessarily explanatory or correct. For example, we might set forth two hypotheses, based on multiple observations that an adversary has deployed weapon-related equipment, that they are getting ready to either (1) test a counter-space weapon, or (2) use a counter-space weapon in anger. The validity of the use-in-anger hypothesis is subsequently rejected since the system was deployed to a known weapon test site. The accepted hypothesis that a test is more likely explains more of the observations, although the other is still plausible. Increasingly, artificial intelligence systems use some form of abductive reasoning to improve their diagnostic accuracy.

When we address the challenge of confidence in SDA assessments, we must understand the nature of the logical reasoning that was employed to arrive at a diagnostic outcome. Careful consideration of the constraints in each of the approaches (deductive, inductive, and abductive) is necessary. In addition, the SDA practitioner must be aware of his or her own experiential biases to ensure the best diagnostic outcomes from what are otherwise highly subjective approaches.

There are several ways to pursue a diagnosis of a given set of observables, and SDA practitioners must develop their tradecraft through personal experience to understand which work and under what circumstances. We address three diagnostic conceptual approaches here that can provide a point of departure for further refinement through real-world experience.

The first diagnostic approach is relatively simple to execute but comes with significant potential for error, especially in scenarios for which little or no experience exists. This approach uses heuristics or experience to drive toward a diagnosis of a situation that appears recognizable; in other words, "we have been here before." Tools that use pattern recognition, pre-constructed decision trees, or checklists and other similar organized means of bringing past experience to the table can be useful and are often executed quickly. For example, the rapid assessment of whether a new foreign satellite launch was successful can be determined with reasonable accuracy from a simple checklist of expected observables. Similarly, templates that reflect prior experience with changes in the patterns of life of a satellite as it ages can yield diagnostics for observed performance degradation in older satellites.

Checklists or decision trees are based on a best effort to establish a priori diagnosis of a set of indicators or observations. Although these approaches can be updated to reflect new foundational understandings, they are rather static in practice and require effort to keep them current. Checklists can be executed relatively rapidly if observables are available to compare with them. Checklists are often based on deductive reasoning, where the observed input conditions and their relationships are presumed to be known. Examples of diagnostic checklist and decision tree approaches can be found in satellite operator procedure manuals and watch officer reference materials. The objective is to provide a first-order or preliminary diagnosis rapidly under the assumption that events for which these tools apply have already been thought through. They often assume that diagnostic possibilities outside of those explicitly considered are exceptionally rare or non-existent. So checklists and decision trees provide rapid, preliminary diagnostics that help us efficiently marshal more detailed diagnostics.

There are several challenges with heuristic, experiential, and intuitive diagnostic approaches. First, the scope and breadth of presumed experience may not be sufficient to diagnose events that are not in the "playbook." Second, the application of experience and intuition in diagnostic processes can lead to unwarranted confidence in the diagnosis; alternative root causes may be overlooked or dismissed without further consideration. Unwarranted confidence in heuristics may also preclude acquiring additional observations or data that could change the substance of the diagnosis. This risk is an example of confirmation bias, which results from the practice of looking at incoming observations and using them to validate one's existing hypotheses to the exclusion of other valid hypotheses.

A second strategic approach to SDA diagnosis is based on the first principles of the underlying systems involved. This approach is most appropriate for physical or engineered systems and the physical space environment. To the extent that institutional, geopolitical, and military operational art reduces to a set of reliable first principles, this approach may yield acceptable results. However, experience suggests that the virtually infinite number of potential factors in the geopolitical environment, for example, precludes the characterization of many first principles to a degree they can be used reliably for diagnosis.

The first-principles approach involves the decomposition of the systems involved into their component elements for which highly accurate (near truth) characterizations are possible. The use of algorithms, modeling, and simulation yields a synthesis of behavior norms, which in turn can help diagnose changes or abnormal behaviors. Observations can be input into such system "models" to determine what conditions or events would be consistent with those observations. Such approaches are useful when sufficient data is available for determining root causes of, for example, satellite anomalies and failures. Complex models of constellations and their replenishment strategies may facilitate diagnosis of observed changes in a foreign constellation of NAVSATs, for example.

This approach is limited to systems that can be analytically characterized with sufficient fidelity that we can diagnose significant changes with a high degree of confidence. The more complex the system, the more complex the modeling involved, and the greater the potential for unknown interrelationships among the model elements to have a significant impact on the diagnostic results.

The last approach we address involves an automated hybrid of the first two. It also involves modeling and simulation but leverages modern data analytic methods rather than exquisite modeling of first principles. With the advent of extremely large volumes of data and related observations being collected throughout the terrestrial, maritime, cyber, air, and space domains, the potential exists to discover diagnostically relevant insights. Known euphemistically as "big data," this ever-increasing torrent of data and information provides an opportunity to contribute significantly to space domain awareness. From the diagnostic perspective, it promises opportunities to link otherwise disparate, difficult to find, and unknown factors to observed changes in the space domain for which we seek explanations.

The experiential nature of heuristics is being captured in artificial intelligence systems for a broad array of applications. Many of these systems foundational to expert diagnostic capabilities are revisiting the application of abductive reasoning to sharpen their diagnoses through extremely large data sets of baseline, foundational, and contextual information. Examining large amounts of historical data and observations under various contexts enables us to search and evaluate more diagnostic outputs. The evolution of the strategic art of SDA will surely be driven by these developments.

We have addressed three approaches to diagnosing observed events and changes in the space domain. The technical expertise and details necessary for success are specific to the nature of the changes, the players involved, the physical environmental conditions, and the geopolitical, military, civil, or commercial context. Real-world diagnosis of events within or affecting the space domain is complex and necessarily requires a combination of approaches. The demands upon the SDA practitioner not only include high confidence diagnosis, but diagnostic results in operationally relevant time frames that may be difficult to achieve. The following are some thoughts and perspectives on strategy and tactics for the diagnostic function within SDA.

- Ensure that all data relevant to the diagnosis of an event are readily discoverable and available.
- Quickly determine whether the confidence required in a diagnosis is achievable with the data available. If not, seek additional information immediately.
- Understand the systems involved in the diagnosis—bring in subject matter experts when the situation involves complex systems and interactions that are beyond the SDA practitioner's expertise.
- Ensure a comprehensive understanding of the prevailing context as it pertains to the space domain. Enlist experts as necessary—do not assume that contexts, especially geopolitical and commercial contexts, are easy to understand by those that are not intimately involved in those sectors.
- Know your enemy or source of risk—exploit foundational intelligence and other insights. Understand your competitors—be aware of their objectives and tactics in the global marketplace.
- Understand the real-world constraints and limitations on acquiring additional information and data to resolve ambiguities.
- Understand the confidence required of the diagnostic results in the context of potential courses of action.
- And lastly, be cognizant of the SDA diagnostic challenges addressed below.

Diagnostic Challenges

There are three special challenges to timely and accurate SDA diagnostics: ambiguities, enigmas, and confidence. Diagnostic processes are often hampered by insufficient data or information, leading to ambiguous results. They can also be flummoxed by enigmatic situations, i.e., combinations of observables that have never been experienced or for which obvious conclusions are elusive. Such situations require us to acquire more data and information until we achieve the needed confidence in the diagnosis. Dire consequences may result from not understanding the limitations of a diagnosis and not adequately informing decision-makers of the confidence of a diagnosis. Although these challenges are interrelated and share common elements, let's look at each in turn.

Ambiguities and enigmas reduce the confidence in SDA diagnostic results. ***Ambiguities*** arise when observations are consistent with two or more potential interpretations or hypotheses. Ambiguous situations may be obvious, or may not be recognized at all, and thus we may overlook them when we don't account for the limitations of inductive reasoning. The risk of the latter increases when there are multiple plausible hypotheses for a set of observations, but only one is ever considered. In contrast, ***enigmas*** arise when observations do not fit any explanatory model nor anything in the historical record. To resolve enigmas typically requires new understanding of heretofore unknown cause-and-effect relationships.

Enigmatic and anomalous events or observations may simply be benign artifacts of space environmental factors, engineering upsets, training, or R&D activities. On the other hand, they may indicate more significant activity of concern to the security or safety of space assets or functions. These latter situations must be detected and pulled out of the noise of the more numerous and likely causes and origins of the former. As an example, current spacecraft anomaly resolution processes tend to favor the expectation (backed by history) that anomalies have engineering or space environment root causes rather than causes of hostile or unauthorized origin. In many failure or anomaly reviews, the last satellite anomaly examined is the one associated with hostile intent or action.

Before addressing concepts to deal with ambiguities and enigmas, we must fully understand what we know and don't know and how cognizant or aware we are of the limits of our knowledge concerning the space domain. One way to look at this challenge is to consider Figure 2-5, where we depict four subsets of knowledge within the superset of all objects and events in the space domain (represented by the ellipse outlined in red). The smallest is the subset of events and objects for which we have highly accurate knowledge. Next is the subset of those events and objects for which our knowledge is imperfect due to inaccurate, biased, and incomplete observations. The third is objects and events that we know exist, but we have virtually no useful information to apply to our SDA processes. We don't know if changes have occurred related to their behaviors and activities and have no basis for conducting useful diagnostic and prognostic analyses. Lastly, the fourth subset is events, objects, and other elements of the space domain that we are not aware of. The first two subsets are known "knowns," the third known "unknowns," and the fourth unknown unknowns.

We can further explore our knowledge of the space domain in terms of "knowns" and "unknowns." In Figure 2-6 we use two dimensions to characterize our knowledge of the space domain across four regimes. The first dimension divides SDA information and insight into either that which we know or we don't know—knowns and unknowns. The second dimension addresses our awareness or cognizance of what we know or don't know. Each regime presents unique challenges and opportunities.

The majority of SDA practice resides in Regime I, the ***known knowns***. Most activities and situations in the space domain follow expected sequences of events where routine space domain awareness strategies for collecting and analyzing information are adequate to support decision making. These include collision avoidance, awareness of new foreign space

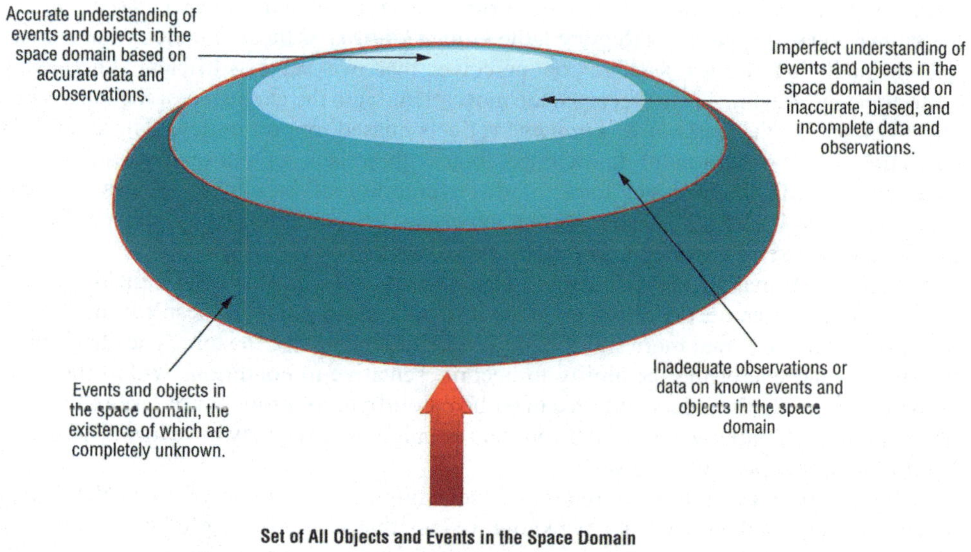

FIG. 2-5. **Subsets of Space Domain Knowledge.** It's important to realize that the bulk of space domain events and conditions lie outside our current understanding.

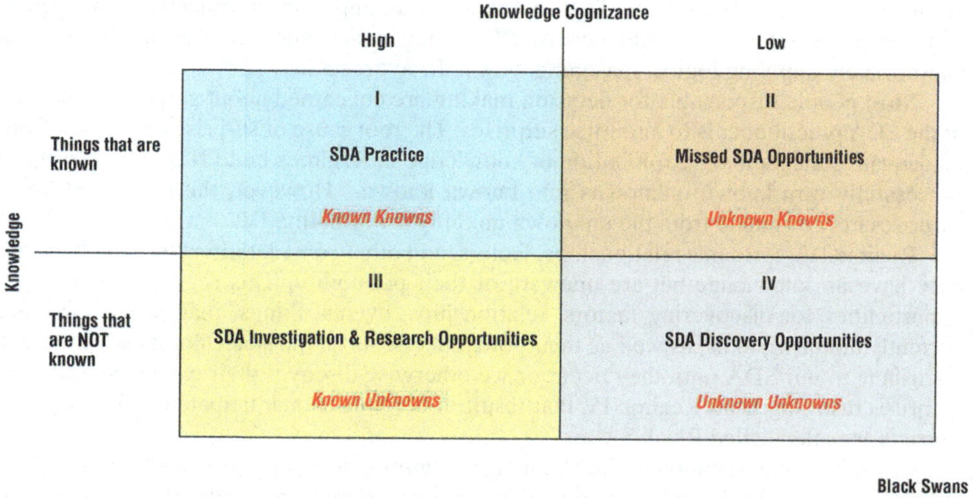

FIG. 2-6. **The Four Regimes of SDA Knowledge.** This matrix comprises both objective information and subjective awareness.

launches, and reaction to the obvious deployment of counter-space weapon systems. Changes in space environmental conditions, routine maneuvers of satellites, routine space launches of new commercial space systems, and even foreign military satellite systems can fit into a space domain focus without much effort. This is particularly true when space events are known and expected. Thus Regime I, the ***known knowns***, is those situations that are well known and that we understand or perceive that we have sufficient knowledge. Unfortunately, Regime I often receives the most attention to the exclusion of the other three.

Regime II, the ***unknown knowns***, reflects missed opportunities. These are the collection and possession of knowledge, but with a lack of subsequent analysis or awareness of that knowledge. Lack of discoverability of baselines, relevant context, foundational knowledge, subject matter expertise, etc., are largely to blame for not knowing what useful knowledge we may have at our disposal.

Regimes III and IV are much more challenging, but can affect SDA diagnoses and prognoses. It's when observations are made or data is collected that doesn't fit the routine or sense of "normal" that more significant developments may be in store. The challenge is developing and exercising the ability to become sensitive to conditions within the space domain that we don't understand. We must be careful not to throw out or ignore data and observations that initially appear to be outliers or enigmatic. We may have to consider new, heretofore unimagined hypotheses.

Regime III reflects our awareness of the limits of our knowledge, i.e., the ***known unknowns***. This is the knowledge we know we lack that would be helpful, even critical, to SDA diagnostic or prognostic processes. The intelligence analyst in the security sector (or perhaps the equivalent in the commercial sector), is often responsible for research to resolve the ***known unknowns***. This might include collecting more data or conducting more intensive analysis. The objective is to convert known unknowns to Regime I ***known knowns***. Recognition of specific known unknowns is important in objectively managing and communicating the confidence of SDA diagnostics and prognostics as well as appropriately applying logical reasoning within those processes.

Most people responsible for decision making are concerned about surprises. The role of the SDA practitioner is to minimize surprises. The root cause of surprise may result from inadequate collection and exploitation of knowledge in Regimes I and II or the inability to successfully turn known unknowns into known knowns. However, the most significant surprises may originate from the unknown unknowns in Regime IV.

Regime IV represents relationships, factors, and other considerations for which we not only have no knowledge but are unaware of their possibility. This regime presents real opportunities for discovering factors, relationships, events, things, places, etc., that are currently unknown to us, as well as their potential effects on the space domain. They can't contribute to our SDA until they occur or we otherwise discover their existence. Extreme surprises resulting from Regime IV that result in heretofore unanticipated and significant impacts are often called Black Swans.

As we turn our attention to the challenges of ambiguities, enigmas, and confidence, it is helpful to reflect upon which knowledge regimes are relevant to the SDA challenge at hand. Being aware of what we know and don't know or even can conceive of knowing is critical to resolving SDA diagnostic and prognostic challenges.

Ambiguities and Enigmas

As noted above, ambiguities involve two or more hypotheses that are consistent with the available observations and foundational knowledge. The approach for dealing with ambiguities generally involves acquiring information that aligns more with one hypothesis (or a smaller set of hypotheses to help narrow the plausible set). Since ambiguities are often the result of small, incomplete, or inaccurate sampling of observables, one of the first tactics is to resample to increase the confidence in the observables. The second tactic is to seek new sources and types of observables that are unique, or nearly so, to a given hypothesis. However, we must take care to ensure that confirmation bias does not drive out other viable hypotheses.

Collecting more data or information may resolve an ambiguity. Alternatively, perhaps more than enough data is available from a given source, but that data is inconclusive with respect to diagnostic interpretations. In these cases, data from independent sources of the same type or an entirely different type of observables may be helpful to support one diagnostic hypothesis over another, and in so doing, increase confidence in the eventual diagnosis. And lastly, focused collection from alternative sources to negate some potential hypotheses or interpretations may be useful in applying abductive reasoning. The objective in resolving any ambiguity is to arrive at a single diagnostic interpretation with high confidence.

Context is important in diagnosing anomalous events, particularly in spacecraft operations. Peacetime anomalies are most likely result from engineering or environmental causes. That said, any unresolved anomalies should be documented and entered into anomaly or enigmatic event logs, something that in practice doesn't happen as frequently as it should. There is a twofold reason for doing this. First, should a similar anomalous event occur in the future, the combination of the historical information with current observations may contribute to a timelier and more accurate resolution. Second, the cause may not be of engineering or environmental origins but a result of unauthorized probing or energy impingement upon a spacecraft. The disciplined keeping of event logs, as well as carefully articulated and documented meta-data that include contextual information associated with such events as spacecraft location, orbital parameters, etc., may help in future diagnostic requirements.

The SDA practitioner should be sensitive to scenarios where significant ambiguities may arise. Efforts should be taken well ahead of these scenarios to ensure that adequate observations and data will be available to diagnose them without ambiguity. As an example, the unambiguous identification of small cubesats launched and released nearly simultaneously from their host launch vehicle can be challenging. The initial separation among them is insufficient to unambiguously identify each object consistently and uniquely. Additional space tracking data along with independent identifying characteristics such as the unique RF command and control frequencies may help minimize ambiguities until the spacecraft naturally drift further apart. Having ready access to information on individual cubesat characteristics and anticipated behaviors before launch can reduce the time to unambiguously identify each satellite.

We now shift to the challenge of enigmas. The above discussion on SDA diagnosis has focused on the logical treatment of observations as they become manifest in the context of baselines, foundational knowledge, and the prevailing context of the domain. The presumption is that observables will be detectable, detected, and recognized as having changed as a result of a specific root cause. In theory, through rigorous integration of observables and supporting SDA information, there will be adequate time and insight to confidently diagnose a change or event. But what if we observe something out of the ordinary, uncorrelated with any set of established norms or baselines? This raises the challenge of dealing with enigmatic, anomalous, and heretofore uncorrelated observations or events.

The strategy for dealing with enigmas is similar to that of ambiguities in that more data or information is needed. However, enigmas require more than just additional data to resolve. In some cases, a new understanding of the relationships among the observations may be necessary. In more challenging cases, we may have to acquire a new understanding of the underlying sources of the observations. In both cases accurate diagnostics are not possible without substantial effort.

The focused collection of additional observations and insights on the apparent parties to the enigma may yield sufficient information to resolve the enigma. Otherwise, analytical approaches that work through a logical series of hypotheses for the root cause may be effective. Using abductive reasoning to refine viable hypotheses by rejecting hypotheses through experimentation, modeling, and simulation or further data collection may be useful. Going back and examining all premises concerning the parties involved to root out unsupported conclusions based on faulty deductive reasoning may also yield sufficiently accurate diagnostic results.

Enigmas are particularly troublesome when they have an apparent relationship to a potentially significant consequence and are not correctly diagnosed in a timely manner. There may not be sufficient time to study and assess all plausible hypotheses or to collect enough additional data to refine an understanding of the relationships among key parameters. Thus, the SDA practitioner should focus diagnostic efforts on those hypotheses associated with the most significant potential consequences. This is one way to avoid the Black Swan syndrome. SDA practitioners and decision-making functions must maintain open communications. Informing decision makers of the potential for a significant diagnosis mitigates the effect of surprise.

Diagnostic Confidence

A fundamental concern for those deciding upon and executing courses of action is the confidence in the SDA insights, including diagnostics, being provided to them. Low-risk courses of action do not require the same level of confidence as those with grave consequences. For example, modest maneuvers executed in a spacecraft in LEO orbit to counter altitude changes due to atmospheric drag are far less risky than an extreme maneuver in a GEO communication satellite to avoid a potential collision. In the former, the consequences of being wrong in the timing or magnitude of the maneuver will likely not permanently affect the mission or mission lifetime. In the case of the latter, a maneuver that was not required may result in lost revenues due not only to a shorter operational life of the satellite, but also due to the out-of-service time during the maneuver and while re-stabilizing the spacecraft at its new location. Clearly, for defense-related courses of action the consequences may be much more severe and thus require even higher levels of confidence. Strategies for achieving SDA must balance the confidence in space domain awareness diagnostics with the gravity of the courses of action under consideration.

Unfortunately, no set of analytical procedures always yields the most accurate diagnosis. SDA practitioners must avail themselves of a broad array of techniques and approaches to understand what is going on in the space domain. This is similar to a medical professional confronted by a set of observations on a patient; the most effective and appropriate treatment depends on the correct diagnosis. Confidence should be an ever-present concern. Dire consequences can result from misunderstanding the limitations of a diagnosis and not adequately informing decision makers of the confidence of the diagnosis.

SDA practitioners must become aware of the limits of observations, baselines, foundational information, and other available data. They also need to understand the risk of three types of potential errors in their diagnostic role. These three error types, classically used in statistical analyses, are useful in assessing and communicating confidence and risk. We state them below in practical terms:

- Type I error: The risk or probability that a hypothesis or observation that is not correct or true will be considered the truth. An example is a false alarm. This is also known as a "false positive."
- Type II error: The risk or probability that a hypothesis or observation that is correct or true will not be considered true. An example is a situation where the warning of a real event is not given. This is also known as a "false negative."
- Type III error: The risk or probability that a hypothesis or observation that is correct or true is recognized as such, but for the wrong reason. An example is incorrectly identifying the root cause of a correctly recognized event.

An important aspect of monitoring the space domain, like the other SDA functions, is the need to minimize errors. Although this may seem obvious, it is a non-trivial aspect of the tradecraft associated with SDA functions. Clearly, missing a significant change in the space domain is of primary concern—a Type II error. Inadequate data collected on the domain is a primary root cause of such shortfalls. This is followed by errors incurred by misinterpreting SDA data wherein significant changes are not recognized or are attributed to a non-significant cause. Examples of Type II errors include failure to identify a significant change in a space object's ephemeris following an intentional maneuver, or incorrectly reporting that an object associated with a space launch is not an operational satellite but rather a piece of debris or the payload shroud.

The flip side is also of concern—Type I errors, often known as false alarms. Examples of Type I errors include the assessment that two space objects have a high probability of collision when in fact their point of closest approach is hundreds of kilometers. Such errors become important when they induce such actions as intentional evasive orbital changes, which incur not only costs and loss of operational availability but also additional risks due to placing the spacecraft in a new orbital regime.

Type III errors may induce inappropriate actions. These are errors involving the wrong diagnosis of a correctly detected change in the space domain. Examples of Type III errors include correctly detecting a change in an adversary's spacecraft ephemeris but attributing the change to preparations for a hostile act rather than a regular orbit maintenance maneuver.

The SDA practitioner may never be able to assign a quantitative or numerical probability of these errors to their diagnostics, but they should be cognizant of their potential to reduce confidence. They should also consider several factors in assessing the confidence of their diagnostics.

Often there are intrinsic uncertainties in the raw observations that are input to the SDA process. The collection of SDA data is generally a sampling process. Rarely, if ever, do we have continuous collection on a space object. In signal theory, there is a well-known theorem which provides us with the limit known as the Nyquist rate, which is the absolute minimum sampling rate at which we can be confident we have captured the salient characteristics of the signal. Sample at a lower rate, and you lose information. The Nyquist rate is generally twice the highest frequency associated with the "signal" or occurrence of events of interest.

Furthermore, the sampling must be uniform across the duration of the observable, such as an RF signal. It may do no good to sample the first thirty seconds of a two-minute signal and ignore the remaining ninety seconds. Finally, there are both short-term and long-term components to almost any observable, and the collection must be long enough to capture the long-term phenomena. An example of this is the small solid earth tides caused by lunar and solar gravitational attraction. These tides affect Earth's rotational speed, producing variations in the length of a day of about a millisecond. They have periods of as long as 18.6 years and other components with periods of days. The standard model used in systems such as GPS incorporates sixty-two separate terms representing short and long term variable components. What does this mean for SDA practitioners? It means that they must understand both collection bias and the sampling rate of the data collection processes that feed their SDA information streams.

To maximize confidence in the diagnostic results, we must understand the veracity of data. This involves the conditions under which the data was and remains valid and conditions under which it may be or become suspect or otherwise inaccurate. The SDA practitioner should have open lines of communication with the sources of SDA observations and be able to acquire the ancillary information on the circumstances and conditions under which the observations were made. Foundational knowledge of potential scenarios where the veracity of the observables may be reduced is an essential component in assessing the confidence in diagnostic results. Thus, users of SDA diagnostic products must clearly articulate their requirements not only in terms of the specific awareness parameters but also the confidence they require to take appropriate action. These requirements may be very difficult to achieve, but they provide SDA practitioners with insights into what they need to strive to obtain.

As we address in the section "Applied SDA Strategy," denial and deception measures in the space domain are a challenge to sustaining confidence in the accuracy of SDA diagnostic results. As space and counter-space capabilities grow globally in number and sophistication, so will their owners' concerns over defensive counters to those capabilities, whether they be security or economic. These concerns span principally across the commercial and security sectors. The objective of denial and deception is to preclude accurate SDA diagnostics by ensuring that few if any observables are detectable or recognized.

The following are some concepts for increasing confidence in SDA diagnostic results:

- Multiple samples of specific SDA data streams over time (relevant to the scenario)
- Multiple, independent sources of SDA data
- Confidence in historical trends (i.e., backed up with lots of data)

- SDA perspectives consistent with the prevailing context (requires good contextual sources)
- Monitoring SDA inputs and assumptions right up to drop-dead decision points
- Understanding the potential for interference or "noise" effects on SDA data and analysis. This included intentional denial and deception efforts as well as unintentional, unrelated interference.

In the next section, we focus on the prognosis of events and situations in the space domain that diagnostic processes have characterized. In many cases the principles associated with diagnostics carry over to prognostics.

2.3.3 Prognosis

Closely following on the heels of diagnostic assessments, which answers the "what just happened and why?" question is the prognosis of the situation, or "what can we expect to happen next?" Prognostic outcomes address tactically relevant future timeframes and provide critical inputs to planning appropriate and effective courses of action. Prognostic perspectives are not long-term forecasts; they are only valid for the combination of current observations and contextual factors. They are subject to change; as soon as things change, so will prognostic perspectives.

Prognostic and diagnostic processes are similar in that they both rely on strong bases of foundational knowledge, historical baselines, contexts, and event-specific observables. But since prognostic processes result in hypothetical expectations but not surety, they inherently can provide no greater confidence than their associated diagnosis. Due to the potentially large number of plausible outcomes of a space domain event, the prognostic expectations cannot generally be inferred with high confidence via deductive reasoning except in scenarios where only extremely well understood physical processes are involved. For example, the high-velocity impact of two space objects will result in a lot of debris. For most events or scenarios, it is more likely that inductive and abductive reasoning will be more appropriate to accommodate the broader spectrum of plausible outcomes. Furthermore, the effects of combining environmental complexity and the nonlinearity of some of the causal factors with sector-specific conditions such as high operational tempos in defense scenarios can affect future outcomes in unanticipated ways.

The prognosis of an observed emergent situation in the space domain is made through either of two basic approaches or a hybrid of the two. In the first, historical baselines and experience drive the prognosis of the future. In the second, we use analytical models and simulations to capture the first principles and known relationships thought to be involved in order to discover potential future outcomes consistent with available information. Hybrid prognostic approaches involve improving the modeling and simulation inputs with historical case information and statistics. The best approach is the one that yields, within the time allotted, the prognosis with the highest confidence. In some cases, this may preclude using time-consuming extreme analytics.

Data-driven prognostic methods require substantial investment in discoverable historical baselines and context. The ability to map a current situation and its diagnosis with something that has occurred in the past can be quite powerful. The experience of subject matter experts is often of significant value to this approach as well. Many routine events and situations in the space domain that become detected and diagnosed through SDA processes can be compared with the historical record to identify a most likely prognosis.

In addition to historical pattern and baseline matching, other techniques leverage the historical record for prognosis. Using Bayesian statistics, we assign and refine the marginal or conditional probabilities of various potential outcomes against the historical baselines. Additional information further refines such probabilities, especially in the case of equally probable outcomes. This is similar to the approach used to resolve ambiguities in diagnostic results.

One shortcoming of data-driven prognostic methods is the risk of not considering possible outcomes. Once Bayesian structures are constructed, it may become difficult to see the need to add additional hypothetical scenarios. The SDA practitioner must always be open to the possibility of alternative, plausible prognostic hypotheses that come late to the process.

The other basic approach to prognostics is modeling and simulation. This approach can be more objective, but doesn't necessarily have higher accuracy or confidence. Modeling and simulation for prognosis has similar challenges to its use in diagnostics. It is only good (accurate) as far as the models represent the real world and relationships among the controlling parameters. Since no single model captures all of the activities, events, phenomenology, etc. in the space domain, the SDA practitioner must fully understand how diagnostic models apply to the prognostic situation at hand. Furthermore, when models are not available to adequately address a given situation, the effect of assumptions (complex or simplistic) made to bridge such modeling gaps on the confidence of prognostic outcomes must be carefully managed.

Good prognostic tradecraft for space domain events requires a good understanding of the timeframe over which subsequent significant events may occur. Some events are instantaneous (e.g., a collision), and others may take place over a period of minutes to hours, even longer. Premature prognosis can occur if it takes place before the "action" is over. A good understanding of the phenomenology, the physics, and system behaviors associated with the diagnosed event or situation is essential to knowing how to judge when a rational prognosis is possible. This is true even in the case of an accurate diagnosis. For example, we accurately diagnose that a new satellite has been launched to GEO, but until the satellite completes its geo-transfer orbit burns we may not know which GEO station slot the satellite will occupy.

For events in space that are associated with environmental conditions or engineering anomalies, we can often reasonably estimate the time horizon for the validity of the prognosis since the root cause remains the root cause. The event occurs, and the prognosis is based on prevailing conditions. For example, we see a foreign (i.e., non-cooperative) satellite on a potential collision trajectory with another satellite. The prognosis is that since there is only a very small chance of collision the satellite will not be hit, and the owner/operator is not likely to take evasive measures. However, for events that may be mitigated or responded to by external parties (those not involved in the SDA prognostic process) we may need to take into account additional considerations to arrive at a more relevant prognosis. Thus, in the previous example, the owner/operator is highly risk-averse and decides not to take a chance and so executes an evasive maneuver. It's important to understand the actions that external parties may take to mitigate (or even exploit) a prognosticated scenario.

The following are some considerations for improving the accuracy and timeliness of prognostic processes:

- Historical baselines and other diagnostic tools should be readily available to support the prognosis of events expected to occur frequently and for which time-critical actions may be necessary. The information within the historical baselines and other foundational material should be readily discoverable.
- A high confidence prognosis may require additional or a wider spectrum of data and information beyond that used in the preceding diagnosis.
- Be willing to reassess a prognosis if additional data becomes available or the contextual assumptions change.
- Use disciplined abductive reasoning to refine the most likely outcomes.
- Recognize the value of all of the data in hand. Understand what is known, what is suspected, and what is not known (the unknowns). Also consider the possibility that unknown unknowns could significantly affect prognostic accuracy.
- Learn from every prognostic activity and refine the tradecraft and update supporting historical baselines for the next time.
- Understand the relative confidence in prognostic alternatives and be able to provide supporting evidence and logic for the prognosis proffered.

2.3.4 Supporting Decisions and Courses of Action

The effective application and exploitation of space domain awareness by the end-user, whether a space operations planner, a diplomat, or someone in a command decision-making position, depends on accurate and timely communication. This consists of communicating current and forecasted conditions as well as anticipated observables; see Figure 2-7. Included should be all relevant aspects of SDA that (1) address current conditions based on observations and their diagnostic interpretation, (2) forecasted conditions and behaviors based on prognostic analysis, and (3) anticipated future observables based on currently available observables and observables consistent with the prognostic analysis. In addition, communication to decision makers must address all observational and analytical limitations as well as their impact on confidence in diagnostic and prognostic conclusions. It may also involve conveying the relevant information and the basis upon which SDA assessments have been made (such as baseline and foundational information, contextual perspectives, and underlying SDA reporting requirements). This transformation and transition of SDA to the end-user is the final process we address.

There are five main objectives to this part of the SDA process framework:

- Accurate and timely notification of significant changes in the space domain to those making decisions regarding potential courses of action.

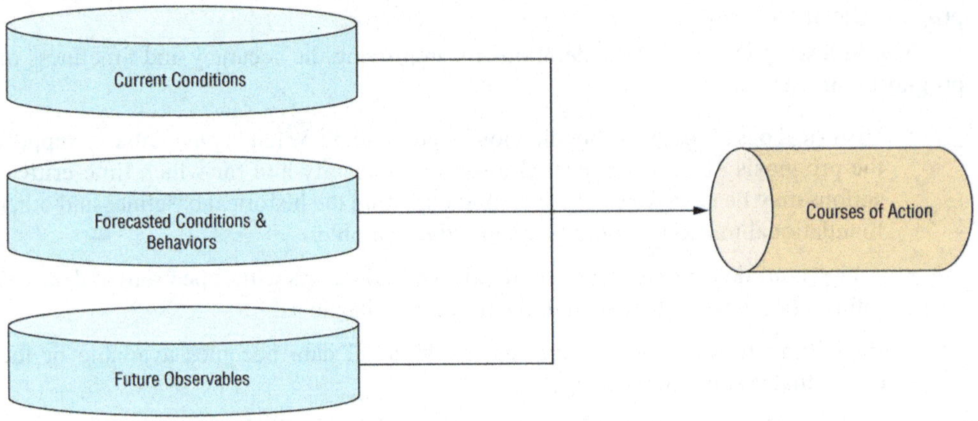

FIG. 2-7. Past, Present, and Future Inputs to Decision-making. Decision makers need all three types of input in their planning.

- Providing the basis or justification for the diagnosis and prognosis of events in the space domain to decision-making processes, including the confidence in those assessments.
- Communicating requirements for additional observations or data to track the evolution of events in space, plus whatever is needed to monitor the effects of any action taken.
- Updating, renewing, or calibrating historical baselines and foundational knowledge from the observations, diagnoses, and prognoses, and lessons learned from an SDA event.
- Providing SDA subject matter expertise to end-users to enable maximum exploitation of the provided SDA.

2.4 Applied SDA Strategy

We now examine specific applications and implementation of the SDA strategy framework for the commercial and security sectors. Each sector has its own mission objectives, performance metrics, and requirements concerning SDA. So tailoring the framework to address those sector-specific needs is important.

This section examines approaches and considerations for achieving more effective space domain awareness within three fundamental applications. It provides some basic concepts to guide SDA practitioners in tailoring the SDA process to their needs and circumstances.

We use the following three fundamental applications for SDA that are generally applicable across both the security and commercial sectors to highlight the SDA process. These applications do not have exactly the same objectives or required results in each sector, but they provide a means for comparing similar requirements and challenges.

- **Monitoring** the space domain for warning and attribution
- **Operations** to be conducted within the space domain
- **Institutional considerations** of the space domain

We treat each of these applications similarly, starting with the fundamental functional objectives of the application; what aspect of space domain awareness do we need to achieve and why? We follow that by working through the elements of the SDA framework. As we proceed through each application, we see how to tailor the SDA process to meet sector mission requirements. Some elements are more critical than others depending on the sector and application involved. Furthermore, aspects of some elements are fairly consistent regardless of where they are used, for example, the need for valid and comprehensive historical baselines and foundational knowledge as well as timely and comprehensive reporting of SDA analytical results and substantiating evidence. We focus less on those common aspects and emphasize the application-specific considerations.

2.4.1 Monitoring the Space Domain

We monitor the space domain for two primary reasons. The first is to pull together all of the current information on the domain to use for such purposes as planning and maintaining historical baselines; this is ***domain characterization***. The second is to look for evidence or indications that a harmful situation within the domain may be imminent; this is ***indications and warning*** (I&W). If we then detect that an important event is likely to occur, we can dig deeper to understand root causes, attribution, and prognosis, and then decide what to do.

Disciplined characterization of the space domain entails continual observation, measurement, and sampling, including understanding the prevailing context. What we observe, measure, and sample depends on the applications involved. Our objective is a pool of current information that we can apply to the routine SDA demands and applications at hand. For example, applications involving planning for space operations, which we discuss later, may need current information on the orbital parameters of satellites that need to be maneuvered and those of other space objects that may pose collision risks.

The following are some perspectives on domain characterization:

- The scope and quality of data should be consistent with the application requirements for accuracy and confidence.
- We should not exceed the "shelf life" of characterization. Refresh and sampling rates should be consistent with expectations concerning the rate of significant change in the domain parameters being sampled.
- We must provide feedback to observational sensors, sources, and processes when requirements for characterization information change.
- Observational data and information, along with their context, should immediately become a component of historical baselines as well as technical, behavioral, and other foundational knowledge.

We now turn to probably the biggest application of monitoring the space domain: indications and warning (I&W). Indications and warning functions alert command and control elements that a situation has occurred (or will soon occur) for which a course of action (COA) may be necessary. Such events may be hostile activities in space or simply changes in the space environment that pose potential hazards to satellite operations.

Both the commercial and security sectors require warnings of potential environmental and unintentional hazards. The security sector, and sometimes the commercial sector, has the added burden of warning of hostile intentions or actions.

Some examples of I&W contingencies important to both commercial and security routine space operations include:

- Impending collisions with other satellites, debris, or other objects in space
- Radiofrequency interference with command, service, or payload links
- Hazardous space or terrestrial environmental conditions
- System degradation, anomalies, and failures

Some examples of I&W contingencies important primarily to the security sector and of occasional commercial sector concern include:

- Pirating of satellite communication bandwidth
- Cyber intrusions or other unauthorized attempts to control space systems
- Indications of counter-space weapon use
- Launching of new adversarial space capabilities
- Maneuvering of adversary satellites to improve support to their forces
- Non-cooperative, close proximity spacecraft operations

Indications and warning is based on the detection of significant change. We focus on changes associated with events, situations, and conditions that pose threats, risks, or hazards to space operations or assets. This involves a priori identification of indicators having high probabilities of association with scenarios of concern. For example, increased activity around an anti-satellite weapon site during a period of heightened tensions might indicate an imminent attack.

Indications and warning requires continuous monitoring of the space domain to ensure adequate time to execute an appropriate and effective COA. This in turn requires a priori insight into the nature and characteristics of the scenarios of concern. Indicators associated with such scenarios are used to highlight and detect important changes, and aid in accurate diagnoses and prognoses. Indications and warning indicators are typically supersets of certain change detection thresholds. In a few cases, an observed change in a single parameter might be enough to trigger an operational warning, but most of the time we require changes across multiple parameters to provide sufficient confidence that a warning is appropriate.

To be effective, I&W must be underwritten by a substantial volume of observations, baselines, context, and foundational knowledge. Furthermore, more serious scenarios are likely to be associated with extremely rapid operational tempos, occurring at extremely high "velocities," both literally and figuratively. Consequently, the information necessary to drive the diagnostic and prognostic processes that underpin I&W must be readily available and discoverable.

Diagnostics for I&W typically have rather short timeframes. Under combat conditions there is little or no time for an extended analytical study of the indicators and other information. This often precludes using modeling and simulation-type tools. In these cases, methods such as pattern recognition to identify behaviors under specific contexts may be more appropriate.

Indications and warnings are forecasts of imminent events or situations; the things have not yet happened. These indicators are often a priori plausible forecasts of an event. As such, indicators may have an inherently implied prognosis based on abductive reasoning.

The information necessary for space domain I&W is broad. It includes not only traditional observables within the outer space portion of the domain (space object tracking data, space environmental conditions, etc.) but also information on adversary defense capabilities and activities. Examples include movement of troops, movement of key technical personnel, changes in launch facility activity, or rocket propellant transport.

The information and knowledge pool must support the exploration of multiple plausible alternative explanations or diagnostic perspectives. Ideally, the SDA practitioner will proactively compile possible diagnostic explanations for the anticipated scope of plausible changes in key indicators. It's sometimes useful to seek negative observables or indicators, i.e., parameters that did NOT change and are therefore not logically consistent with a need for warning.

Indications and warning processes are driven by an acute need for very current observations that are relevant and mapped to the specific indicators that trigger warnings. Some considerations include:

- Observations should occur at sufficient frequency (sampling rate) to detect fleeting or short-duration conditions associated with the indicators being sought.
- Observations should take place across the portions of the domain where we anticipate the events, phenomena, and conditions.
- Observations should be consistent with the established baselines underwriting the indicators.
- Multiple sources of observations may be necessary to reduce ambiguity, increase sensitivity, and improve subsequent diagnostic and prognostic processes.
- Observations that negate some plausible diagnostic alternatives may be helpful and improve the efficiency of diagnostic processes. Negative indicators may contribute significantly in this regard.
- Understanding the prevailing context is important for shaping the collection of relevant observables as well as their accurate interpretation.

Indications and warnings associated with defense scenarios are a serious matter, and the confidence in the accuracy of a warning is critical. Sufficient and appropriate observations, along with relevant change detection thresholds, contribute confidence in warning. Avoiding Type I errors (false alarms) is important not only because they waste resources but also because of the risk of reducing warning sensitivity through repeated false alarms; such situations contribute to Type II errors—the missed warning of a real event.

A required outcome of I&W diagnosis is often the attribution of an observed event or situation to a particular root cause or actor. Generally, it is hard (but not impossible) to conduct a conventional military action in the terrestrial, maritime, or air domains without (1) detection and (2) attribution. Conversely, for events in the cyber and space domains, attribution can often be denied (or at least delayed) relatively easily. A ground-based high-energy laser, for example, could disable a LEO imagery satellite, and all the owner-operator might know is that suddenly their spacecraft can no longer take imagery. They might know where the satellite was when it failed but not whom to blame.

Relevant information for attribution is anything that connects back to the root cause of an event. One of several challenges for attribution is that it involves initiating analytical efforts after the fact and is based in part on evidence that may not last long. Current observations of the aftermath (post-event) will probably provide only part of the evidence necessary for accurate attribution.

The assignment of attribution may also be driven by very short time constraints, especially if associated with combat operations. We may not have time for exhaustive diagnostic analyses involving modeling and simulation or incorporation of subject matter expertise. The SDA practitioner should anticipate circumstances where short diagnostic and attribution timelines prevail.

The gravity and severe implications of some security sector options for courses of action against a party to whom attribution has been assigned demand a high degree of confidence. The ability to assign attribution with high confidence is underwritten by the quality and comprehensiveness of the underlying SDA data, information, and processes. Some events in space, such as on-orbit collisions or RF interference, may not be hostile in intent, but attribution is necessary to enable effective mitigation and subsequent legal action.

Achieving confident diagnostics for attribution can be very difficult. The diagnosis must be of extremely high confidence, and we must have objectively ruled out alternative root causes for the observed event or situation. Like the supporting SDA data, the logic behind the diagnosis may be subject to rigorous examination and challenge. Thus, we must carefully document the logical reasoning behind the diagnostic conclusions and ensure that alternative hypotheses for root causes or parties can be refuted.

The quality and validity of information required for accurate attribution are at least as stringent as those for triggering warnings. In some cases, the information must provide a credible, legal basis for claims for damage, or reactive, defensive, or offensive measures. Some considerations for ensuring adequate information and knowledge for ascribing attribution include:

- The provenance and chain of custody of the applicable SDA data and their sources
- Information and observables that sufficiently negate hypotheses of alternative root causes, sources, or parties to a space domain event or situation
- Contextual insights and foundational knowledge that are consistent with the assigned attribution

- The potential need for releasable (unclassified or non-sensitive) evidence to support attribution claims
- Observations of all known directly involved systems, targets, facilities, sites, locations, and associated intangibles such as communications, RF and IR emissions, etc., as soon after the event as possible
- Observations of all suspected ancillary systems, facilities, sites, locations, personnel, etc., that may have been involved and that could help in assigning attribution
- Observations of entities known or suspected of NOT being involved, to provide counter-evidence or a basis for comparison in the diagnostic processes
- The relevant contexts, to help in the diagnostic processes.

Indications and warning diagnostic insights provide valuable decision support in selecting appropriate courses of action based on the attribution of an event. Decisions depend on a number of considerations, including the severity of the event, the responsible party, and the confidence in attributing the event to that party. Diagnostic assessments must address not only the logical reasoning behind the attribution but also the validity of the underlying evidence.

Some examples of the types of SDA information that are relevant to decision support include:

- The record of contextual and background information leading to the warning, including all observations, assessments, context perspectives, indicators, and their thresholds
- Sources of data and observations (and their assessed veracity) that triggered the warning
- The diagnostic reasoning behind a warning along with the assessed confidence in the resulting diagnosis
- The prognoses of anticipated evolution of the situation, reactions, next steps, etc., and the confidence in those prognoses
- Indicators that might be useful for monitoring the progression of the situation and the sources and nature of the necessary observables.

The SDA practitioner must know the evidence behind the initial indications and warning and subsequent assignment of attribution. The supporting calibrations and analytical confidence metrics or approach to validation include:

- The rationale behind refuting alternative hypotheses with its associated analytical confidence
- The foundational information, including baselines, foundational intelligence, and other information that support diplomatic or other courses of action
- The ability to release potentially sensitive data, observations, and reasoning behind the attribution. This may be critical, particularly to diplomatic actions.

2.4.2 Operations

Being comprehensively aware of the space domain is essential for routine space mission operations, such as safely inserting a satellite into orbit or accurately diagnosing the effects of the space environment on satellite electronics. It is also critical for adequately protecting a spacecraft against interference or attack and ensuring effective space support to defense and security operations. We consider here two important dimensions of operations. The first is routine operation of space systems and services in support of their missions, both for the commercial and security sectors. The second, which is largely (but not entirely) a security sector concern, are space operations that support defense and security requirements. The two overlap, but we address them separately to highlight some of the important differences.

Routine Space Operations

We define for the purposes of this discussion routine space operations as activities associated with launching and keeping spacecraft in service throughout their lifetimes regardless of mission, plus the final steps at end of life. In the next section, we address defense and security-related space operations. Routine operations include the launch of a new satellite, maneuvering a satellite to avoid collision, or orbital rendezvous for docking or on-orbit servicing. We also include non-hostile or defense contingency scenarios such as engineering anomalies, electronic upsets caused by the space environment, and even events on the ground that affect space system command and control. All of these operations require awareness of the space domain for mission success. In the following we address how the SDA strategy framework applies to the three operational phases of planning, execution, and assessment.

Planning

Planning of space operations either for direct, immediate execution, or for future execution in a contingency requires acknowledging and incorporating space domain insights. A key to successful operational planning is the ability to accurately project future conditions, environment, and other factors that may shape the execution of a plan. Space domain awareness provides the foundation for such projections to minimize risks and to exploit potential opportunities. The following are examples of operational activities that incorporate some degree of SDA:

- Collision avoidance planning
- Geo-transfer orbit planning
- Launch collision avoidance planning
- End-of-life deorbit or disposal orbit insertion planning
- Contingency planning for engineering and environmental anomalies
- Efficient orbital insertion
- Efficient station-keeping maneuvers
- Proximity operations for servicing or inspection (future capabilities)

SDA support for planning begins with understanding the time horizon for executing the plan. This establishes the boundary conditions for the historical baselines used to project future operational conditions and relevant context. The further into the future plan execution is anticipated the less relevant some baselines become. For example, commercial satellite operators planning near term orbital maneuvers may use current space catalogs to plan the timing and magnitude of such maneuvers. The longer they wait to execute the maneuver, the more likely planning assumptions based on the catalog will be incorrect.

Foundational knowledge, historical baselines, and current space domain conditions are all essential inputs to operational plans. Since planning is often conducted well in advance of execution, the currency of the information may not be as critical as that necessary to support immediate operational decisions. However, operational planning for space also involves anticipating any SDA data and information that must be collected, processed, and delivered where and when needed. For example, operational plans may need to specify requirements for orbital change detection. This might apply to planning satellite end-of-life disposal, where we need to monitor the final changes in the satellite's orbit before it burns up in the atmosphere. We may need to incorporate requirements for tracking observations into operational plans.

Examples of space domain information that may be necessary for planning routine space operations include:

- Space catalog information to support launch insertions and associated collision avoidance plans
- Seasonal and current terrestrial weather forecasts affecting launches
- Status of the operational health of current spacecraft and
- the potential need for replacement
- RF frequency allocations among red, blue, and gray space systems
- RF frequency interference sources and characteristics

Execution

We almost always execute operational plans (as well as routine non-planned operational activities) for space systems to minimize risk to spacecraft, which are expensive and nearly impossible to fix if damaged (to say nothing of the lengthy time to replace). Also, the commercial sector is driven to ensure efficient delivery of space services and capabilities with the highest economic returns at minimal costs. Security sector space operations must also avoid unnecessary risks to provide required services and capabilities to protect and serve national equities.

Every operational activity, whether continuous or periodic, is affected by conditions within the space domain. Understanding the prevailing conditions within the domain helps optimize the delivery of services and minimize the risks to space systems. The following are some examples of operational activities and associated space domain considerations:

- Relocation of a geosynchronous communication satellite to a new orbital slot
  - Orbital parameters, object tracking during relocation, collision avoidance

- Drag compensating maneuver of a LEO satellite
  - Monitoring altitude, collision avoidance, reestablishing orbital parameters
- Collision avoidance maneuver
  - Monitoring catalog for potential orbital conjunctions
- Increasing transponder power to compensate for RF interference
  - Monitoring RF environment, geolocation of RF emitters.

Observations on current conditions and activities within the domain that support change detection and other analytical functions necessary for efficient and safe operations must be collected in a timely manner. In some cases, the satellites themselves provide indications through changes in their telemetry or external characteristics of their communications signals that something has occurred or changed in the space domain. The concept of satellite-as-a-sensor is particularly useful in providing information on changes in the space domain.

Change detection concerns usually involve one of two things: (1) ensuring that spacecraft have not suffered some externally induced upset or damage, or (2) warning of conditions conducive to degradation or damage, such as changes in the space environment or pending collisions. Appropriate thresholds associated with changes in parameters monitored on board a satellite can provide the basis for detecting changes associated with the space domain. Not all changes relate to the space domain, however; some are simply changes in routine onboard operations or engineering failures. The changes induced via the space domain that could be manifest in telemetry include:

- The apparent loss of solar panel elements from collision with another space object (e.g., debris)
- The apparent loss of attitude control with concurrent indication that a change in momentum was experienced prior to the loss of control. This may indicate an impingement with another space object.
- Transponder power levels changing due to unauthorized signal transponding or jamming
- Onboard computer upset requiring a reset of the operating system or other software remediation measures as a result of a digital circuit impacted by a high energy particle.

Changes in external (to the spacecraft) parameters that are of concern to the commercial sector operator include:

- Changes in orbital parameters of another space object that may now pose a collision risk
- Reports of RF interference by other space operators at frequencies and from locations relevant to the operator's spacecraft
- Terrestrial and space environmental conditions that may affect space services such as communications

- Changes in geopolitical or competitive market dynamics that may result in competing signals interfering with existing space communication services.

Timely detection of changes in space that could substantially affect the outcome during operations is imperative. The SDA practitioner should have already identified the potential risks and threats in the domain that must be detected during an operation. Since the timelines involved in some operations preclude exhaustive examination of all SDA data and observations, it's important to have a priori identified those parameters and situations that could alter operational outcomes. Efficient data fusion and management systems are essential as SDA data and observations flood change detection, diagnostic, and prognostic processes during an operation.

The timelines for routine space operational SDA diagnostic and prognostic processes are driven largely by the events and circumstances surrounding just the space systems involved (unlike the multitude of external factors driving the pace of defense operations). However, the objective is to rapidly identify domain changes that need immediate attention.

Diagnostics during operations are likely to be much abbreviated due to the time constraints and ops tempos of defense and security operations. The window for decisions to execute many techniques, tactics, and procedures (TTPs) may only be sufficient for very limited assessments of the root cause. This precludes diagnostic analyses involving time-consuming modeling and simulation or group consensus approaches.

SDA practitioners who anticipate becoming involved in security sector operational activities should strive to ensure that the necessary SDA data and other inputs will be collected and made readily available. They should also train and practice against plausible scenarios using the available tools and techniques.

Assessment

We usually associate the role of SDA in the assessment function for routine space operations with verifying whether the operation has indeed proceeded as planned and that the results are consistent with the planned objectives. If not, SDA may help determine the root cause of any operational failure, or any unexpected or undesired outcome. The focus may be limited to considering just those observables necessary for verification. For many operations, this may be as simple as re-establishing the orbital track of a satellite that has undergone a maneuver or relocation. However, if there have been anomalies or deviations in the planned operation, then the assessment function provides for the detection of such anomalies and contributes to the diagnosis and prognosis of the situation.

The assessment of the results and impacts of routine operations relies on SDA to focus on the relevant observable changes in the domain. The fundamental SDA information supporting assessment tends to focus on a limited set of systems, behaviors, and locations; this depends on the scope of the operation. The information requirements for assessment focus heavily on change detection for verifying or detecting anomalies. The diagnosis of the health and status of systems experiencing an anomaly may require additional foundational information. We may need historical baselines and foundational knowledge concerning spacecraft and support systems, for example, to make such assessments. Thresholds and other change detection requirements should help determine relevant health and status. For example, the functional capability of a satellite may be more important to ascertain than a modest change in its lifetime. Contextual insights other than those related to the engineering and environmental conditions may not be relevant, since the proximate root cause of any change is presumed to be understood.

Current observations are key to assessment. Without current observations of potentially affected systems, assessments may rely on hypothetical impacts based on the nature of the operation. The observations should focus on supporting the diagnostic and prognostic processes concerned with assessing the health and status of systems and their continued ability to function. For routine space operations where we don't anticipate damage or other impacts, we may just need observations to verify that the executed plan has resulted in the desired new condition. An example is verifying the change in orbital parameters following a station-keeping maneuver. We match the anticipated orbital parameters against the observations to check that any changes are within the expected range or required tolerance.

For operations involving a high risk of unanticipated consequences, the purpose of change detection is to rapidly determine if any significant changes have occurred in a system. Change detection in this context focuses on observables related to the continued viability of a system. If no changes are detected, then we might skip subsequent diagnostic and prognostic functions. However, if the confidence in this finding is low, we might have to resample, or collect additional observations. We must always remember the adage, "absence of evidence is not evidence of absence" when we don't see expected changes based on the operational circumstances.

The diagnostics associated with the assessment function are first concerned with verifying that any observed changes during the operation are, in fact, likely to have been due to the operation. The second role is to provide insight into the impact or damage sustained by a system during an operation. This may be useful in the subsequent prognosis of an affected system on its potential to be repaired or returned to service. In some such cases, the diagnostics are very specific with respect to the damages or impact incurred where the root cause is known.

For routine space operations where the validation of anticipated changes is consistent with expectations and operational plans, there is no need for a prognosis. That has already been accommodated in the requirements for the operation. On the other hand, for operations that have not gone according to plan, prognoses can be important to subsequent courses of action.

The prognostic approaches for assessing unknown or unanticipated effects focus on forecasting the continued ability of an affected system to function at some level for some time. This involves understanding the consequences of damage or degradation on essential components as well as understanding the potential for repair or self-healing (e.g., reset). For prognosis of capability (for example, satellite communications) rather than individual systems per se, the focus may be on the ability to reconstitute or replenish lost or degraded capabilities.

Defense Operations

Defense operations in the space domain include both defensive and offensive activities. Defensive considerations involve protecting space assets from hostile intent. Offensive considerations involve both the denial of space capabilities to others and using one's space capabilities to support military operations in all warfighting domains—maritime, terrestrial, air, cyber, and space. As with routine space operations, space domain awareness is critical for successful defense operations involving space. We address the application of SDA to each of the three operational phases of planning, execution, and assessment.

Planning

Space domain awareness for planning defense operations includes both defensive and offensive considerations. Defensive considerations involve protection plans for one's military assets and capabilities writ large (for example, overflight warning of hostile ISR space capabilities) as well as one's space assets and capabilities (for example, warning of potential counter-space attack preparations). Offensive planning requires SDA for optimal performance of space assets providing force enhancement capabilities (for example, optimizing satellite communication services under adverse space environmental conditions such as solar events) or supporting space control functions (for example, geolocation of the source of SATCOM jamming signals to provide weapon targeting inputs).

For planning both defensive and offensive operations, the SDA of most value is the current characterization of the domain along with insights into historical behavioral norms. The latter have a bearing on anticipated adversarial activities and reactions to our operations. The objective is to maximize the effectiveness of the military options with minimal chance of surprise and risk to one's own assets and capabilities.

Plans for security operations must reflect as much insight into the current domain conditions as possible. Although we may not need to change the core tactics and strategy, the relevance of any operational plan decreases sharply with time—far faster than a plan for routine space operations where many of the domain conditions don't change quickly. The foundation on which the viability and efficacy of a defense operations plan are based should be verified and validated through current observational data. Similarly, if behavioral assumptions were based on a contextual environment that is now no longer valid, a reassessment of those assumptions may be necessary.

Planning for military operations does not typically involve change detection processes in a first-order manner. However, embedding within plans the need to monitor the space domain for changes that affect the outcome of an operation may be appropriate. In such cases, specific thresholds that, if exceeded, would indicate the need to change plans may be important.

The following are some examples of contributions of SDA to operational planning:

- Preparation of the battlespace—a fundamental element of operational planning that involves a holistic perspective of all red forces as well as relevant civil and other non-combatant entities. In the case of SDA, this means the red space order of battle as well as the equivalent in the civil and commercial sectors.
- Space environmental conditions—a backdrop against which planned space operations will take place. For example, understanding the anticipated changes in solar activity over the planned mission period or the status of annual micrometeorite activity depends on SDA data sources.

- Space launch activity—representative of activities for which deconfliction measures may be necessary. This includes SDA of the status of global launch sites, launch manifests, etc.
- Status of adversarial counterspace forces and capabilities—this provides the basis for planning defensive countermeasures. Examples include direct ascent and orbital ASATs, directed energy weapons, RF jammers, and cyber weapons.
- Red, gray, and commercial order of battle. This should be as comprehensive as possible in the context of anticipated or potential engagements or interactions with other space objects.
- Baseline operational behaviors for the aforementioned orders of battle. Examples include all of the orbital behaviors manifested in satellite catalogs along with similar historical information on terrestrial space assets to support the characterization of normal.
- Radiofrequency assignments and usage for space systems (on-orbit and terrestrial) relevant to the planned mission
- Planned behavior of other parties' space systems, e.g., the planned or anticipated changes in GEO orbital slots of other communication satellites.
- New commercial or foreign space launch plans, or other activities (such as high energy terrestrial RF emissions) that we may need to accommodate.

Execution

Execution of defense operations in the space domain requires substantial SDA for mission success. Although a significant amount of SDA may have been incorporated into operational plans, its relevance likely will quickly fade upon contact or engagement with the adversary. What sets the SDA requirements for defense operations apart from routine space operations is the uncertainty caused by an adversary's actions as well as those taken to counter them. This uncertainty is temporal, spatial, and behavioral. SDA must reduce that uncertainty for effective operational command and control.

The temporal aspects of defense operations may be exceedingly short, thus placing challenging demands on SDA. Frequency changes in the domain must be recognized and communicated quickly. That includes the full scope of domain awareness of the battlespace, including both space and relevant terrestrial elements. Global space domain awareness may become of secondary importance relative to awareness of domain activities and events directly relevant to the battlespace.

During the execution of operational plans, the currency of SDA inputs to command and control elements is critical. Knowledge of significant changes in the domain that require immediate courses of action as well as tracking operational plan progress requires current awareness of the domain. The high ops tempo involved in real-world operations requires that SDA inputs be available immediately. The SDA processes associated with monitoring, diagnostics, and prognostics must align with the operational decision-making timelines. Issues or questions regarding ambiguities and enigmatic situations must be communicated in terms that command elements can understand quickly.

The velocity of data and information from all sources must match the velocity of conflict; in other words, the operational tempo. If observations are not frequent enough, their value to decision-making may vanish. For example, SDA supporting a find-fix-track function for a specific weapon system may be needed in real time.

Timely detection of relevant changes in space during operations is imperative. The SDA practitioner should have already identified the potential risks and threats in the domain that must be detected. Since the timelines involved in some operations may preclude exhaustive examination of all SDA data and observations, it's important to have a priori identified those parameters and situations that can affect operational outcomes. Efficient data fusion and management systems are critical as SDA data and observations flood change detection, diagnostic, and prognostic processes during an operation.

Diagnostics during operations are likely to be much abbreviated due to the time constraints and ops tempos of defense and security operations. The window for decisions to execute many techniques, tactics, and procedures (TTPs) may only be sufficient for very limited assessments of the root cause. This precludes diagnostic analyses involving time-consuming modeling and simulation or group consensus approaches. Prognostic functions take place under much the same constraints. Furthermore, the period over which prognostic assessments are reasonably valid may be short compared to pre-conflict strategic warnings.

SDA practitioners anticipating involvement in security sector operational activities should strive to ensure that the necessary SDA data and other inputs are collected and made readily available. They should also train and practice the core SDA processes against plausible scenarios using the available tools and techniques.

Assessment

The assessment of the results and impacts of defense space operations relies substantially on SDA, often referred to in this context as battle damage assessment (BDA). For example, SDA characterizes the damage sustained by an attack upon either one's own or the adversary's spacecraft or terrestrial space support facilities. Assessments in a defense environment must be done in a manner consistent with the prevailing battle rhythm to ensure effective inputs to succeeding operations.

To a first order, the SDA required for operational assessment tends to focus on a limited set of affected targets, activities, and locations; that is a function of the scope of the operation. Second-order effects and unintended consequences may require a larger SDA scope.

The important difference between the assessment function for defense operations in space and routine space operations is the additional burden of assessing the operational impact on systems other than only one's own. Operational impacts on space systems belonging to an adversary, ally, or an uninvolved party must be considered. This places significant importance on foundational knowledge and historical baselines for those systems, which are often not as comprehensive and detailed as for one's own systems.

The diagnostic and prognostic processes are much the same in principle as those for routine space operations. The objective is to detect damage, diagnosis the root cause of any damage inferred from observational insights, and to provide a prognosis of the continued viability of the system.

The following are some considerations relative to SDA and assessment of defense operations in the space domain:

- Undertake advanced preparation (via good operational planning) of the volume and requirements for observations needed after the fact (operation) to support timely assessment.
- Diagnostic and prognostic functions may need to look beyond the effects on individual space systems elements (e.g., satellites) and consider the impact at the constellation or enterprise level.
- If attribution (or its denial) is important, collection and assessment of SDA observables available to external parties may be necessary.
- Understand the plausibility of alternative diagnostic and prognostic assessments. As appropriate, consider the potential effect of denial and deception measures on available observables. Consider alternative corroborating, independent sources of SDA.
- Anticipate assessments and ensure that operational plans manifest pre-event requirements.
- Assessment of defense operations in space should link to the indications and warning monitoring of the domain for possible reactions to the operations that may affect the validity of prognostic assessments.

2.4.3 Diplomacy

The role of diplomacy in space operations is typically quite limited. Diplomacy associated with the space domain usually involves much longer timeframes than those associated with security operations. However, geopolitical institutions, both governmental and non-governmental, are increasingly attempting to influence the norms of behaviors in the space domain. In addition, their actions may be instrumental in both long- and short-term deterrence messaging. For diplomatic actions, such as demarches, to be effective, especially during periods of tension, they must be based on an informed awareness of relevant developments, activities, and changes in the space domain. There are three interrelated diplomatic areas of concern relative to the space domain for which SDA is essential:

- Compliance—compliance with international agreements, including treaties, that concern the space domain either directly (e.g., space debris) or that space capabilities or services enable (e.g., treaty monitoring from space)
- Messaging—contributing to the content or basis for diplomatic messaging on space domain affairs (e.g., "we are concerned about your activities..." messages)
- Demarches—contributing the evidence and basis for diplomatic demarches addressing unacceptable situations involving the space domain (e.g., putting space-to-ground weapons in space).

The information for diplomatic SDA application must be sufficient to enable the monitoring of international behaviors in space that indicate possible inimical shifts in motivation or intentions. Some of these shifts may be more detectable from sources unrelated in the first order to SDA, such as intelligence sources. However, some changes in operational behavior may only be manifested in traditional SDA data sources, for example, intentional operations of a satellite in close proximity to another's. Another important constraint on SDA data and observations for diplomatic purposes is that some of it may need to be releasable to foreign, potentially hostile states, especially if such data is offered up as evidence of some accusation or attribution. Such data should be as clean as possible, free from real or perceived biases, calibrated, and, in some cases, be subject to verification by independent players.

Other information needed to support diplomatic applications of SDA is not typically associated with the other security sector applications. Understanding the details or parameters surrounding a treaty or other established "red lines" may be necessary to detect important changes indicative of violations. Foundational knowledge of the equities and behaviors in the domain of potentially affected non-aligned states and parties may be important in prognostic analyses.

The SDA practitioner supporting diplomatic functions must ensure that the changes that are significant from a diplomatic perspective are plausibly detectable in the data and observations routinely collected via traditional defense and security sector sensors and sources.

Changes having diplomatic significance within the space domain could be quite small, such as slight changes in the apparent position of a foreign government relative to global space security. However, since these changes don't necessarily require an immediate response, we don't dwell on them here. More important change detections are those for which an immediate course of action is necessary to preclude further escalation or behaviors considered inimical to national interests. SDA change detection for diplomatic applications is both quantitative and qualitative. The quantitative (including binary) changes of significance relate to such things as movements of military equipment, on-orbit behaviors (e.g., close approaches), in-space testing of debris-producing weapons, etc. Detection of significant changes in these situations can be supported by historical baselines, foundational knowledge, and pre-established thresholds or red lines of concern.

Significant changes in qualitative factors are more difficult to ascertain since they rely on the subjective interpretation of observables and apparent changes relative to qualitative thresholds of concern. For our purposes, we focus on the quantitative changes of diplomatic concern and subsequent diagnostic and prognostic processes.

The diagnostic processes associated with supporting diplomatic applications of SDA focus more on the institutional root causes for an observed change than on the physical or systems level. Changes due to the latter are still important to understand, especially when they affect the attribution of an observed event. The question diplomatic elements may seek answers for is whether the change observed in the domain was the result of an intentional political or military decision on the part of the other party or an unintentional (albeit perhaps reckless or careless) action. The diplomat's ultimate attention is on those situations where courses of action involving persuasive changes in behavior may be necessary.

SDA supporting diplomatic applications can be challenged by the need to communicate what may involve technically complex topics that are not easily understood by those outside of the space community. The SDA practitioner supporting diplomatic activities must be able to provide and portray SDA information in objective, easy to comprehend terms. The SDA practitioner should be able to provide materials that support demarches and other diplomatic instruments with releasable SDA information. This may be challenging when the sensors or sources of critical SDA information are sensitive or classified.

2.5 Summary

The objective of this chapter is to provide a structured and disciplined approach to achieving space domain awareness across virtually all dimensions of the domain and to support the needs of all space applications, services, and activities. The essence of awareness is the ability not only to observe the elements of the domain relevant to one's needs but also to be able to detect significant changes, correctly draw conclusions from such changes, and then forecast in a relevant time frame anticipated subsequent events. This chapter provides an analytical framework for working through the sometimes daunting task of observing, analyzing, and acting upon significant events in the space domain.

We present a basic model for considering the flow of data and information collected on the domain and how to process it to detect significant changes and support diagnostic and prognostic analyses. The model is adaptable to virtually all aspects of space domain awareness and provides a foundation for developing effective operational strategies for SDA, regardless of the application—civil, commercial, or national security.

A robust strategy is essential for observing, capturing, and managing information on the space domain. Our awareness of the domain relies on two types of information. The first is the current state of the domain derived from up-to-date observations and measurements and the prevailing context or big-picture backdrop affecting the domain. The second is the foundational basis for awareness of the domain. It consists of historical baselines of behaviors and knowledge of such things as the physical environment and space systems in orbit. It also includes standards and thresholds that define our requirements for reacting to changes in space.

Synthesis and analysis of observations made in the context of foundational information and knowledge enable actionable awareness of significant events and conditions in space. Three fundamental processes are involved. First is the detection of significant changes in the domain that necessitate a reaction. The significance of a detected change is determined by prescribed thresholds of concern and its cause. Thus, the second analytical process is the diagnosis of the detected event and its root cause; in other words, what happened and why? The third process is the prognosis of the developing situation. Knowing what and why something happened leads to the need to understand what we can reasonably expect to happen next. Throughout all three of these SDA processes it is critical to understand and communicate the level of confidence in the analytical results and to objectively consider alternative explanations when faced with ambiguities and enigmatic situations.

The final function of SDA is support to the planning and execution of courses of action following the detection of a significant change in the space domain. The last part of the chapter addresses some of the considerations needed when tailoring the SDA analytical model for monitoring the space domain (for indications and warning), routine and defense-related space operations, and diplomacy.

Reference

U.S. Department of Defense. 29 May 2013. Joint Publication 3-14 Space Operations.

Chapter 3

Characterization of the Space Environment

Ryan M. Weisman, Ph.D.,
Air Force Research Laboratory, Space Vehicles Directorate

This chapter describes how we characterize the space environment and its influences on space object motion, space object characterization, and its impact on space domain awareness. We intend to demonstrate what interactions are happening, how we conceive of them, how we model them mathematically, and how they can lead one astray if neglected or incorrectly modeled. The latter is crucial for attribution of events and implications for establishing and maintaining custody of space objects over time in a fluctuating space weather environment. We examine gravity, neutrals, plasma, radiation, and particulate environments with their potential impacts on the state of space objects (dynamical, sensing, communication, and prediction uncertainty). The five areas are coupled via the sun's emissions yielding combined effects on space objects, sensors, and communication. We can predict the likelihood of some effects but not others, since solar emissions arrive at Earth in a little more than eight minutes. The impacts of their local and geographically diverse effects last for days.

3.1 Introduction

This chapter describes how we characterize the space environment and its influences on space object motion, space object characterization, and the impact on space domain awareness. With the increased use of electronics and the need for more timely reporting from smaller, lighter, and more capable platforms, we need to understand the effects of the natural and man-made space environment. There are many space environment books

[Hastings and Garrett, 1996], [Tribble, 2003], [Pisacane, 2005a], and book chapters [Pisacane, 2005b], as well as space vehicle and constellation design methodology books [Wertz et al., 2015; Wertz, 2009], and books on dynamical interaction [Wertz, 2002], [Milani et al., 1987], [Hughes, 2004].

3.1.1 Lead-Off Example

On April 5, 2010, the Galaxy 15 satellite, owned and operated by Intelsat, stopped responding to commands. Galaxy 15 was at the time in a five-year-old geosynchronous Earth orbit (GEO), and carried C-band transponders and an L-band payload for the U.S. Federal Aviation Administration's GPS Wide Area Augmentation System (WAAS) [Ferster, 2010]. The manufacturer, who had produced similar units, attributed the behavior to a solar storm upset [de Selding, 2010a]. Even though the systems appeared to operate normally, the lack of response to commands made stationkeeping for the 133-degree west space vehicle impossible. With physics taking over, easterly drift made Galaxy 15 a large conjunction threat to its GEO neighbors and a source of C-band communication disruption [de Selding, 2010b]. The lack of stationkeeping control lasted approximately eight months and required 15 maneuvers by other spacecraft to avoid conjunctions and service disruptions. Control of Galaxy 15 was re-established on January 7, 2011 [de Selding, 2011a]; the failure review board attributed the cause to electrostatic charging and not to solar activity [de Selding, 2011b]. This example demonstrates that passive effects, even when known, can have large consequences not only on mission operations of a single spacecraft but on those of other space users. Executing unplanned maneuvers can degrade the mission life of spacecraft, requiring replacement or augmentation that costs significant sums of money in additional manpower, new spacecraft manufacture, service disruption, and launch costs.

3.1.2 Overview

The space environment comprises multiple coupled processes that possess different time-scales and manifest differently based on orbit regime, object type and construction, and solar condition. The Sun is the dominant driver for most space environment processes, but Earth's non-spherical shape significantly affects spacecraft pointing. Moreover, humans have populated the environment with debris ranging in size from thruster expulsion particulates to defunct spacecraft, with speeds ranging from 3 km/s (GEO) to 10 km/s (GEO transfer orbit). Given the increasing dependence on space for communications and navigation, designers and operators need to be aware of the complex space interactions, so as to mitigate or reduce the risk of on-orbit anomalies.

The Sun is responsible for the dynamic nature of the space environment [Tribble, 2003]. It has an 11-year electromagnetic radiation cycle that drives the long-term periodic behavior of the atmosphere, especially the ionosphere. Additional radiation is transmitted by the solar wind, a magnetized plasma moving at speeds between 300 and 800 km/s [Pisacane, 2005a]. Shorter time-scale events also occur, such as solar flares and coronal mass ejections. Solar flares emit extreme ultra-violet (EUV) and X-rays, which reach the Earth in minutes and can last hours, causing atmospheric heating, notably in the ionosphere. Sometimes particle events occur, especially during solar maximum. Solar flares also produce radio emissions that may temporarily interrupt communications [Tribble, 2003]. Coronal mass ejections (CMEs) are extremely energetic events whose shock waves reach Earth in hours, with the actual CME arriving days later. The leading wave of the CME is composed of energetic particles leading to radiation effects; the CME itself creates a geomagnetic storm that can reconfigure Earth's magnetosphere and interfere with power transmission on Earth's surface [Hastings and Garrett, 1996; Pisacane, 2005a].

Besides the influence of the Sun, designers and operators need to be aware of Earth's geopotential effects and the particulate environment stemming from micrometeoroids and man-made hazards. Gravity gradient torque effects impact flexible spacecraft operations, and orbital debris represents a significant hazard to spacecraft and mission operations.

3.1.3 Observe-Orient-Decide-Act Loop

Space environment modeling accuracy and implementation are important to the Observe-Orient-Decide-Act (OODA) loop, due to the feedforward nature of the Observe and Orient steps to the Act and Decide steps [Boyd, 2018]. Figure 3-1 displays space weather effects on space vehicles and signals. Incorrect modeling or neglect of the effects of the space environment can lead to wrong decisions on anomaly resolution and mitigation strategies. In the other direction, misuse of the feedback from the Act and Decide steps to the Observe and Orient steps may entail wrongfully assimilating data, leading to prediction inaccuracy and imprecision. Understanding what the fundamental drivers are and how they manifest is crucial in making sure that the appropriate level of information is gathered and forward fed to the Orient, Act, and Decide steps. During the Act step more observations and information may be needed to feed back to the Orient step.

Understanding the space environment and its impacts on varying time scales of object motion and vehicle systems is critical to attributing telemetry and observation of recorded events, as well as predicting the evolution of a large population of space objects and debris. We need to bring to bear lessons learned in space vehicle design, space environment physics, orbit and attitude physics, and estimation theory. These have implications for object tracking (force and torque perturbations for trajectory augmentation), asset fault detection and mitigation (hardware upset events, hardware degradation, signal transmission alteration, sensor occultation), and control (thermal control, attitude control, guidance law, and trajectory design).

With proper physical modeling and feedback between the Observe-Orient-Decide-Act steps, we can operate within an event's OODA loop, operate through the event, and own the outcome. For example, properly understanding and accounting for the South Atlantic Anomaly allows attitude control system designers to use alternative control means (e.g., reaction wheels) during the encounter to maintain proper attitude control. Additionally, even though we don't know when hardware upset events will occur, accounting for such losses makes the system more resilient to operate through events without additional input from ground operators.

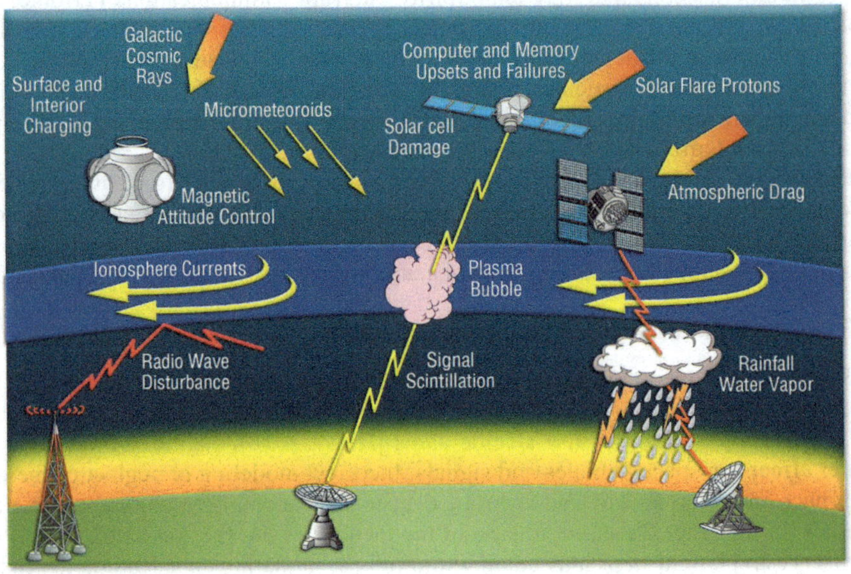

FIG. 3-1. Effects of Space Weather Environment on Space Vehicles and Signals [Lanzerotti et al., 1997]. Accurate modeling of effects and correct use of data are vital to the success of the Observe-Orient-Decide-Act (OODA) loop.

3.1.4 Other Examples

The United States' first satellite, Explorer I, whose initial science results led to the Van Allen Belt discovery, is probably the first case study in resolving an on-orbit anomaly due to the space environment and its impact on object dynamics. Figure 3-2 displays a picture of Explorer I, in which a critical aspect of environmental effects on object attitude motion was overlooked—dissipation of rotational energy. The vehicle was designed to spin about its axis of least inertia while transmitting information from its on-board cosmic ray instrument. The antennas dissipated the rotational energy about the axis of least inertia, causing the satellite to seek its lowest energy state, which resulted in a tumble about its axes of largest inertia, as predicted by Euler's equation. Unfortunately, this effect was not accounted for until after the "anomaly" was observed via communication irregularities.

The South Atlantic Anomaly (SAA) is a dip in the magnetic field of the Earth that allows higher energy particles and radiation to penetrate closer to Earth. This corrupts extremely sensitive scientific instruments and vehicles utilizing magnetic torque rods for attitude control. For example, on the Hubble Space Telescope, certain scientific instruments cannot be used and its magnetic torque bars don't function for attitude control when transiting this region.

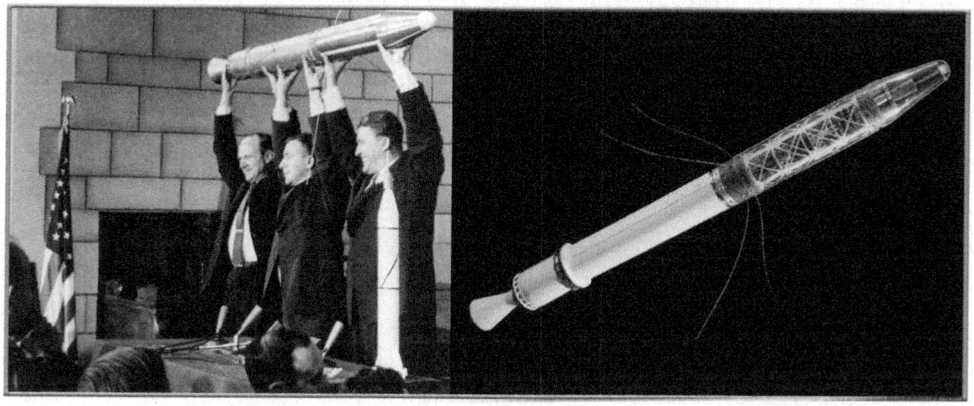

FIG. 3-2. **(Left) Explorer I held by (left-to-right) William H. Pickering, James A. Van Allen, and Wernher von Braun; (Right) Explorer I.** This satellite, the first launched by the U.S., delivered data that led to the discovery of the Van Allen Belts.

3.2 Dynamical Modeling

3.2.1 Force, Torque, and Attitude Equations

Here we give the general dynamic equations for simulation and prediction of space object motion in the space environment. This section is mainly concerned with modeling and attributing mismatches between predicted dynamics and observations, but another point is the coupling between attitude and trajectory. Figure 3-3 shows the inertial and body frames, denoted by "n" and "b" respectively, of an arbitrary object that is spinning with a given angular velocity, $\boldsymbol{\omega}^{\mathbf{b/n}}$. In the figure, "**R**" is a vector from the origin of the inertial frame to the origin of the body frame (assumed to be at the center of mass of the object and oriented along its principal inertia axes) while "ρ" is a vector from the body frame's origin to an arbitrary location on the body (in this case a differential mass element dm), and "$\mathbf{r}_1$" is a vector from the origin of the inertial frame to the arbitrary location on the body related by Equation (3-1).

$$\boldsymbol{r}_1 = \boldsymbol{R} + \boldsymbol{\rho} \tag{3-1}$$

$$\frac{d}{dt}(\boldsymbol{r}_1) = \frac{d}{dt}(\boldsymbol{R}) + \frac{d}{dt}(\boldsymbol{\rho})$$

Equation (3-2) summarizes the transformation of vectors between the body and inertial frames using the direction cosine matrix (DCM) as well as their time rate of change. The rotation between frames and the time rate of change are critical because some perturbing force and torque effects are more easily expressed in a body frame and rotated into the inertial frame. For notation, $\boldsymbol{a}$ represents a vector quantity, ${}^{b}\boldsymbol{a}$ represents the vector $\boldsymbol{a}$ expressed in the body frame bases, $\boldsymbol{b}$, ${}^{n}\boldsymbol{a}$ represents the vector $\boldsymbol{a}$ expressed in the inertial frame bases, $\hat{\boldsymbol{n}}$. The DCM elements are computed from the dot products of the body bases and inertial bases expressed in the inertial frame. For example, the first row, second column element of the DCM is given by the dot product of the first body frame basis and the second inertial frame basis, $C_{12} = \hat{\boldsymbol{b}}_1^T \hat{\boldsymbol{n}}_2$, with both bases expressed in the inertial frame. For the time rate change of vectors in the inertial frame, $\boldsymbol{\omega}^{b/n}$ is the angular rate of the body frame with respect to the inertial frame and is used for computing the rate of change of the DCM when the chain rule is applied to the position level.

$$\hat{b} = C\hat{n} \tag{3-2}$$

$$C_{ij} = \hat{b}_i^T \hat{n}_j$$

$$^b a = C\, ^n a$$

$$\frac{d}{dt}C = -[\omega^{b/n} \times]C$$

$$\frac{^n d}{dt}(^b a) = C^T\left[\frac{^b d}{dt}(^b a) + \omega^{b/n} \times {}^b a\right]$$

Equation (3-2) makes use of the cross-product skew-symmetric matrix given by Equation (3-3), but $\boldsymbol{a}$ is replaced by $\boldsymbol{\omega}^{b/n}$.

$$[a \times] = \begin{bmatrix} 0 & -a_3 & a_2 \\ a_3 & 0 & -a_1 \\ -a_2 & a_1 & 0 \end{bmatrix} \tag{3-3}$$

The general equation for accelerations experienced by a space object can be given according to Newton's Second Law or Gauss' variational equations. See Chapter 4: Astrodynamics for more information on how different models are generated, derived, and implemented. Equation (3-4) is expressed in the inertial frame (e.g., J2000).

$$\ddot{r} = a_{two\text{-}body} + a_{gravitational} + a_{neutrals} + a_{plasma} + a_{radiation} \\ a_{maneuver} + a_{impact} + a_{dist} \tag{3-4}$$

$a_{two\text{-}body}$ $= a_{two\text{-}body} = -\dfrac{\mu}{\|r\|^3}r$

$a_{gravitational}$ $= \ddot{r}_{Earth\ geopotential} + \ddot{r}_{Sun} + \ddot{r}_{Moon} + \ddot{r}_{planetary\ bodies}$

$a_{neutrals}$ = atmospheric drag and particle interaction

a_{plasma} = spacecraft charging and interaction with magnetic field

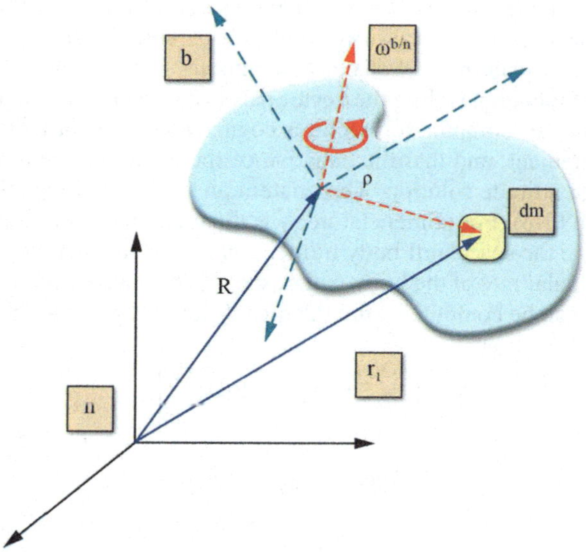

FIG. 3-3. Reference and Body Frame.

$a_{radiation}$	= Earth re-radiation pressure and solar radiation pressure
$a_{maneuver}$	= orbit maintenance and stationkeeping
a_{impact}	= micrometeoroid and orbital debris impact
a_{dist}	= unmodeled disturbances

The general rigid body equations for torques experienced by a space object about the center of mass are given by Equation (3-5) according to Euler's law [Wertz, 2002]. Here $\boldsymbol{\omega}^{b/n}$ is represented by $\boldsymbol{\omega}$ to allow for better readability. For nomenclature, $\boldsymbol{H}_{sys}$ is the total angular momentum of the spacecraft and controls (reaction wheels, control moment gyroscopes, magnetic torque rods, etc.), $\boldsymbol{L}$ is the summation of all external torques (i = [1, n]) on the spacecraft, $\boldsymbol{\rho}$ is the vector from the center of mass of the spacecraft to the applied external force, $\boldsymbol{F}$ is the external force on a particular spacecraft panel or facet, and $\boldsymbol{I}$ is the inertia tensor of the spacecraft.

$$\frac{d}{dt}(H_{sys}) = I\dot{\omega} + \omega + I\omega = L_{external} \tag{3-5}$$

$$L_{external} = \sum_{i=1}^{n} \rho_i \times F_{external,i}$$

$$L_{external} = L_{gravity\ gradient} + L_{drag} + L_{magnetic} + L_{particulate}$$

The torque affects the forces by rotating the exposed area to the sun or local velocity vector, which in turn affects the torques by changing the forces experienced by the panels and facets of the vehicle. Common singular representations are Euler angles (roll, pitch, yaw) as well as the modified or classical Rodriguez parameters; the common non-singular representation is the quaternion [Schaub and Junkins, 2003]. The scalar element is the cosine of half the Euler angle while the vector part is the product of the Euler axis and sine of half the Euler angle. Computation requires cognizance of the order of the elements, the unit normal requirement, and that the negative of the quaternion and original quaternion will give the same attitude solution. The quaternion differential equation is linear and is given by Equation (3-6). For nomenclature, q is the quaternion for attitude mapping from the inertial frame to the spacecraft body frame, $Im \times m$ is the identity matrix of size m by m, and ω is the angular rate of the body with respect to the inertial frame expressed in body frame. The DCM can be computed from the quaternion using Equation (3-7).

$$\frac{d}{dt}(q) = \Xi(\omega)q = \Psi(q)\begin{bmatrix} 0 \\ \omega \end{bmatrix} \tag{3-6}$$

$$\text{subject to: } \|q\| = 1$$

$$q = [q_s, q_\upsilon]^T$$

$$q_s = q_0 \, , q_\upsilon = [q_1, q_2, q_3]$$

$$\Xi(\omega) = \frac{1}{2}\begin{bmatrix} 0 & -\omega^T \\ \omega & [\omega \times] \end{bmatrix}$$

$$\Psi(q) = \frac{1}{2}\begin{bmatrix} q & \begin{matrix} q_\upsilon^T \\ q_s I_{3 \times 3} + [q_\upsilon \times] \end{matrix} \end{bmatrix}$$

$$C = I_{3 \times 3} - 2q_s[q_\upsilon \times] + 2[q_\upsilon \times][q_\upsilon \times] \tag{3-7}$$

From the above, we see that the torque directly affects the force and vice-versa. Often the effect of imposed torques takes a long time to manifest, and dominance of the gravity force may mask effects of the smaller conservative and nonconservative forces. So it is important to understand how, when, or even if one can separate out effects of force and torque of various perturbations.

Due to the speed at which the force states (position, velocity) and torque states (angular velocity, attitude) tend to change, numerical stiffness does arise when attempting to simultaneously solve both sets of equations. Integrator choice, coordinate changes, and state normalization can combat problems with differential equation stiffness, but one must be aware of these nuances; otherwise the prediction may be inaccurate and observations inappropriately timed.

In the above, we assumed rigid body behavior for modeling. This is not always the case with flexible pieces of insulation which may have separated from rocket bodies (with or without other large pieces of debris) but whose orbital motion is more in line with objects whose area to mass ratio is much greater than normal (0.01–0.02 m^2/kg). These are commonly known as high area to mass ratio (HAMR) objects [Schildknecht et al., 2005]. Flexibility of the body also has implications when on-board accelerometers are used to characterize non-conservative forces, such as those experienced by the CHAMP, GRACE, and Mars Reconnaissance Orbiter. In such cases, flexibility can introduce additional measured effects or noise.

3.2.2 Gravitational Environment

Modeling of Earth's gravitational environment is important for high accuracy trajectory prediction as well as gravity gradient effects on large space vehicles such as Hubble Space Telescope. Complete gravity force modeling incorporates the effects of additional bodies or "third bodies" (Sun, Moon, Jupiter, etc.) along with the geopotential modeling of a non-spherical Earth. Gravitational models for Earth are highly accurate but are sometimes computationally burdensome when solid tide or high order and degree evaluations are required [Vallado, 2007]. Accounting for the Sun and Moon is critical for relativity, as shown by Montenbruck and Gill, 2000. Their modeling techniques will be discussed more in Chapter 4: Astrodynamics. Here we concentrate on the torque on the spacecraft due to Earth's non-spherical shape.

The gravity gradient torque derivation for a non-spherical object subject only to two-body acceleration is in Montenbruck and Gill, 2000. To account for a non-spherical Earth the differential force equation is modified slightly [Gottleieb, 1993]. Using matrix C from Equation (3-7) to map inertial frame differential force to body frame differential force gives Equation (3-8), and the overall torque on the rigid body gives Equation (3-9). Higher order geopotential force is commonly modeled in the Earth-Centered-Earth-Fixed (ECEF) frame, so we must rotate from ECEF to Earth-Centered-Inertial (ECI) and then from ECI to the spacecraft body frame to form the direction cosine matrix. For nomenclature, $a_{earth\text{-}geopotential}$ is the acceleration resulting from higher degree and order gravity terms, G is the first gradient of the potential where is the gradient of the higher order gravity vector (theoretically the trace of (i.e., the Laplacian) equals zero due to the gravity potential field, but numerically this is not always the case). If $\boldsymbol{\rho}$ is taken from the center of mass of the spacecraft, then the first integral in the body frame torque equation is zero by definition.

$$L_{gravitygradient} = \int_{Body} \rho \times \, dF_{gravitygradient} \tag{3-8}$$

$${}^{b}dF_{gravitygradient} = C({}^{n}a_{gravity} + \nabla_r {}^{n}a_{gravity} C^T \rho)dm$$

$${}^{b}L_{gravitygradient} = \underbrace{\int_{Body} \rho \times C^{n} a_{gravity} dm}_{0 \text{ if } \rho \text{ from Mass Center}} + \int_{Body} \rho \times C \nabla_r a_{gravity} C^T p dm$$

$$^{n}a_{gravity} = a_{two\text{-}body} + a_{earth\text{-}geopotential}$$

$$^{b}L_{gravitygradient} = \int_{Body} \rho \times G\rho dm \tag{3-9}$$

$$^{b}L_{gravitygradient} = \begin{bmatrix} G_{23}(I_{33}-I_{22}) - G_{13}I_{12} + G_{12}I_{13} - I_{23}(G_{33}-G_{22}) \\ G_{13}(I_{11}-I_{33}) + G_{23}I_{12} - G_{12}I_{23} - I_{13}(G_{11}-G_{33}) \\ G_{12}(I_{22}-I_{11}) - G_{23}I_{13} + G_{13}I_{23} - I_{12}(G_{22}-G_{11}) \end{bmatrix}$$

$$G = C\nabla_r {}^{n}a_{gravity} C^T$$

Equation (3-9) is different from that given in Gottlieb [Gottlieb, 1993] because there Equations 8-16 and 8-30 use the nomenclature of

$$I_{i,j} = \int_{body} \rho_i \rho_j dm$$

which is counter to the traditional definition

$$I_{i,j} = -\int_{body} \rho_i \rho_j dm$$

for the off-diagonals of the inertia matrix [Schaub and Junkins, 2003]. Equation (3-9) reduces to Equation (3-10), expressed in body frame, when we consider only two-body motion.

$$^{b}L_{gravitygradient} = C\left(3\frac{\mu}{\|R\|^5}R \times iR\right) \tag{3-10}$$

Figure 3-4 plots the norm difference of the torque vectors computed by the higher order geopotential model and the two-body gravity model at an altitude of 400 km. The difference between the 2 × 0 gravity gradient torque and two-body is approximately the same order of magnitude as that shown in the figure. The inertia tensor is [477 63 0; 63 770 0; 0 0 821] Nm, the direction cosine matrix is a 2-3-1 rotation with a pitch angle of 20°, a yaw angle of 30°, and a roll angle of 40°.

3.2.3 Particulate Environment

The particulate environment comprises both man-made and natural objects. The man-made objects are uncontrolled debris (inoperable satellites, rocket bodies, paint flecks, etc.) along with active objects (under control of operators). Natural objects include micrometeoroids, which pose hazards not only to space objects but to manned missions. The

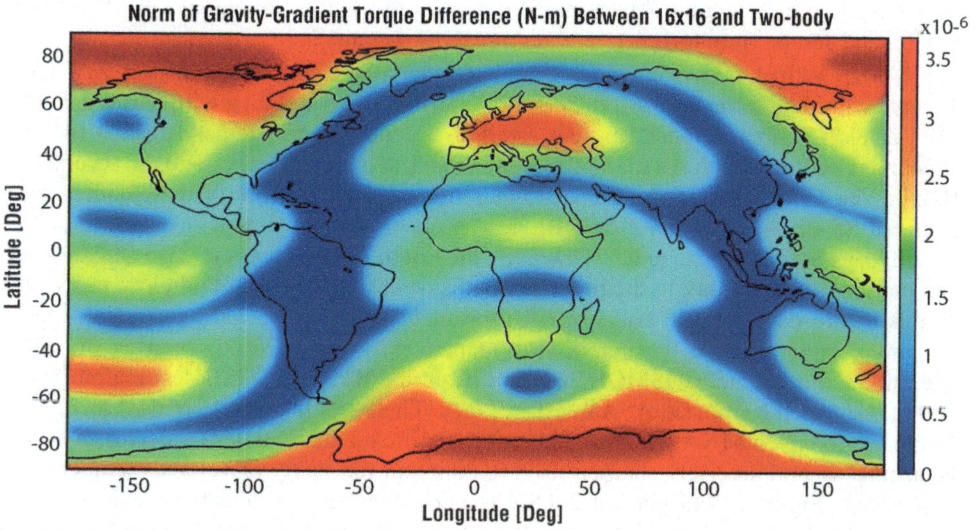

FIG. 3-4. **Difference of Gravity Gradient Torque Computations at 400 km Between Higher Order Geopotential and Two-Body.**

Long Duration Exposure Facility (LDEF) mission sought to characterize micrometeoroid and space debris impacts. Post-analysis found that 30% of the impacts on the trailing surfaces were from orbital debris attributed to highly eccentric orbits [Stuckey, 1993]. The Space Shuttle and International Space Station have been impacted by micrometeoroids; at least two shuttle windows per mission had to be replaced [National Research Council, 2011]. In 1996, the French CERISE satellite had its gravity-gradient boom severed by an untracked piece of debris later attributed to an Ariane rocket and considered to be the first verified case of an on-orbit collision of two objects [Alby et al., 1997; Surrey Satellite Technology LTD, 2016].

Orbital debris stems from inactive payloads, rocket bodies, break-ups, fuel particulates, paint flecks or particulates from surface degradation, or separation of thermal blankets or protective shields, giving rise to the HAMR object population.

Prediction and compensation methods for collision mitigation, estimation of the debris population dispersion and evolution, and reconstruction of past events are on-going areas of research [Akella and Alfriend, 2000; Jenkin, 2004; Klinkrad, 2006; Romgrens et al., 2013]. Current estimates are that there are approximately 18,000–23,000 payloads, rocket bodies, and debris fragments on-orbit, with about 94% being inoperative [NASA, 2016; ESA, undated]. For conjunction screening, a common practice is to use a scaled control volume around the object in question to compensate for the lack of knowledge about other objects [Klinkrad, 2006]. Communication and exchange of uncertainty information are key to being able to confidently assess the conjunction probability and predict the evolution of the debris environment. Chapter 4 discusses these matters in detail. Both NASA and ESA have debris environment models, ORbital Debris Engineering Model (ORDEM) and Meteoroid And Space debris Terrestrial Environment Reference Model (MASTER) respectively, which predict object spatial density and the flux onto a spacecraft for large objects and small (1mm to 1cm) particulates [Krisko et al., 2015].

In March 2014, post-event analysis showed that a South African CubeSat, ZACUBE-1, came within an estimated 200 meters of COSMOS 2151 and then within 100 meters of METEOR 2-5, both defunct Russian satellites [Republic of South Africa, 2014]. The encounters occurred within a span of 24 hours, the first over Antarctica and the second over Brazil. The United States Joint Space Operations Center did notify the South African National Space Agency that ZACUBE-1 was on close approach with COSMOS 2151 several days before the encounter. However, the South African press release noted, "There is a margin of uncertainty in the predicted paths of the satellites, and all close encounters of this nature can result in collision." Unfortunately, not all scenarios result in objects remaining intact, due to the inability to maneuver [NASA, 2009], such as the active Iridium-33 and defunct COSMOS 2251 collision in 2009, or planned destruction [Kelso, 2007], such as the Chinese ASAT test in 2007. These conjunctions, as well as on-orbit fragmentation events [NASA, 2008], produce significant debris whose trajectories need to be accurately and precisely characterized to maintain the safety of vehicles with crossing trajectories. Figure 3-5 displays a 2008 estimate of the spatial density of objects in low Earth orbit (LEO) and Figure 3-6 shows the same for the near-GEO altitudes (about 35,800 km). The disposal orbit for GEO objects is approximately 300 km above GEO [Jehn et al., 2005]. Debris has also been observed in highly eccentric and GEO transfer orbits, but their low perigee allows for their removal at a much faster rate than the 100,000 years estimated of the GEO graveyard orbits.

More recently, DigitalGlobe's WorldView-2 Satellite in low Earth orbit suffered a debris creating event where an impact with an untracked object caused at least eight pieces of additional debris [Gruss, 2016]. In August 2016, an estimated 5-millimeter-sized particle struck a European Space Agency Earth environment monitoring satellite's solar array. At the time of impact, power reduction was noticed as well as small changes in the attitude and orbit of the spacecraft [ESA, 2016]. Cameras on board the spacecraft identified an approximately 40-cm area of deformation on one of the 2 m × 10 m solar arrays of the 2300-kg vehicle in a sun-synchronous orbit at 693 km. Most spacecraft do not possess onboard structural health monitoring devices, so it will be interesting to see what conclusion would be generated given present analysis methodologies using relayed telemetry data or ground sensor tasking follow-up, as were available for Galaxy-15 and WorldView-2.

3.2.4 Neutrals Environment

The neutral environment interacts with objects having low perigees. Collision with neutrals at LEO speeds causes degradation of the surface material. The presence of highly reactive atomic oxygen in this regime leads to surface erosion and causes visible glow effects [Tribble, 2003]. Because of potential surface contamination from thruster firings and atomic oxygen, care is necessary with the placement of on-orbit remote sensing equipment. Delineation between upper atmospheric and outer atmospheric environment varies, but the Von Karman altitude of 100 kilometers is typically used as the transition point where astrodynamics forces dominate aerodynamic forces.[41] Other boundary altitudes exist depending upon basic science or human safety factor purposes. The altitude

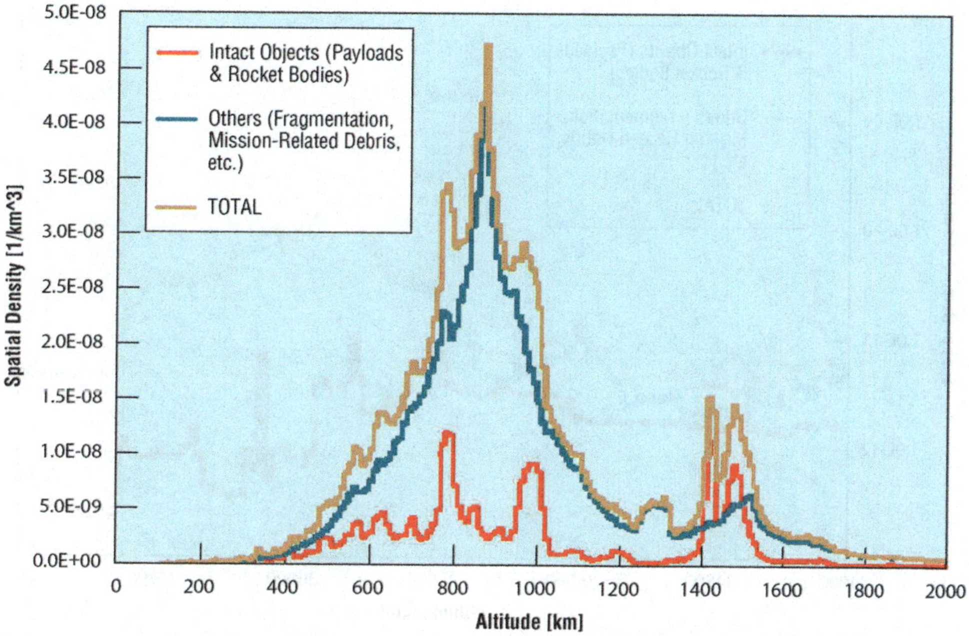

FIG. 3-5. **Spatial Density per Cubic Kilometer for Altitudes Less than 2000 km [NASA, 2008].** Even small debris fragments pose serious hazards to orbiting spacecraft and other space missions.

at which Earth's atmosphere attains its minimum temperature is 85 kilometers (mesopause) [Pisacane, 2005a] and NASA used 400,000 feet (~120 kilometers) for Apollo's entry interface for insertion maneuvers [Woods, 2008].

The outer atmospheric environment (greater than 100 km) is composed of neutral gases whose concentration depends on Earth's gravity and solar radiation. Forecasting the atmospheric environment is extremely complicated due to the sparsity, low accuracy, and lack of availability of high-altitude atmospheric monitoring systems as well as having to account for the impact of the solar cycle. Solar and geomagnetic behavior drive the exospheric temperature, which drives the atmospheric density, but precisely how the exospheric temperature and density are related is not well known, which is largely why there are so many different models [Pisacane, 2005b].

Modeling of the environment composition and behavior is crucial for prediction of objects with perigee below 1000 km and is critical for objects re-entering the atmosphere. The exosphere (~ 400 km to 3 Earth radii) is the outer area of the atmosphere, where the mean free path of molecules is large enough to not require a continuum model for interaction with space objects. The mean free path is the average distance a gas molecule travels between collisions and defines the flow regime through the Knudsen number (Kn) [Hastings and Garret, 1996], Equation 3-11. If the Knudsen number is $>>1$, the flow is considered free, with the molecules on ballistic trajectories; $Kn < 0.1$ is deemed continuum

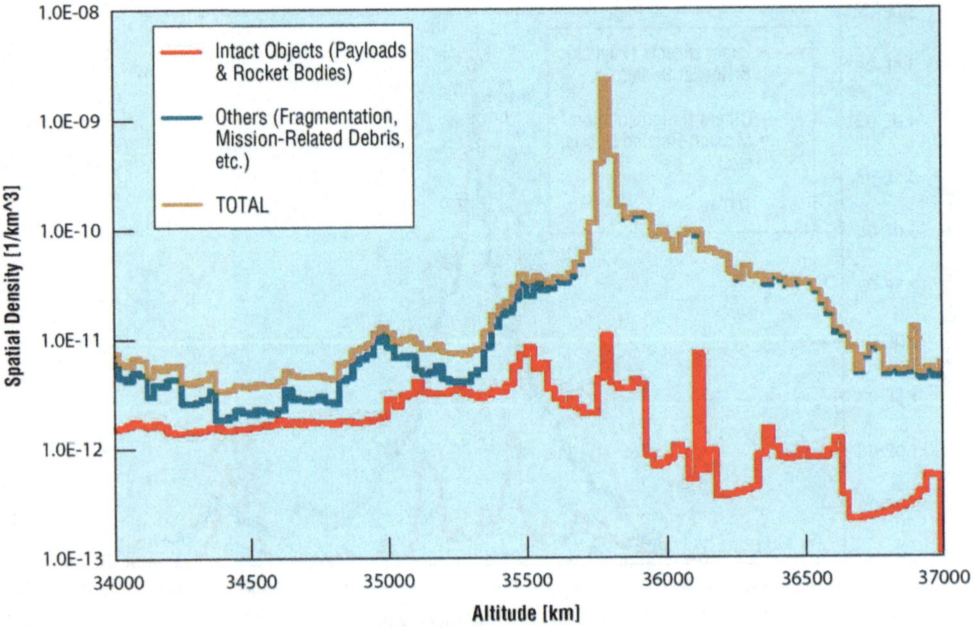

FIG. 3-6. Spatial Density per Cubic Kilometer for Near Geosynchronous Altitude [NASA, 2008]. Debris at near-GEO altitudes can last up to 100,000 years.

flow and Kn ~1 is deemed transitional flow. By defining the flow regime through the Knudsen number, a proxy is established for determining when we need to consider computational fluid dynamics and thermodynamics interactions in computing forces and torques on the object body. Prediction of a re-entering object's trajectory and its break-up point requires analysis that can hand off information between the free molecular and transition regimes as well as the transition and continuum regimes. Simulation analysis should consider uncertainty of the object's initial attitude and trajectory to provide a sensitivity analysis with respect to state and parameter uncertainty and a confidence measure of the solutions.

$$Kn = \frac{\lambda}{L_b} \tag{3-11}$$

λ = mean free path

L_b = length of body (space object)

The key point for the effects of the neutral environment is the amount of uncertainty associated with the characterization of its constituent components for prediction of drag effects. Table 3-1 reports the estimated uncertainty for the parameters, with respect to nominal, for objects in low Earth orbit. The table demonstrates that solar fluctuation is the predominant driver for variations. Measures of solar emission, such as F10.7, are commonly used to address temperature fluctuation, but these indices are typically indirect and time-lagged. The F10.7 index is the measurement of solar flux at the 10.7-cm wavelength and is highly correlated with sunspot number and solar flare activity [Pisacane, 2005b]. A swing of 300 K can vary the density by up to a factor of six, depending on solar and geomagnetic behavior. The ballistic coefficient is the product of the drag coefficient and area exposed to the flow divided by mass. For a cannonball, the ballistic coefficient is constant, but it varies with orientation for faceted models of spacecraft that are composed of different materials or project a different cross-sectional area antiparallel to the velocity. For objects whose drag coefficient, area-to-mass ratio, and trajectory are known extremely well, the modeling uncertainty is due to atmospheric density, but fluctuations in solar activity due to solar storms can severely increase error in estimating atmospheric density. Objects whose vehicle models are not known or characterized prior to orbit require the ballistic coefficient to be estimated in-flight, thereby creating a separation problem between the product of atmospheric density and ballistic coefficient and increasing uncertainty for the parameters. For non-spherical objects that are slewing or rotating such that their projected areas vary, uncertainties for the ballistic coefficient are much higher than the table.

TABLE 3-1. Cannonball Drag Acceleration Parameter Uncertainties [Pardini and Anselmo, 2001; [Pisacane, 2005b; Marcos et al., 2001].

Parameter	Estimated Uncertainty
Atmospheric Density	5% (well-known object, quiescent solar conditions) 20% (unknown object, quiescent solar conditions) Up to 100% (solar storms)
Ballistic Coefficient (C_dA/m)	5%–60% (depending upon solar condition)
Relative Velocity (Wind Uncertainty)	5%

The analytical drag coefficient, C_d, for a sphere is 2.0 with 2.2 commonly used. Pisacane, 2005b; Sutton et al., 2007; Doornbos, 2010 survey other methods for computing the lift and drag coefficients. Atmospheric models vary from empirical [Standard Atmosphere, Harris Priester; see Montenbruck and Gill, 2000 for comparison] to computational fluid dynamic models (Thermosphere-Ionosphere General Circular Model, Global Ionosphere Thermosphere Model) and approximations thereof [Jacchia, Jacchia-Bowman, MSIS; see Akins et al., 2003 and Bowman et al., 2008 for comparison]. The empirical models entail less computational burden, but the physics-based hydrodynamic circulation models are more accurate and less susceptible to biased data. However, model accuracy depends highly on the amount of data available and the initial uncertainty of the input parameters, which may or may not be readily available. Severe fluctuations of the environment can severely alter orbit trajectory and orbital lifetime.

The vacuum of space leads to out-gassing of materials. Out-gassing in turn can lead to inadvertent coating of vehicle and sensors as well as acting as a thrust force. Atomic oxygen is highly reactive, degrading sensor coatings and eroding surface materials. The LDEF experienced contamination on optical surfaces, causing degradation of performance; removal of the contamination restored performance in some cases with some contamination attributed to out-gassing effects [Stuckey, 1993].

The ionosphere is generated from the same solar radiation that drives the exospheric temperature and causes neutrals to become ionized. It is due to the ionosphere that propagation of amplitude modulated (AM) radio waves is better at sunset [Pisacane, 2005a]. The ionosphere and its perturbations absorb or alter the transmission path of communication signals and signals used for navigation [Bothmer and Daglis, 2007] above 80 km, where the atmospheric density is no longer able to absorb all ionizing wavelengths. Accurate forecasting is critical for mitigation and investigation of these effects, which led the Air Force Research Laboratory to conduct the Communication/Navigation Outage Forecasting System (C/NOFS) mission [377th Air Base Wing, 2008; AFRL, 2008]. Solar radiation also causes material degradation, making surface materials redden over time. This is important for space object characterization, which uses light reflectance time histories and models.

Drag is caused by interaction with neutral molecule collisions. Examination of the drag effect on an object's orbital elements using variation of parameters shows secular deterioration in semi-major axis and eccentricity as well as periodic behavior for semi-major axis, eccentricity, argument of perigee, and mean anomaly [Schaub and Junkins, 2003]. The overall effect of drag is to reduce altitude and circularize the orbit. Figure 3-7 compares the drag force experienced by a spherical object with orbits of varying eccentricity and a ballistic coefficient of 0.22 m^2/kg to the J_2 perturbation (see Chapter 4: Astrodynamics) and the two-body gravity term.

As an example of the variability of orbit lifetime as a function of solar condition, the lifetime at an initial altitude of 400 kilometers ranges from four years at solar minimum to seven months at solar maximum [Lean et al., 2006]. Because of the large density variability in altitude and the dependence upon solar parameters, a spacecraft may experience more than an order of magnitude difference in atmospheric drag due to strong solar or geomagnetic activity [Pisacane, 2005a]. Figure 3-8 displays the time for a spacecraft's initial perigee altitude to decay to 180 kilometers depending upon the solar condition and the estimated density with a ballistic coefficient of $0.022m^2/kg$. The density estimates vary by up to 30% based upon the differences observed in literature; see Table 3-1. Neglecting the effect of the environment can lead to severe misconceptions of object trajectory, which reduces our ability to predict the re-entry path and uncertainty envelope.

Accurate and reliable state uncertainty prediction depends on understanding motion models and characterizing uncertainty in initial conditions. If the motion model possesses a degree of uncertainty, we can use data assimilation techniques to update model parameters based upon system observability, availability of new measurements, and the measurement uncertainty. Yurasov et al., 2008, demonstrated that data assimilation methodologies can improve re-entry predictions for spherical satellites with known physical characteristics by updating parameters within the atmospheric density model. However, non-spherical satellites still pose a significant problem due to the time-varying ballistic coefficient. Additionally, problems arise as to the appropriateness of the amount of parameter change in the density models, since the changes may only reflect local or short-term behavior instead of global or long-term behavior. It is thus important to

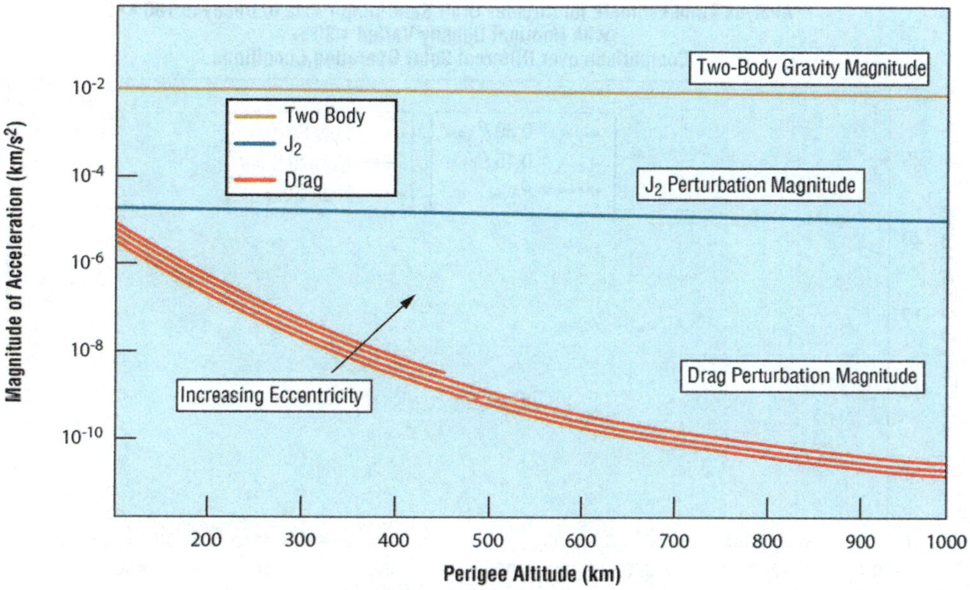

FIG. 3-7. Comparison of Dominant Forces for Low Earth Orbits.

understand the general motion of the object, the effects of the space environment, and what the information from sensing resources is telling us about the object and its motion.

The effects of lift and side forces, Equations (3-12) and (3-13), are minimal but the lift formulation is included for completeness [Sutton et al., 2007; Doornbos, 2010]. For nomenclature, is the velocity of the spacecraft relative to the atmosphere, $\hat{\boldsymbol{n}}_{Facet,i}$ represents the unit normal of the facet that is in the flow, $\rho_{density}$ is the atmospheric density as a function of altitude and solar conditions, $\boldsymbol{\rho}_i$ is the distance from the vehicle center of mass to the facet considered, $C_{d,i}$ is the coefficient of drag of the facet, $C_{l,i}$ is the coefficient of lift of the facet, *i* represents the index of the particular facet, and N is the total number of facets in the spacecraft model). If a summation of only one term is used, the equations reduce to the conventional "cannonball drag" assumption of the drag force. The dot product of the facet normal and relative velocity represents the percentage of the facet's area exposed to the flow. The differential form of the equations is with respect to the differential area exposed to the flow.

$$F_{Drag} = -0.5\sum_{i=1}^{N} \rho_{density}\frac{C_{d,i}A_i}{m}(\hat{n}^T_{facet,i}V_{rel,atmo})V_{rel,atmo} \tag{3-12}$$

$$L_{Drag} = -0.5\sum_{i=1}^{N} \rho_{density}\frac{C_{d,i}A_i}{m}(\hat{n}^T_{facet,i}V_{rel,atmo})(\rho_i \times V_{rel,atmo})$$

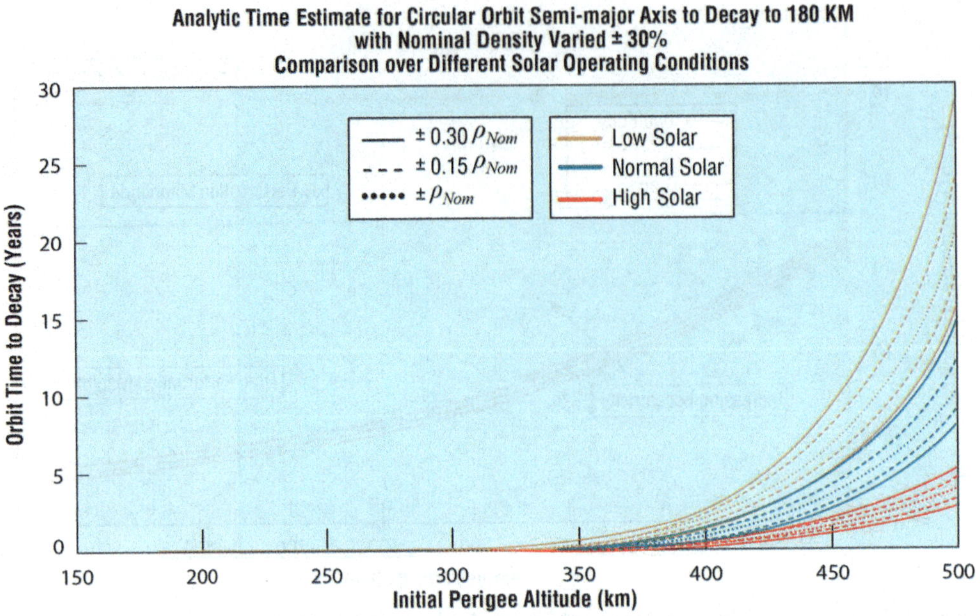

FIG. 3-8. **Time to Decay to 180 Kilometers from Initial Altitude Depending Upon Solar Conditions.** Solar conditions can make an order of magnitude difference in spacecraft lifetime; density variations can make a difference of a factor of two.

$$F_{Lift} = -0.5\sum_{i-1}^{N}\rho_{density}\frac{C_{l,i}Ai}{m}(V^T_{rel,atmo}V_{rel,atmo})\frac{(V_{rel,atmo}\times\hat{n}_{facet,i}\times V_{rel,atmo})}{\|V_{rel,atmo}\times\hat{n}_{facet,i}\times V_{rel,atmo}\|} \quad (3\text{-}13)$$

$$L_{Lift} = -0.5\sum_{i-1}^{N}\rho_{density}\frac{C_{l,i}Ai}{m}(V^T_{rel,atmo}V_{rel,atmo})\rho_i\times\frac{(V_{rel,atmo}\times\hat{n}_{facet,i}\times V_{rel,atmo})}{\|V_{rel,atmo}\times\hat{n}_{facet,i}\times V_{rel,atmo}\|}$$

3.2.5 Plasma Environment

Plasma is composed of charged particles and interacts with magnetic fields. Earth's natural magnetic field, the magnetosphere, is condensed by the solar wind composed of ions and electrons and gives rise to the magnetopause (stagnation point between the solar wind and Earth's magnetosphere). The magnetopause is typically around 10 to 15 Earth radii from the center of Earth on the sunlit side, but coronal mass ejections or solar storms can cause the magnetopause to be pushed inward to around three Earth radii. The solar wind speed and electron density are quite variable (300 to 900 km/s and 5 to 40 cm^{-3} respectively), making the magnetosphere a highly dynamic system with response times on the order of 10 minutes. Figure 3-9 plots the three-day time history of the magnetopause and the instant magnetic field line conditions reported from the GOES spacecraft [CCMC, 2016].

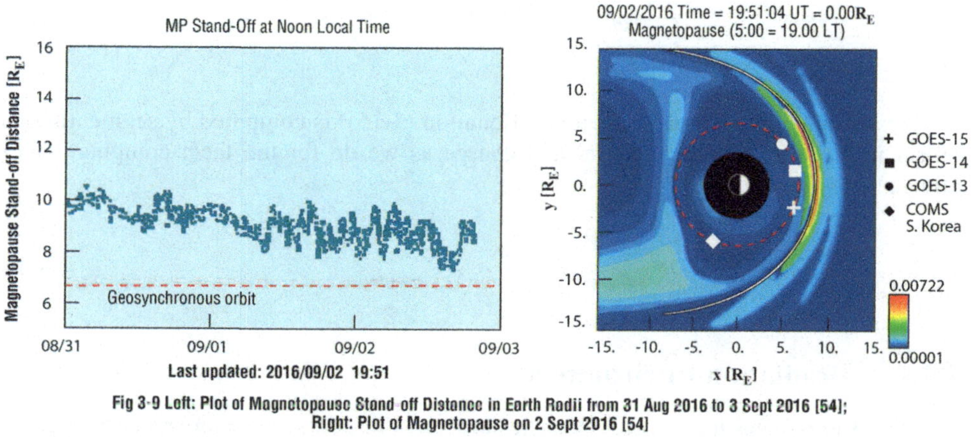

FIG. 3-9. **Magnetopause Data from Aug-Sep 2016.** Left: Plot of magnetopause stand-off distance in Earth radii from 31 Aug 2016 to 3 Sept 2016 [CCMC, 2016]; Right: plot of magnetopause on 2 Sept 2016 [CCMC, 2016].

Magnetosphere compression often causes unanticipated charging of spacecraft, leading to arcing across different materials, damaging solar arrays, electronics, or instruments because the spacecraft collects a positive charge when sunlit and a slightly negative charge in eclipse. Besides the direct implications to medium Earth orbit (MEO) and GEO spacecraft, plasma effects also occur for polar orbiting LEO spacecraft, where negative charging is sometimes several thousand volts. The highly dynamic nature of the magnetosphere can inject large errors into the attitude determination system for spacecraft using vector magnetometers. Modeling the magnetic field is important for on-board attitude determination algorithms that use vector magnetometers and missions using magnetic torque rods for attitude control, such as Hubble Space Telescope.

References CCMC, 2016; Frueh et al., 2014; and Paul and Frueh, 2016 discuss differential charging and the incorporation into orbital evolution for debris populations at GEO and LEO. The force due to the magnetic (Lorenz) fields is given by Equation (3-14). A straight forward spherical harmonic model for the magnetic field direction, $\boldsymbol{B}$, prediction in spherical and Earth-Centered-Earth-Fixed coordinates is given by Wertz, 2002, with further implementation discussion given by Davis, 2004. In Equation (3-14), is the velocity of the spacecraft relative to the magnetic field. For nomenclature, *CAP* is the body capacitance and φ is the body voltage. Magnetometers measure the local magnetic field and can be coupled with other sensors (such as a coarse sun sensor) for full determination of an object's attitude. However, magnetometer sensitivity becomes an issue because the field varies as R^{-3} and solar storms can cause the magnetic field to rotate. Higher-fidelity models that include solar wind and in-situ measurements such as the Tsyganenko model [Community Coordinated Modeling Center, 2007] are useful for transient conditions and sensitivity analysis.

$$F_{magnetic} = q(V_{rel,b\,field} \times B) \tag{3-14}$$

$q \quad = CAP \times \phi$

The torque imparted on the vehicle (Equation (3-15)) is computed by segmenting the object in facets of similar material and charge as we do for the facet computation for atmospheric drag modeling.

$$L_{magnetic} = \sum_{i=1}^{N} \rho_i \times F_{magnetic} \tag{3-15}$$

3.2.6 Radiation Environment

The Sun radiates high-energy particles and photons across the entire electro-magnetic spectrum, which displace or ionize materials in their path [Tribble, 2003]. Radiation from the Sun causes a shift in material properties (e.g., "reddening" in wavelength), degrades materials, produces upsets and failures in electronics, causes thermal fluctuation, and alters momentum on space objects. Absorbed radiation leads to hardware malfunction if thermal balancing is not considered during on-orbit operations. Earth exerts radiation pressure in the optical and longwave infrared bands on the order of 10% of the radiation pressure exerted by the Sun [Gottlieb, 1993]. Solar and thermal radiation pressure forces will be discussed in Chapter 4: Astrodynamics, so here we give only the torque equations of motion. Trapped particles are responsible for the Van Allen radiation belts; precise modeling and prediction of their behavior as well as how to protect sensitive on-board electronics is an on-going area of research.

The LDEF experiment found that in LEO ultra-violet (UV) radiation made the heaviest deposits on the vehicle. For spacecraft in MEO, the Air Force Research Laboratory is undertaking an experiment for operation in harsh radiation environments by studying the slot region of the Van Allen radiation belts and collecting data on degradation of electronics, thermal, optical, and mechanical structures, surfaces, and coatings [Scherbarth et al., 2009]. The Van Allen belts consist of trapped radiation plus protons and electrons in two different belts: a low altitude (up to 6000 km) belt composed of high energy protons and electrons, and a high altitude (up to 60,000 km) belt composed of high energy electrons. The inner belt's electron concentration correlates closely with geomagnetic storms while the outer belt is more dynamic due to the geomagnetic tail and can experience fluctuations of up to a factor of 1,000. The inner belt extends into the LEO orbit regime at the poles and in the area known as the "South Atlantic Anomaly," where Earth's magnetic field is weak. Mapping of the occurrences of the single event upsets of the TMUS4416 RAM for the UOSAT2 mission (700 km altitude and 98° inclination) demonstrates these effects [DeWitt et al., 1993].

Spacecraft electronic component design typically uses cumulative dose instead of instantaneous radiation dose for measuring radiation exposure and estimating performance degradation effects. Some short-duration cubesat or amateur satellites use non-space radiation rated components since they cost much less and the designers are willing to assume the risk. Solar arrays are susceptible to solar flares and damage from displaced atoms or ionization due to trapped electrons or protons and manifested as drops in voltage and current [Hastings and Garrett, 1996]. Hardware events result when high energy particles penetrate electronic circuits or sensors and produce unwanted ionization or displacement resulting in burned-out pixels, shorted circuits, changes in memory elements, or flipped logic states [Wertz et al., 2015]. Radiation also leads to internal charging with component effects similar to those of surface charging.

Abercromby et al., 2007, demonstrated that even with pre-flight measurements of materials and known composition, adjustments are necessary for on-orbit models to compensate for space weathering effects ("reddening") as early as one month into the mission and on a continuing basis. Weather effects include degraded surface paint and protective glass, which are crucial for thermal control. Understanding the reflectance properties of the observed space object is critical for high precision orbit prediction [Grey and Ziebart, 2014; Ziebart, 2004; McMahon, 2013] and object characterization [Wetterer et al., 2014]. For objects with a high area to mass ratio, accounting for the effect of radiation pressure is critical for maintaining space object trajectory custody and performing data association [Kelecy et al., 2014]. One area of on-going research is how to properly and in a physically realistic sense, incorporate space weather effects, changes in surface curvature (e.g., "wrinkling"), object flexibility, and bi-directional reflectance distribution function modeling. This must be done while accounting for objects with different size faceted surfaces that can rotate so that one does not wind up using cannonball models with negative ballistic coefficients or models with uncertainty bounds that include negative ballistic coefficients.

Torques produced by solar and terrestrial radiation pressure are computed by accounting for the absorption and the reflection (specular or diffuse) of the incident photons emitted from solar radiation or Earth-albedo and Earth-infrared radiation. Specular reflection is at an angle with respect to the facet normal while diffuse reflection is uniform in all directions with respect to the facet normal. For higher fidelity reflection models, bi-directional reflection distribution functions (BRDFs) are commonly used [Wetterer et al., 2014], which account for illumination angle and material properties.

Equations (3-16) give a simplified solar radiation pressure force solution, $\boldsymbol{F}_{srp}$, using a specular mirror and an averaged diffusion reflection model, and Equation (3-17) gives the torque computation. The object is assumed to be modeled as a collection of facets, each with their own area A_i, illuminated by the sun with the unit vector to the sun given by $\hat{\boldsymbol{s}}$ and the normal of the facet given by $\hat{\boldsymbol{n}}$. With the constraint that the specular reflection, absorption, and diffuse coefficients sum to one for each facet, Equations (3-16) reduce to Equation (3-18). If higher fidelity BRDFs are available that are physically consistent with the energy balance (absorbed versus reflected) of the system, they can be switched out for the diffusion model. For nomenclature, $C_{abso,i}$, $C_{diff,i}$, $C_{spec,i}$ are the coefficients of absorption, diffusion, and specular reflection respectively, $\boldsymbol{P}$ is the mean momentum flux from the Sun, and $\boldsymbol{P} = \boldsymbol{F}_e/c$ where $\boldsymbol{F}_e$ is the solar energy flux on the spacecraft (1360 W/m^2) and c is the speed of light in vacuum (3×10^8 m/s).

$$F_{srp} = \sum_{i=1}^{N} F_{srp,absorptions,i} + F_{srp,specular,i} + F_{srp,diffusion,i} \tag{3-16}$$

$$F_{srp,absorption,i} = -PC_{abso,i}(\hat{n}^T_{facet,i}\hat{s})\hat{s}A_i$$

$$F_{srp,specular,i} = -2PC_{spec,i}(\hat{n}^T_{facet,i}\hat{s})^2\hat{n}_{facet,i}A_i$$

$$F_{srp,diffusion,i} = -PC_{diff,i}(\hat{n}^T_{facet,i}\hat{s})\left(\frac{2}{3}\hat{n}_{facet,i} + \hat{s}\right)A_i$$

subject to: $C_{abso,i} + C_{diff,i} + C_{spec,i} = 1$, For All i

$$L_{srp} = \sum_{i=1}^{N} \rho_i \times F_{srp,i} \tag{3-17}$$

$$F_{srp} = \sum_{i=1}^{N} -P\left\{(1-C_{spec,i})\hat{s} + 2\left[C_{spec,i}(\hat{n}^T_{facet,i}\hat{s}) + \frac{1}{3}C_{diff,i}\right]\hat{n}_{facet,i}\right\}(\hat{n}^T_{facet,i}\hat{s})A_i \tag{3-18}$$

3.3 Summary

Understanding space environment effects, time-scales, and coupling on object attitude and orbital dynamics, hardware, instrumentation, and materials is vital for mission success, whether conducting basic science (Hubble Space Telescope), delivering navigation information (Global Positioning System), or sending television broadcasts. Efficient and effective feedforward and feedback of the OODA loop components require us to understand the fundamental drivers of the different regimes, the time-scales of the drivers, and driver coupling, and continue to update hypotheses on the interaction since only partial observations are available. Care must be taken during this cycle of prediction and refinement because observations are distilled versions of the true interaction with the local environment.

The previous sections demonstrated the fundamentals required for modeling the space environment, the impact of its characterization accuracy on predictions of space object motion, and how real anomalies have been attributed to environment aspects. Neglect of such interactions affects the ability to predict object motion and conduct on-board attitude determination and mission operations. Importantly, it is not just the combined and multiple time-scale effect of the natural environment that we must be concerned about but also the effect of man on the space environment, which adds complication to detection and prediction of artificial space objects. This is already a challenge due to sensor observation types and cadences. Table 3-2 summarizes the orbit types with respect to the dominant space weather effects and examples of missions where the space environment has played a significant role during operations. For a summary of spacecraft system anomalies attributed to space environment interactions, see Bedingfield et al., 1996.

TABLE 3-2. Summary of Orbit Regime and Space Environment Effects to Consider ("debris" comprises both natural and man-made hazards).

Orbit Regime	Example Mission	Space Environment Drivers	Manifested Effects
LEO	International Space Station, Hubble Space Telescope, Space Shuttle, Long Duration Exposure Facility, CubeSats	Gravity gradient, neutrals, magnetic (South Atlantic Anomaly), plasma (near polar), debris	Flexible body torques, drag, material erosion, signal effects, instrumentation upsets, charging, object conjunction
MEO	Global Positioning System	Radiation, plasma, debris	Electronic upsets, material reddening, surface charging and arcing, object conjunction
GEO	Galaxy 15	Radiation, plasma, debris	Electronic upsets, material reddening, surface charging and arcing, object conjunction
GTO	Debris	All of the above	All of the above

References

377th Air Base Wing Public Affairs. 2008. "Communication/Navigation Outage Forecasting System Fact Sheet" Available: http://www.kirtland.af.mil/library/factsheets/factsheet.asp?id=12776.

Abercromby, K. et al. 2007. "Remote and Ground Truth Spectral Measurement Comparisons of FORMOSAT III." In *Advanced Maui Optical and Space Surveillance Technologies Conference.* Kihei, HI.

Air Force Research Laboratory (AFRL) Space Vehicles Directorate. 2008. "C/NOFS: A Mission to Forecast Scintillations." Available: http://www.kirtland.af.mil/shared/media/document/afd-080718-072.pdf.

Akella, M. and K. Alfriend. September - October 2000. "Probability of Collision Between Space Objects." *Journal of Guidance, Control, and Dynamics,* Vol. 23, No. 5, pp. 769-772.

Akins, K. et al. 2003. "Comparison of MSIS and Jacchia Atmospheric Density Models for Orbit Determination and Propagation." In *AAS/AIAA Spaceflight Mechanics Meeting*. Ponce, Puerto Rico.

Alby, F., E. Lansard, and T. Michal. March 17-19, 1997. "Collision of CERISE with Space Debris." In *Second European Conference on Space Debris*, Darmstadt, Germany.

Bedingfield, K., R. Leach, and M. Alexander. 1996. "Spacecraft System Failures and Anomalies Attributed to the Natural Space Environment." National Aeronautics and Space Administration. Marshall Space Flight Center, MSFC, Alabama.

Bothmer, V. and I. Daglis. 2007. "Space Weather - Physics and Effects." Chichester, UK: Praxis Publishing.

Bowman, B. et al. 2008. "A New Empirical Thermospheric Density Model JB2008 Using New Solar and Geomagnetic Indices." In *AIAA/AAS Astrodynamics Specialist Conference and Exhibit.* Honolulu, HI.

Boyd, J.R. March 2018. "A Discourse on Winning and Losing." Edited by G.T. Hammond, Curtis E. LeMay Center for Doctrine, Development, and Education, Maxwell AFB, AL. Pp. 383–385.

Community Coordinated Modeling Center (CCMC). 2 September 2016. "Magnetopause Position and Conditions at GOES Satellites." Goddard Spaceflight Center, National Aeronautics and Space Administration. Available: http://ccmc.gsfc.nasa.gov/cgi-bin/display/RT_t.cgi?page=mpause.

Davis, J. May 12, 2004. "Mathematical Modeling of Earth's Magnetic Field: Technical Note." Virginia Institute of Technology, Blacksburg, VA.

De Selding, P. 20 April 2010. "Orbital Blames Galaxy 15 Failure on Solar Storm." *News.*

De Selding, P. 21 May 2010. "SES, Intelsat Prepare Evasion Tactics as Threat from Failed Satellite Looms." *Space News.*

De Selding, P. 7 January 2011. "Intelsat Moving Recovered Galaxy 15 to Test Location." *Space News.*

De Selding, P. 13 January 2011. "Electrostatic Discharge Crippled Galaxy 15, Intelsat Says." *Space News.*

DeWitt, R., D. Duston, and A. Hyder. 1993. The Behaviour of Systems in the Space Environment. New York, NY: Kluwer Academic Publishers.

Doornbos, E. December 2010. *Thermospheric Density and Wind Determination from Satellite Dynamics.* TU Delft.

European Space Agency (ESA). 31 August 2016. "Copernicus Sentinel-1A Satellite Hit by Space Particle." Available: http://www.esa.int/Our_Activities/Observing_the_Earth/Copernicus/Sentinel-1/Copernicus_Sentinel-1A_satellite_hit_by_space_particle.

European Space Agency (ESA). Undated. "About Space Debris." Available: http://www.esa.int/Our_Activities/Operations/Space_Debris/About_space_debris.

Ferster, W. 8 April 2010. "Intelsat Losses Contact with Galaxy 15 Satellite." *Space News.*

Frueh, C., D. Ferguson, C. Lin, and M. Jah. 2014. "Passive Electrostatic Charging of Near-Geosynchronous Space Debris HAMR Objects and its Effects on Coupled Dynamics." In *AAS/AIAA Spaceflight Mechanics Meeting*. Santa Fe, NM.

Gottlieb, R. February 1993. "Fast Gravity, Gravity Partials, Normalized Gravity, Gravity Gradient Torque and Magnetic Field: Derivation, Code, and Data." National Aeronautics and Space Administration: Lyndon B. Johnson Space Center, Houston, TX.

Grey, S. and M. Ziebart. 2014. "Development in High Fidelity Surface Force Modeling and their Relativistic Effects on Orbit Prediction." In *2014 AIAA SPACE Conference and Exposition*. San Diego, CA.

Gruss, M. 19 July 2016. "DigitalGlobe Says WorldView-2 Operational After Debris Causing Event." *Space News.*

Hastings, D. and H. Garrett. 1996. Spacecraft Environment Interactions. New York, NY: Cambridge University Press.

Hughes, P. 2004. Spacecraft Attitude Dynamics. Mineola, NY: Dover.

Jehn, R., V. Agapov, and C. Hernandez. 2005. "End-Of-Life Disposal of Geostationary Satellites." In *4th European Conference on Space Debris*. Darmstadt, Germany.

Jenkin, A. July – August 2004. "Effect of Orbit Data Quality on the Feasibility of Collision Risk Management." *Journal of Guidance, Control, and Dynamics,* Vol. 41, No. 4, pp. 677-683.

Kelecy, T., M. Jah, J. Baldwin, and J. Stauch. 2014. "High Area-to-Mass Ratio Object Population Assessment from Data/Track Association." *Acta Astronautica,* Vol. 96, pp. 166–174,

Kelso, T. 2007. "Analysis of 2007 Chinese ASAT Test and the Impact of Its Debris on the Space Environment." in *Advanced Maui Optical and Space Conference.* Wailea, HI.

Klinkrad, H. 2006. Space Debris: Models and Risk Analysis. Chichester, UK: Praxis.

Krisko, P., S. Flegel, M. Matney, D. Jarkey, and V. Braun. 2015. "ORDEM 3.0 and MASTER-2009 Modeled Debris Population Comparison." *Acta Astronautica,* Vol. 113, pp. 204–211.

Lanzerotti, L., D. Thomson, and C. Maclennan. Summer 1997. "Wireless at High Altitudes - Environmental Effects on Space-Based Assets." *Bell Labs Technical Journal,* Vol. 2, No. 3, pp. 5–19.

Lean, J. et al. April 2006. "Thermospheric Densities Derived from Spacecraft Orbits: Application to Starshine Satellites." *Journal of Geophysical Research,* Vol. 111, No. A4.

Marcos, F., W. Burke, and S. Lai. 2007. "Thermospheric Space Weather Modeling." In *AIAA Plasmadynamics and Lasers Conference.* Miami, FL.

McMahon, J. 2013. "Improving Orbit Determination with Non-Cannonball Solar Radiation Pressure Models." In *AAS/AIAA Spaceflight Mechanics Meeting.* Kuaui, HI.

Milani, A., A. Nobili, and P. Farinella. 1987. Non-gravitational Perturbations and Satellite Geodesy. Bristol, England: IOP Publishing.

Montenbruck, O. and E. Gill. 2000. Satellite Orbits: Models, Methods, and Applications. New York, NY.: Springer-Verlag.

NASA Orbital Debris Program Office (NASA). 2008. "History of On-Orbit Satellite Fragmentations, 14th Ed."

NASA Orbital Debris Program Office (NASA). July 2016. "Satellite Box Score," *Orbital Debris Quarterly News.*

National Aeronautics and Space Administration (NASA). April 2009. "Satellite Collision Leaves Significant Debris Clouds." *Orbital Debris Quarterly News,* Vol. 13, No. 2, pp. 1–2.

National Research Council. 2011. "Limiting Future Collision Risk to Spacecraft: An Assessment of NASA's Meteoroid and Orbital Debris Programs." National Academies Press, Washington, D.C.

Pardini, C. and L. Anselmo. June – April 2001. "Comparison and Accuracy Assessment of Semi-Empirical Atmosphere Models Through the Orbital Decay of Spherical Satellites." *Journal of the Astronautical Sciences,* pp. 255–268.

Paul, S. and C. Frueh. 2016. "Space Debris Charging and Its Effects on Orbit Evolution." In *AIAA SPACE Forum.* Long Beach, CA.

Pisacane, V. 2005a. The Space Environment and Its Effects on Space Systems. Reston, VA: AIAA.

Pisacane, V. 2005b. Fundamentals of Space Systems, 2nd ed. New York, NY: Oxford University Press.

Republic of South Africa, South African Government News Agency. 2014. "SA's First Nano-satellite Encounters Space Debris." Pretoria.

Romgrens, B., E. Mooij, and M. Naeije. May–June 2013. "Satellite Collision Avoidance Prediction Using Verified Interval Orbit Propagation." *Journal of Guidance, Control, and Dynamics,* Vol. 36, No. 3, pp. 821–832.

Schaub, H. and J. L. Junkins. 2003Analytical Mechanics of Space Systems, 2nd ed. Reston, VA: AIAA.

Scherbarth, M. et al. 2009. "AFRL's Demonstration and Science Experiments (DSX) Mission." In *Advanced Maui Optical and Space Surveillance Technologies Conference.* Kihei, HI.

Schildknecht, T. et al. 2005. "Optical Observations of Space Debris in High-Altitude Orbits." in *Fourth European Conference on Space Debris*, Darmstadt, Germany.

Stuckey, W. 1993. "Lessons Learned from the Long Duration Exposure Facility." Defense Technical Information Center.

Surrey Satellite Technology LTD. [Accessed 2016]."CERISE Satellite Data," [Online]. Available: https://www.sstl.co.uk/Missions/CERISE-Launched-1995/CERISE/CERISE-The-Mission.

Sutton, E., R. Nerem, and J. Forbes. 2007. "Density and Winds in the Thermosphere Deduced from Accelerometer Data." *Journal of Spacecraft and Rockets,* Vol. 44, No. 6, pp. 1210–1219.

Tribble, A. 2003. The Space Environment: Implications for Spacecraft Design. Princeton, NJ: Princeton University Press.

Vallado, D. 2007. Fundamentals of Astrodynamics and Applications, 3rd ed. New York, NY: Springer.

Wertz, J. 2002. Spacecraft Attitude Determination and Control. Hingham, MA: Reidel Publishing Company.

Wertz, J. 2009. Orbit & Constellation Design & Management. New York, NY: Springer.

Wertz, J., D. Everett, and J. Puschell. 2015. Space Mission Engineering: The New SMAD. Hawthorne, CA: Microcosm Press.

Wetterer, C. et al. Jan.–Feb. 2014. "Refining Space Object Radiation Pressure Modeling with Bidirectional Reflectance Distribution Functions." *Journal of Guidance, Control, and Dynamics,* Vol. 37, No. 1, pp. 185–196.

Yurasov, V. et al. March–April 2008. "Reentry Time Prediction Using Atmospheric Density Corrections." *Journal of Guidance, Control, and Dynamics,* Vol. 31, No. 2, pp. 282–289.

Ziebart, M. Sept. - Oct. 2004. "Generalized Analytical Solar Radiation Pressure Modeling Algorithm for Spacecraft of Complex Shape." *Journal of Spacecraft and Rockets,* Vol. 41, No. 5, pp. 840–848.

Chapter 4

Space Object Motion and Events

Marek Ziebart, Ph.D., *University College London*
Carolin Frueh, Ph.D., *Purdue University*
Matthew D. Hejduk, Ph.D., *Astrorum Consulting*

Previous chapters describe space's physical environment and the types of human-made objects resident therein. This chapter discusses the form of trajectory resulting from the interactions between the environment and resident space objects (RSOs).

We introduce concepts and characteristics, relevant to SDA practitioners, of both the astrodynamics and the motion of objects in the space environment. We summarize key forcing effects and RSO characteristics. We introduce the reference frames required to describe orbital motion, and explain the principal geometric characteristics of orbital trajectories. We do this by explaining the first-order model of Keplerian motion, and subsequently the impact of perturbing forces. We then present various trajectory models used to represent and communicate the time-varying position and velocity of RSOs. We emphasize the limitations of the representations and quantify the accuracies achieved. Finally, we deal with aspects related to object orientation and rotation.

4.1 Causes of Actual Motion

4.1.1 External Sources (Four Fields)

Isaac Newton's first law of motion states: an object will remain at rest or will travel in a straight line with uniform motion unless an external force acts upon it. This is pertinent to understanding the trajectories of objects in the space environment, which exerts—by virtue of the various fields permeating it—forces and torques upon space vehicles and space debris. These forces and torques systematically change the rotational and translational states of space objects. How the resident space objects (RSOs) respond to those forces and torques depends upon several factors such as mass, moment of inertia, area-to-mass ratio, and charge-to-mass ratio. A second consideration is the design of a space vehicle to respond to forces and torques via attitude stabilization and even feedback loops derived from onboard accelerometers. This reduces the impact of surface forces [Canuto et al., 2010]. A reasonably elegant, concise description of the environmentally induced force and torque characteristics considers four 'fields':

- Gravitational potential
- Electro-magnetic radiation
- Particulates (neutral gas and plasmas)
- The geomagnetic field

These fields change characteristically in near-Earth space and one can plan for the consequences of the induced forces and torques for any given space vehicle based on an understanding of the fields. A few examples will illustrate the concept:

- Geopotential 'wells' at geostationary altitudes cause station-keeping problems for communications satellites, which tend to slip into the wells seeking a minimum gradient in the gravitational potential.
- Solar radiation pressure affects the orbits of GPS satellites, changing the satellite trajectories by about 200 meters along track per orbit—if ignored in orbit prediction routines, precision navigation of weapon systems becomes unreliable.
- Mis-modeling of the coupling of solar FUV and EUV radiation fluxes with the geomagnetic field accounts for significant variations in the modeling of atmospheric drag coefficients for RSOs [Bowman et al., 2008]. This last point illustrates the nature of the forcing via the environment for both atmospheric density and geomagnetic field characteristics.

Hence, we arrive at a preliminary conclusion: the physics of orbital motion is described for the most part by the physics of the space environment itself and its coupling with space object characteristics to create forces and torques. While we do not have a complete understanding of the four fields described above, considerable advances have been made during the space age primarily because of the availability of space-borne sensors designed to measure, *in situ*, the parameters of the space environment. In a sense, the problem can be stated simply: we need to understand the physics of the space environment, and the physics of the interaction of that environment with the objects we deploy. That in turn allows us to model the respective forces and torques mathematically.

4.1.2 Body-generated Sources

Some RSOs absorb and re-radiate energy, and this re-radiation tends to manifest itself as a thermal response. Thermal imbalance forces can also be generated by the anisotropic emission of heat via both passive and active thermal control of spacecraft payload. These control mechanisms are required to maintain specific temperature regimes for the spacecraft instrumentation. The downside being that systematic forces are caused, driven by the thermal maintenance cycles. Moreover, some RSOs also experience outgassing (trapped materials surface in the absence of atmospheric pressure), and these effects are observable, particularly in the early phases of mission lifetime (such as over the first year on orbit). Many active RSOs also have propulsion systems, which impart a force or torque to the space objects. In order to maintain specific ground tracks thruster events are required to counter-act naturally occurring perturbations that cause satellites to drift away from their nominal or design trajectories. All these effects induce accelerations on the respective RSOs, and, depending on the required accuracy of the tracking, may need to be modeled. Such modeling tasks require knowledge of the timing, direction and magnitude of the forces.

4.2 Reference Frames and Coordinate Systems

To describe RSO motion we need 'reference frames'—such reference frames enable the user to describe the positions of points in space and time. The simplest of these are Cartesian frames consisting of three orthonormal basis vectors that meet at a common origin. SDA requires four key Cartesian frames.

The Earth Centered Inertial (ECI) frame has its origin at the center of mass of Earth, with the Z-axis coincident with Earth's spin axis (defined at a particular moment in time—known as the frame epoch). The XY plane aligns with Earth's equatorial plane at epoch. The X-axis is defined by the intersection of the ecliptic and the equatorial plane at the frame's epoch. An inertial frame is effectively non-accelerating and non-rotating and one in which we can invoke Newton's Principle of Determinacy. It is the frame of choice for describing the motion of space vehicles. By contrast, an Earth-centered, Earth-fixed (ECEF) frame co-rotates with Earth. Its Z-axis is the same as the ECI Z-axis, and the XY-plane is equatorial, but the X- and Y-axes are fixed to Earth and rotate in inertial space. The ECEF frame is useful for describing the locations of tracking stations on Earth's surface. A topocentric basis or frame is one fixed, as a rule, to a tracking station. The axes point along local north, east, and up (NEU) directions. Finally, there are a number of frames fixed to and moving with space vehicles. Probably the most important of these is the height, cross track, and along track (HCL) frame. The height vector is parallel to the radial vector (from the origin of the ECI or ECEF frame) to the spacecraft center of mass. The cross track vector is normal to the instantaneous orbital plane, and the along track vector completes the right-handed set, and aligns loosely with the velocity vector. Figure 4-1 depicts these four reference frames.

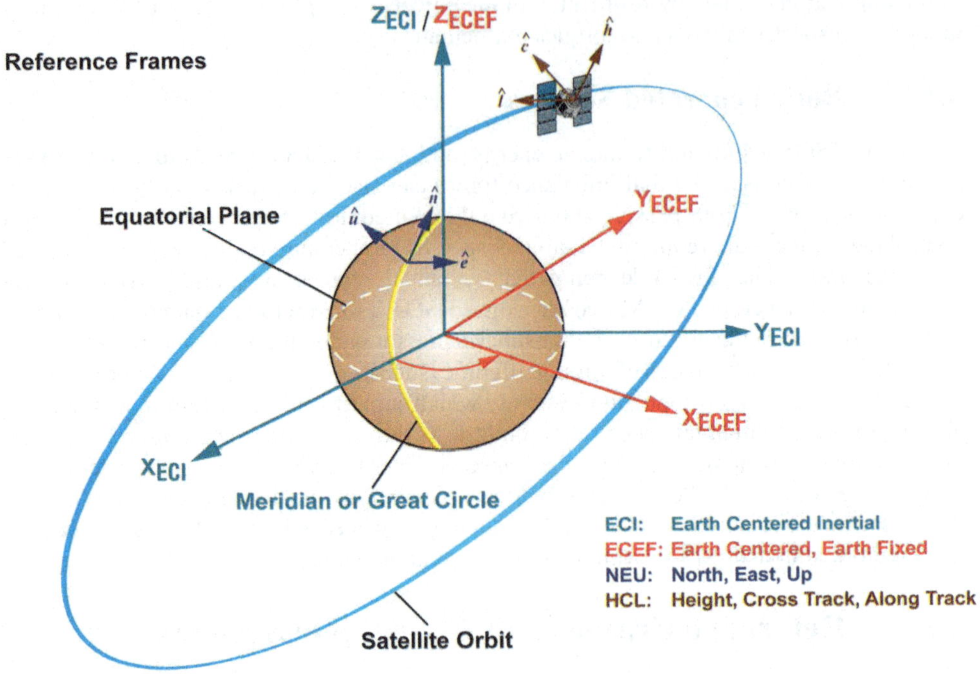

FIG. 4-1. Reference Frames. These are the main reference frames used in astrodynamics.

There are standard formula to transform between these frames. McCarthy [1996] and Vallado [2001] are useful references for formula and background theory.

Before we turn to coordinate systems it is useful to distinguish between 'reference systems' and 'reference frames'. The former are sets of definitions that describe all the requisite numerical and physical data types used to model the frame. For example, one requires a value for the speed of light in vacuum, a method for defining the orientation of Earth's spin axis in space and its temporal evolution, and a method to then calculate the position of 'fixed' stars on the celestial sphere. These ideas create a reference system. The realization of that process is a reference frame. Reference systems provide standards and clear definitions of process. Reference frames are the bases within which events are described and measured; they have associated practical data sets (e.g., star catalogs, tracking station coordinates, and velocities) provided to users to calculate parameters like position and attitude. Many of these standards are maintained and published by the IERS (International Earth Orientation and Reference Systems Service [Souchay and Feissel-Vernier, 2005]).

Finally, a coordinate system is purely a mathematical formulation to describe, conveniently, a point in space with respect to a reference frame. The most widely used coordinate system is Cartesian, being simply the projection of the range vector of a point onto the reference frame axes. However, depending on the positioning problem to be solved, more intuitive systems may be used. Spherical and ellipsoidal systems describe space in terms of latitude, longitude and height above a reference surface. Mapping projections extend this concept to create coordinate systems used to describe positions that can be mapped onto 2D surfaces to represent ground tracks or other collections of spatial data.

4.2.1 Reference Frames for Orbital Motion

Historically speaking there has been a plethora of reference frames, many of which had no formal mathematical relationship with each other. That was unimportant while there were very few missions on orbit, and the agencies that operated such missions had little need to exchange data or manage traffic. Each agency could define and maintain its own reference frame for its own operational requirements. It is very different today. Present challenges involve a contested and congested space environment. The need to exchange data has become far greater.

The International Celestial Reference Frame (ICRF) is defined by the positions (right ascension and declination) of some 1,500 extragalactic radio sources. This is analogous to defining a celestial frame using a star catalogue. The primary measurement type used to calculate the frame parameters is very long baseline interferometry using a worldwide network of astronomical radio observatories.

J2000 is the most widely used scientific inertial frame. It is defined at 12 hours Universal Time on January 1st 2000. Standard planetary ephemerides published by NASA's Jet Propulsion Laboratory use J2000 as the preferred inertial frame for calculating and modelling planetary trajectories.

US Strategic Command (USSTRATCOM)'s two-line elements (TLEs) use the true equator, mean equinox (TEME) ECI frame. This differs from J2000 in that the ECI frame is defined for the epoch of the mission described using the TLEs. That is, the frame is defined specifically for each mission. In other respects it is similar—tri-axial right-handed basis, with the Z-axis aligned with the instantaneous spin axis at the frame epoch.

Just as there are multiple inertial frames, the same is true for Earth-Centered, Earth-fixed frames.

There are many ECEF frames, but the fundamental realization for science applications is the ITRF. The International Terrestrial Reference Frame (ITRF) is derived from the International Terrestrial Reference System. It consists largely of a catalogue of positions and velocities of tracking stations at a particular epoch. Successive realisations of the ITRF come out every few years. Their production ingests measurement data from four techniques: GNSS (Global Navigation Satellite Systems); SLR (Satellite Laser Ranging); VLBI (Very Long Baseline Interferometry) and DORIS (the French satellite tracking system—Doppler Orbitography and Radiopositioning Integrated by Satellite). Tracking station velocities account for the relatively slow (mm per year) motion caused by plate tectonics, as well discontinuities caused by earthquakes. The scientific community primarily uses the ITRF to monitor and model planet-scale processes such as global sea level rise and the earthquake cycle. Other notable examples of ECEF systems include WGS-84 (USA) and PZ-90 (former Soviet Union, now Russia).

For the purposes of SDA it is vital that ancillary information is supplied identifying the relevant frame within which data (e.g. RSO positions and trajectories) is oriented.

4.2.2 Reference Frame Transformations

Provided a sub-set of physical points known in two different reference frames, we can compute a reference frame transformation between them. The most common, and classical, transformation is the Helmert 7 parameter transformation. This might be computed to transform between two ECEF frames. It is common in comparisons between computed orbits of the same space vehicle by different agencies to first carry out a Helmert transform between the ephemerides prior to comparing their trajectories because of weakness in the consistency of frame definitions. A Helmert transform typically involves three shifts, three rotations, and a scale parameter. A second form of transformation involves the problem of relating ECI to ECEF frames (and vice versa). This is a complex task involving models of precession, nutation, polar motion and Earth rotation about the spin axis (usually related to a time argument such as the Greenwich Apparent Sidereal Time). Detailed standards exist to describe the required procedures [Petit and Luzum, 2010].

4.2.3 Description of Rotational Motion

In addition to the orbital motion of a space object, attitude motion is of prime interest. The term attitude describes the orientation of a spacecraft or space object relative to inertial space. The attitudes of all objects are perturbed by naturally induced torques. The same perturbations that act on the orbit also induce attitude changes. Forces that are not uniform over the object introduce torques that lead, if not counteracted, to attitude changes. Non-uniformity arises via non-spherical non-convex shapes, different material properties, and non-uniform mass distributions.

Most operational satellites feature active or passive attitude control in order to comply with mission requirements. Active attitude control includes measurement of the attitude and a means to correct for it by introducing additional torques. For the attitude measurements we use either relative or absolute attitude reference systems; absolute ones include, for example, star- or sun- trackers and magnetometers. Relative one—those systems that measure only a change of attitude—include inertial measurement units (IMU) and gyroscopes. Mechanisms to introduce torques that lead to active changes or stabilization of the attitude include reaction wheels and thrusters. Passive attitude stabilization in contrast relies on mechanisms that keep the attitude roughly stable over time by relying on natural torques, such as gravity or magnetic torque.

An RSO's orientation in space affects its dynamics, which are in turn required in the process of orbit prediction and orbit estimation. In reverse, attitude models usually require knowledge of the position and velocity of the RSO. Hence, orbit and attitude states are coupled—one depends on the other and vice-versa. This occurs most strongly where surface/non-conservative forces are concerned, although second order gravity torques are also possible effects for extended bodies. Space object characterization requires at least some understanding of the structure of an RSO and its orientation in space. Orbit and attitude states are central to SDA. [Calef, 2006; Linares, 2014].

Local-Vertical, Local-Horizontal (LVHV) Reference Frame

One often used reference frame is the so-called local-vertical, local-horizontal reference frame. Two similar but not identical definitions of this frame exist [Vallado, 2001; Sauceda, 2001]. In the first definition, it is identical to the Gaussian reference frame or to the radial-transverse-normal (RNT) reference frame. In this definition, the local horizontal direction is in the radially outward direction and is denoted as the X-axis, and the Y-axis is perpendicular to the radial direction in the orbital plane. For circular orbits, it coincides exactly with the velocity direction. It is called the along-track direction. The Z-axis is perpendicular to the orbital plane and completes the right-handed system. Figure 4-2 illustrates these two definitions.

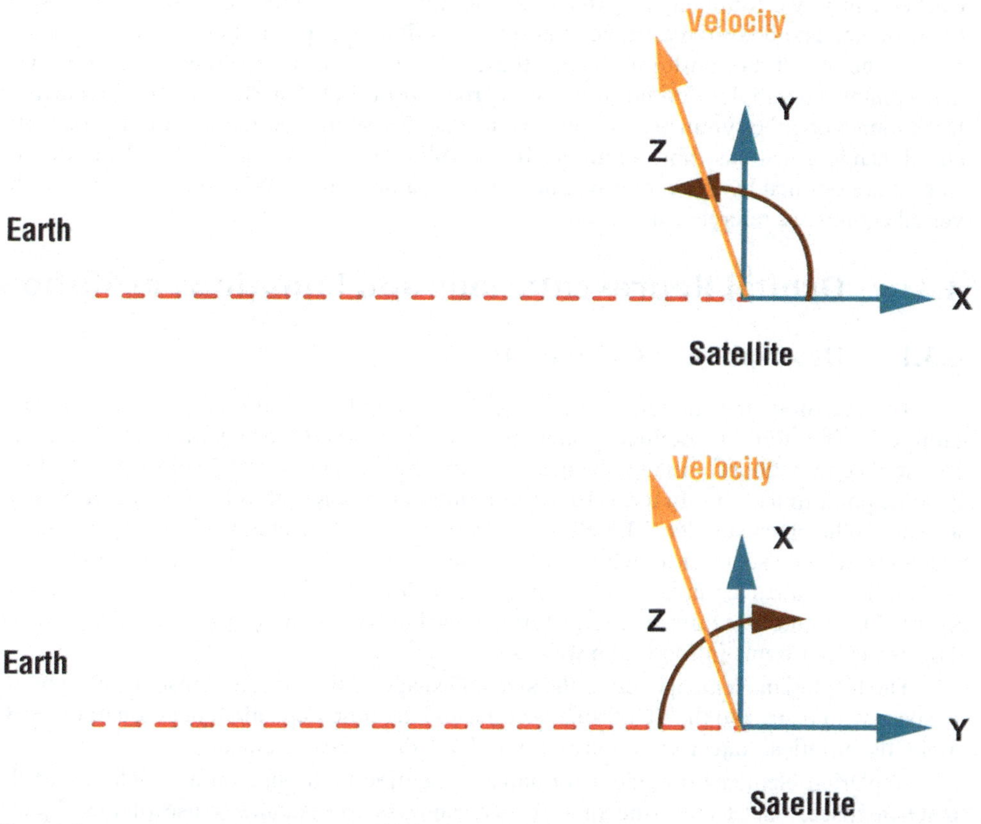

FIG. 4-2. **Body-Centered Reference Frames, Both Denoted LVLH.** Left: Gaussian reference frame Right: Roll-pitch-yaw reference frame.

The conflicting definitions of the LVLH reference frames stem from the wish to use the same frame for satellites as the one used for airplanes and which is identical to the roll-pitch-yaw frame. The roll-pitch-yaw reference frame is identical to the RNT frame, rotated by 90 degrees around the Z- or out of orbital plane axis, followed by a rotation around the along-track direction. This means the new X-axis is pointing in the along track direction (roll), the new Y-axis is orthogonal to the orbital plane (pitch), opposite to the former Z-axis, and the Z-axis is pointing from the satellite to the center of the earth (yaw), opposite to the former radial direction.

Body-Fixed Reference Frame

Unlike the case of an airplane, neither of the above defined LVHV reference frames is a true body-fixed frame, as the satellite can and does move relative to the velocity direction. Three-axis stabilized satellites orient their panels towards the sun throughout the orbit. Objects without active or passive attitude control, including space debris, tumble freely. Consequently, we formulate any dynamic and kinematic attitude equations in the body-fixed frame. The body-fixed frame is normally defined along symmetry and main inertia axes of the object. For uniform objects, these coincide. Definition along the axis of main inertia allows a simpler definition of the inertia tensor, but that is only for convenience. Definition along the symmetry axis makes the results obtained from the attitude kinematic and dynamic equations more intuitive. In the following, the body-fixed and the inertial frames are defined by the center of Earth, the equator as the fundamental plane, and the vernal equinox as reference direction.

4.3 Orbital Representations and Equations of Motion

4.3.1 Description of Orbital Motion

To first order the motion of RSOs (objects bound gravitationally to the Earth) is elliptical. The form is defined completely by a two-body problem treatment given Newton's gravitational constant, the mass of Earth and the position and velocity of the RSO at some point in its orbit. In terms of acceleration magnitudes, all other forcing effects are at least 1000 times smaller. The elliptical form is described classically using Keplerian elements, which use, in turn, two Cartesian reference frames: one aligned with the orbital plane (the so-called Gaussian vectors) and one with its Z-axis parallel to the Earth's spin axis and its X and Y vectors in the equatorial plane but fixed with respect to the distant stars. This is the ECI frame introduced previously.

The Keplerian elements define the size and shape of the orbital ellipse, its orientation in space with respect to the ECI frame vectors, and the time-dependent position of the RSO along the elliptical trajectory. Figures 4-3 and 4-4 show these elements.

Keplerian elements provide a formalism to define both position and velocity of the RSO in space, but at the same time to give an easy to visualize sense of the object's trajectory around the planet.

Looking directly down upon the orbit (that is, along the orbit normal) gives a clear sense of the shape of the trajectory, and its relationship to the position of the Earth.

The six Keplerian elements are:

a semi-major axis of the orbital ellipse (length, usually in km)

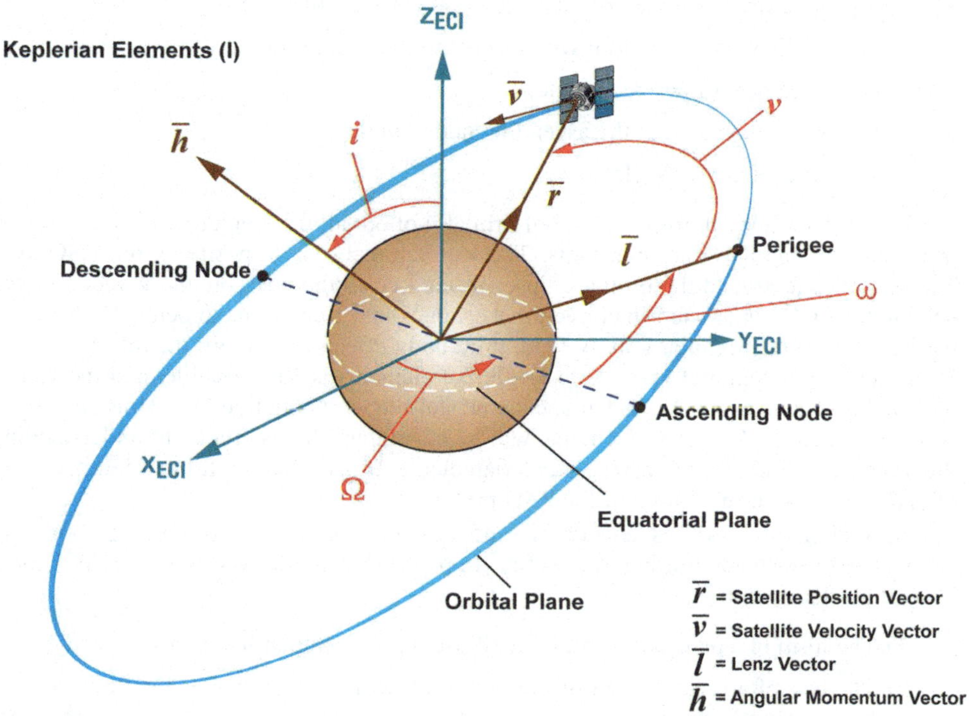

FIG. 4-3. Keplerian Elements Used in Defining First-Order Orbital Motion.

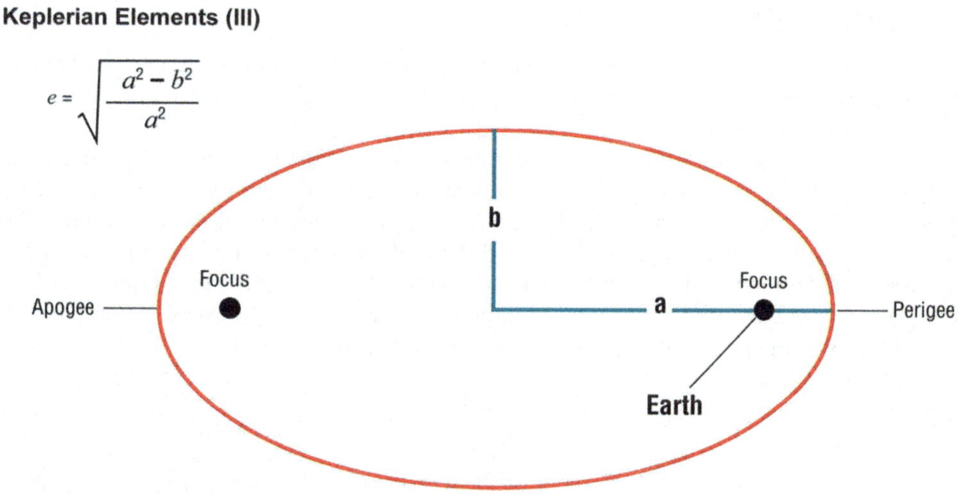

FIG. 4-4. Geometry of an Orbital Ellipse.

e eccentricity of the orbital ellipse ($e = 0 \Rightarrow$ circular orbit)

i inclination of the orbit with respect to the equator (angle)

ω argument of perigee (angle)

Ω right ascension of the ascending node (angle)

ν true anomaly (angle)

For the two-body problem model of orbital motion, all the Keplerian elements (apart from the true anomaly) are constants. The true anomaly value positions the RSO as a function of time around the orbital ellipse. If one knows the position and velocity of the RSO in the ECI frame, one can convert these values into Keplerian elements. Under more realistic dynamics (irregular gravity field, third body effects and so on) the transformation from inertial position and velocity to Keplerian elements is still possible, and the values obtained give a sense of the instantaneous orbital plane, inclination, etc. However, from epoch to epoch the Keplerian elements will be changing. This is crucial to understanding the motion in space—time-varying Keplerian elements give a powerful and intuitive way of understanding the evolution of the RSO motion.

Several helpful aspects concerning the visualization of an orbit using Keplerian elements are worth committing to memory. In part, these are derived from Kepler's three laws of planetary motion:

1) The form of a planetary orbit is an ellipse with the Sun at one focus.
2) The law of areas: the area of the sector traced by the radius vector from the Sun between any two positions of a planet in its orbit is proportional to the time occupied in passing from one position to the other
3) The square of the orbital period, T, is proportional to the cube of the semi-major axis: $T^2 \propto a^3$.

Kepler derived his laws empirically from large amounts of observational data, but they can be proved analytically for the two-body problem. This gives us a series of first-order characteristics of RSO motion.

The semi-major axis gives us insight into orbital altitude and orbital period (from Kepler's third law). For near-circular orbits, subtracting the mean radius of Earth gives us a measure of the height of the RSO above Earth's surface. Equatorial radius is about 6378 km, and the polar radius is approximately 6356 km. GPS satellites have a semi-major axis of around 26,600 km, and the orbits are near circular. Taking Earth's mean radius as 6371 km (say 6400 km as a first-order approximation), we get an orbital altitude of ~20,200 km.

We can compute the orbital period (seconds) via Kepler's third law:

$$T = 2\pi\sqrt{\frac{a^3}{GM_E}} \tag{4-1}$$

GM_E is Newton's gravitational constant times the mass of the Earth, with a standard value of $3.986004405 \times 10^{14}\ m^3/s^2$

Hence, a GPS satellite's orbital period is around 11h 58m. Figure 4-5 shows altitudes and periods for several types of satellite.

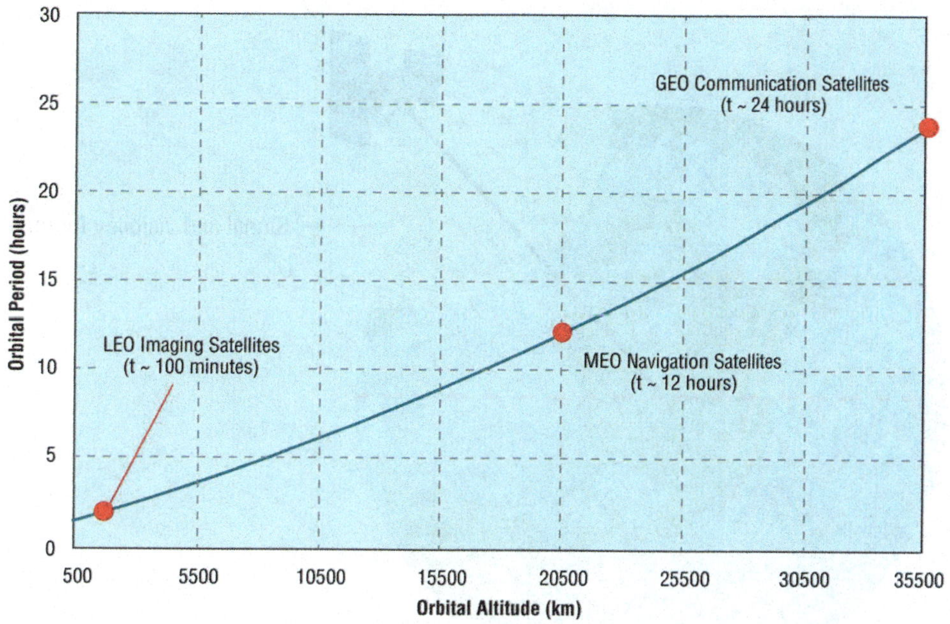

FIG. 4-5. Orbital Periods and Semi-major Axis.

There is a simple, but dependable, first order relationship between the orbital semi-major axis and the orbital period. The vast majority of satellite orbits are also near circular, which means the orbital period also defines the speed of motion to first order. Practitioners tend to have a clear sense of these numbers as they are characteristic of operating in particular parts of space.

Orbital eccentricity yields useful insights into orbital motion, too. For a non-circular orbit ($0 < e < 1$), the satellite moves with non-uniform speed. Kepler's second law implies that at perigee the satellite will move fastest, and at apogee the slowest.

$$V_{perigee} = \sqrt{\frac{GM_E}{a}}\sqrt{\frac{1+e}{1-e}} \tag{4-2}$$

$$V_{apogee} = \sqrt{\frac{GM_E}{a}}\sqrt{\frac{1-e}{1+e}} \tag{4-3}$$

For a Molniya orbit ($a = 26{,}500$ km, $e = 0.7$) the perigee speed is 9.2 km/s, and the apogee speed is 1.6 km/s. Contrast that with a near-circular GPS orbit, where the speed stays close to 3.9 km/s throughout.

The combination of orbital inclination and instantaneous altitude controls the area of the earth visible to the satellite, and this in turn limits the visibility of the satellite from ground tracking stations. Figure 4-6 illustrates this situation.

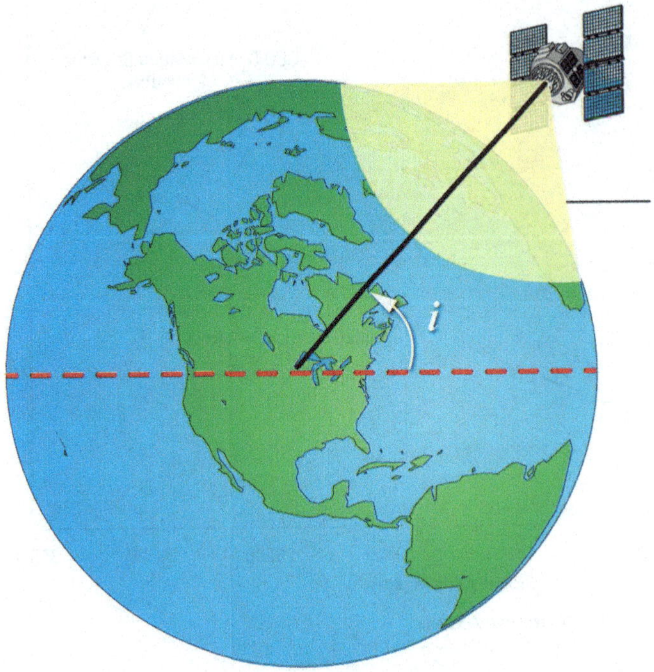

FIG. 4-6. **Area of Earth Visible to Satellite.** Orbital inclination and altitude affect the area of Earth visible to the satellite and vice versa.

The right ascension of the ascending node (Ω or RAAN) is the point at which an RSO moves from the southern to the northern hemisphere in inertial space. In Figure 4-3 it's the angle in the equatorial plane between the inertial X-axis and the line of nodes. For a two-body problem model this angle is a constant. However, due to the oblateness of Earth (Earth's polar radius is about 22 km shorter than the equatorial radius)—and here we begin to consider the impact of more realistic models of Earth's gravity field and its effect upon orbital motion—the RAAN changes slowly and systematically (i.e., secularly) moving westwards or eastwards, depending upon the orbital inclination:

$$\dot{\Omega} = -\frac{3nR_E^2}{2(a(1-e^2))^2}J_2\cos(i) \tag{4-4}$$

where

$\dot{\Omega}$ = time rate of change of RAAN

n = mean motion (2π /orbital period)
R_E = equatorial radius of the earth (6378.137 km)
J_2 = $1.08262668 \times 10^{-3}$

Figure 4-7 shows this dependence graphically. Table 4-1 gives the Keplerian elements for several satellites.

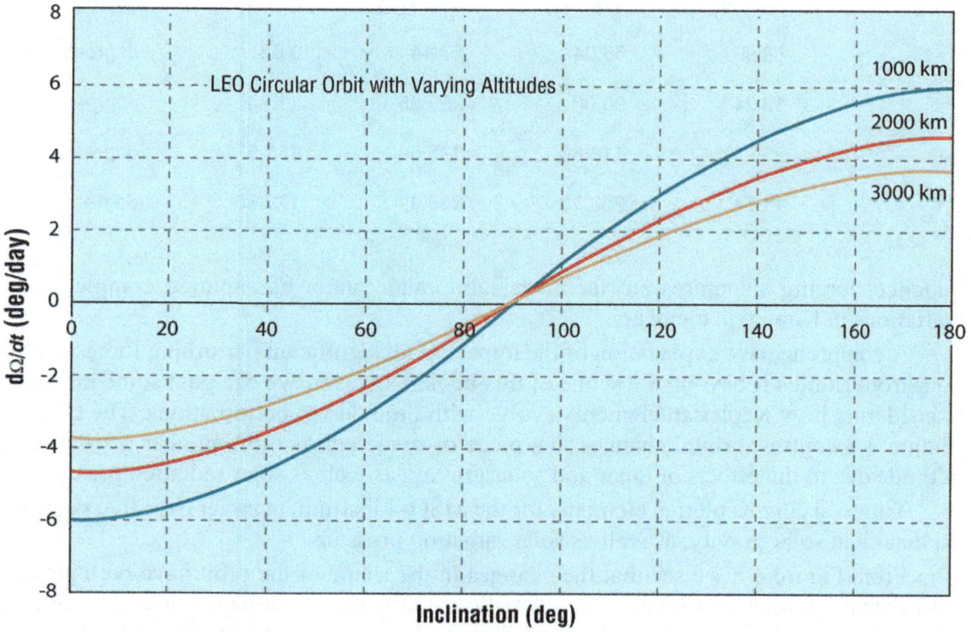

FIG. 4-7. Rate of Change of Ascending Node as a Function of Orbital Altitude and Inclination.

Typical Keplerian elements for a selection of missions. The first two are LEO RSOs, the first in a polar orbit, the second inclined to provide coverage of most of the world's oceans. The GPS satellite is in a medium earth orbit (MEO) and GALAXY 25 is at geostationary altitude.

4.3.2 Beyond Keplerian elements

The assertion above that all other forcing effects are a factor of a thousand smaller than gravitational acceleration can be misleading. The so-called 'perturbing forces' (gravitation, radiation pressure, atmospheric drag, and Lorentz forces due to coupling of surface charge and magnetic fields) are generally systematic (as opposed to truly random processes). They create wide amplitude periodic effects, secular trends, and even shock-like discontinuities. At the simplest level these perturbations cause linear variations in Keplerian elements (e.g., the drift of the right ascension of the ascending node due to the oblateness of the Earth). In

TABLE 4-1. Sample Keplerian Elements for LEO, MEO and GEO Missions.

	LEO		ME	GEO	
	GRACE-A	Jason-1	GPS	GALAXY 25	units
a	6857	7714	26561	42148	km
e	0.0008	0.0001	0.0231	0.0003	no unit
i	88.99	66.04	53.44	0.02	degrees
ω	13.24	90.00	259.65	11.90	degrees
Ω	255.97	116.56	175.54	168.53	degrees
ν	11.74	253.13	239.11	179.56	degrees

practice, forcing signatures are far more subtle and cannot be captured completely via variations in Keplerian elements.

A comprehensive explanation of the impact of all significant perturbing forces applied in astrodynamics is beyond the scope of this book. However, we can gain some insight by considering how Keplerian elements evolve with time due to perturbations. The charts in Figure 4-8 contrast orbital changes to a piece of space debris orbiting near geostationary altitude due to the effects of lunar and solar gravity, as well as solar radiation pressure.

Time-varying Keplerian elements for the MSG-2 instrument cover over five years due to lunar and solar gravity, as well as solar radiation pressure.

From Figure 4-8 we see that the changes in the shape of the orbit have both periodic and secular components. These characteristic patterns can often be captured in trajectory models expressed as a base set of Keplerian elements along with a set of functions describing the time-variation of the elements. This method is used for the so-called osculating Keplerian elements used in the GPS navigation message (and for other satellite systems). Further readings on orbit perturbations are in Battin, 1987; Montenbruck, 2000; and B. Tapley, G. Born, and R. Schultz, 2004.

4.3.3 Representations of Orbital Motion

Simplistic though they are, Keplerian elements are useful as a first-order model of RSO motion. Moreover, an analytical solution exists so that given a set of Keplerian elements one can quickly compute the motion of the RSO in space at any future time (i.e., Newton's Principle of Determinacy applies). When considering more realistic force models (with a view to calculating position and velocity in an orbit) analytical solutions of the equations of motion do not exist, and we must resort to numerical methods. In the latter case, we calculate the trajectory from some initial conditions of position and velocity (usually in an inertial reference frame) and a numerical integration of differential equations describing forces acting on the RSO. We determine the initial conditions from some form of observation of the RSO (e.g., optical measurements made with a telescope or telemetered

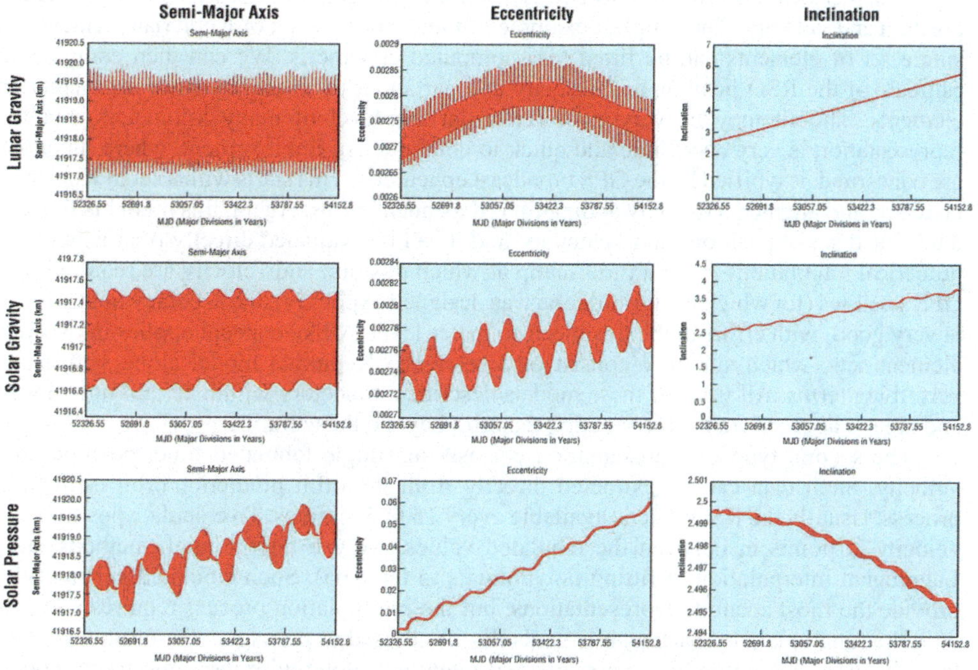

FIG. 4-8. Osculating Keplerian Elements by Differing Force Effects.

GPS position and velocity). The accuracy of the resulting trajectory depends upon how well we determine the initial conditions, and how accurately the force models describe the actual force field in space.

It is worth considering the difference between *orbit prediction* and *orbit estimation*. In the prediction problem, the analyst usually has a set of initial conditions and some tracking data for the RSO collected over the past few hours or days. These data are combined, along with the available force models, to provide a best estimate of the initial conditions (as well as additional parameters estimated to augment the force models if they seem insufficient). The orbit is then computed by propagating forward this initial state using numerical integration of the differential equations. The resulting arc length can be a few seconds to several years depending on the application. Any errors in the force models then appear as a systematic bias in the computed trajectory. *Orbit estimation*, on the other hand, fits a trajectory to a set of tracking measurements (such as laser ranging observations or Doppler data) in a least squares sense. Hence, orbit estimation always occurs after the fact—that is, once we have acquired the measurements. The process still requires the numerical integration of differential equations, but has the added constraint that the resulting trajectory must 'fit' the tracking data in some quasi-optimal sense. Both processes (orbit prediction and orbit estimation result in a table of positions and velocities time-tagged at regular intervals. Classically, this is known as an ephemeris. The last element of the problems addressed here deals with the communication of the trajectory data to a user.

Ephemerides fall broadly into two types: functions and tabulated lists of time-tagged position and velocity. The simplest example of the former is a set of Keplerian elements. A single set of elements can be fitted to a computed ephemeris. We can then compute an estimate of the RSO position and velocity at a particular time directly from the Keplerian elements. The fit may be very poor (errors at the level of many kilometers) but the representation is very accessible and quick to compute. An improvement, where functions are concerned, is typified by the GPS broadcast ephemeris. This starts with a set of Keplerian elements and augments them by adding time-dependent terms. Again, this model is fitted to a tabulated set of positions and velocities, and it can be evaluated directly (that is, without numerical integration) given a time stamp at which position and velocity are required. For GPS satellites (for which the formulation was designed explicitly), the accuracy of the model is very good, with errors at the decimeter to meter level. A like concept applies in two-line element sets, which similarly consist of a reference Keplerian model along with some perturbing terms. All three of these models describe a trajectory within certain time limits, and are not appropriate for propagating the orbit beyond those limits.

The second type of representation consists of simple tabulated time, position, and velocity. Such data can be extracted directly from the orbit prediction/orbit estimation process. Usually the RSO state is available every 15 to 5 minutes. To calculate position and velocity at points in between the tabulated values, we use interpolation methods (e.g., Lagrangian interpolation or fitting polynomials to the data). Such tabulated ephemerides provide the most accurate representations, but the interpolation process requires care. As with the functional approaches, it is generally ill-advised to propagate the orbit outside of the tabulated data unless using a numerical integrator along with the same force models used to derive the trajectory. Orbit accuracy in a tabulated form has no theoretical limit. In practice, accuracy is limited by the quality of tracking data and the accuracy of the force models used. NASA JPL has demonstrated orbit determination accuracy of better than 1 cm for the Jason-class LEO satellite altimetry missions, and GPS orbit accuracy has been demonstrated at the 2 cm level [Haines, 2004; Ziebart et al., 2007].

The methods above describe RSO motion in the space environment. From the perspective of an observer or analyst this motion is more usefully expressed in a topocentric frame (with North, East and Up unit vectors at the observer's site). Finally, we can visualize the motion of the sub-satellite point using a ground track—usually a table of time-tagged latitudes and longitudes of the sub-satellite point. Figure 4-9 illustrates this concept.

Satellite ground tracks help visualize the motion and coverage of a satellite mission with respect to the Earth's surface. In this case note the extent of the footprint as a function of latitude—driven in part by the RSO inclination (circa 66°).

4.3.4 Gravitation

Newton hypothesized a universal force called gravity acting according to an inverse square law such that the force between a planet of mass m_1 and a body of mass m_2 in orbit around that planet obeys the following equation:

$$m_2 \ddot{\boldsymbol{r}} = -\frac{G m_1 m_2}{r^3} \boldsymbol{r} \tag{4-5}$$

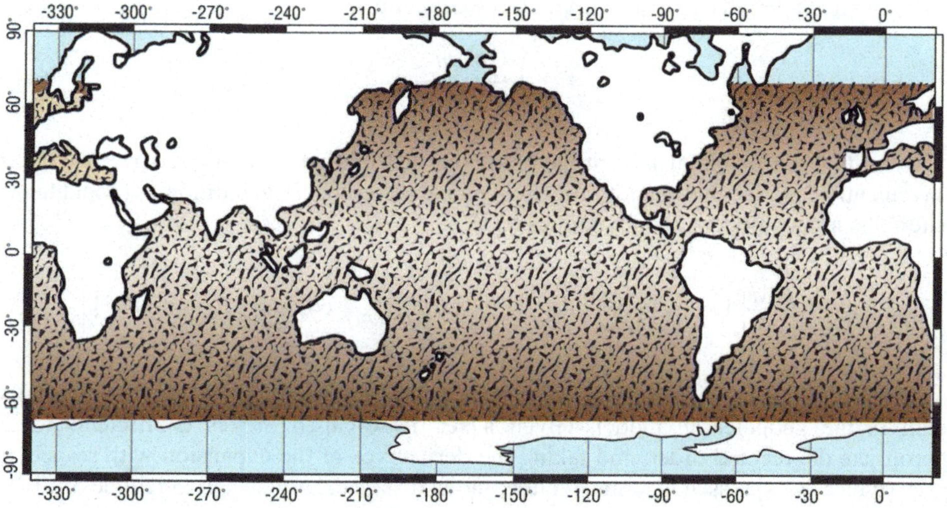

FIG. 4-9. Ground Track of the TOPEX-POSEIDON Mission Over the Oceans for a Ten-Day Period.

where $\boldsymbol{r}$ is the position vector of the orbiting body with respect to the center of mass of the planet, and G is Newton's gravitational constant. Newton's law explained the motion of the Moon about Earth. However, we cannot prove the law—at least not in the sense of deducing it from axioms. Instead it makes useful predictions about the motion of objects. Precise measurements of orbital motion show quickly that Newton's law as stated above is insufficient to describe how objects truly move. Keplerian elements are predicated upon such a simplistic two-body problem.

However, Newton's law yields some very useful ideas—and for our purposes here the concept of 'gravitational potential' is key. Say we define some quantity *U, gravitational potential energy* to be:

$$U = -\frac{Gm_1}{r^2} \tag{4-6}$$

This defines a property of near-Earth space, a potential energy field varying as a function of distance from the center of mass of the planet. Taking the first derivative of U with respect to the radial distance r we get:

$$\left|\frac{\partial U}{\partial r}\right| = -\frac{Gm_1}{r^2} \tag{4-7}$$

Which is the acceleration due to gravity, expressed as a vector quantity. Thus:

$$\ddot{\boldsymbol{r}} = -\frac{Gm_1}{r^3}r \tag{4-8}$$

From which we derive easily the simple force equation:

$$m_2\ddot{r} = -\frac{Gm_1m_2}{r^3}r \qquad (4\text{-}9)$$

Gravitational potential is a property of the distribution of mass in space. We can conveniently describe this quantity for a real planet using spherical harmonics (explanatory material is in any modern astrodynamics text):

$$U(\varphi, \lambda, r) = \frac{GM}{r}\left\{1 + \sum_{n=2}^{\infty}\sum_{m=0}^{n}\left(\frac{a}{r}\right)^n P_{n,m}(\sin\varphi)[C_{n,m}\cos m\lambda + S_{n,m}\sin m\lambda]\right\} \qquad (4\text{-}10)$$

For our purposes it is the *spherical harmonic coefficients* ($C_{n,m}$, $S_{n,m}$) that provide the detail in the geopotential model. Given a set of so-called Stokes coefficients to an appropriate degree and order, and taking the derivatives of the expansion with respect to space (typically expressed in terms of either radial distance, latitude and longitude or ECEF XYZ), we compute a commensurate gravitational acceleration field. The successful development, launch, and analysis of novel forms of measurement of geopotential models by dedicated gravity field missions such as GRACE, CHAMP, and GOCE have revolutionized knowledge of Earth's gravity (e.g., Tapley et al., 2004). This led to an understanding of the time variation of many of the geopotential coefficients allowing for time-variable gravity. Tidal variations in the field are easily computed. Current high-resolution geopotential models are complete to degree and order 2159. To put that in perspective, gravity fields from the 1990s were complete only to degree and order 360.

To summarize—the modeling of the Earth's gravity field involves taking the spatial derivatives of the Earth's gravitational potential expressed in spherical harmonics. The accuracy of such geopotential models has improved dramatically since the 1990s and provides a tried and tested resource for astrodynamics today.

Current and future mission requirements for position and orientation make strenuous demands for accuracy in the modeling of physical effects and in the realization of appropriate orbital and planetary reference frames.

4.4 Attitude Representations and Equations of Rotational Motion

4.4.1 Euler Angles

Euler's theorem states that three angles are sufficient to describe an arbitrary reference frame transformation between two reference frames that share the same origin. Euler angles hence describe the transformation from the inertial reference frame to the body frame and back at any point in time, using projections or the so-called direction cosine matrix [Goldstein, 1950]. The order of the rotations is not commutative. In the following sequence, 3-1-3 (φ,θ,ψ) is used. This means a sequence of rotations first around the body z-axis (by angle φ), then around the new x-axis (by angle θ), followed by a rotation around the new z-axis (by angle ψ) to complete the transformation from the body frame to the inertial frame. The direction cosine matrix is:

$$\begin{bmatrix} x_{body} \\ y_{body} \\ z_{body} \end{bmatrix} = \begin{bmatrix} \cos\psi\cos\phi - \cos\theta\sin\theta\sin\phi & \cos\psi\sin\phi + \cos\theta\sin\psi\cos\phi & \sin\theta\sin\psi \\ -\sin\psi\cos\phi - \cos\theta\cos\psi\sin\phi & -\sin\psi\sin(\phi + \cos\theta\cos\psi\cos\phi) & \sin\theta\cos\psi \\ \sin\theta\sin\phi & -\sin\theta\cos\phi & \cos\theta \end{bmatrix} \begin{bmatrix} x_{inertial} \\ y_{inertial} \\ z_{inertial} \end{bmatrix} \quad (4\text{-}11)$$

as illustrated in Figure 4-10. The attitude equations of motion can hence be formulated as the following. The kinematic equations are:

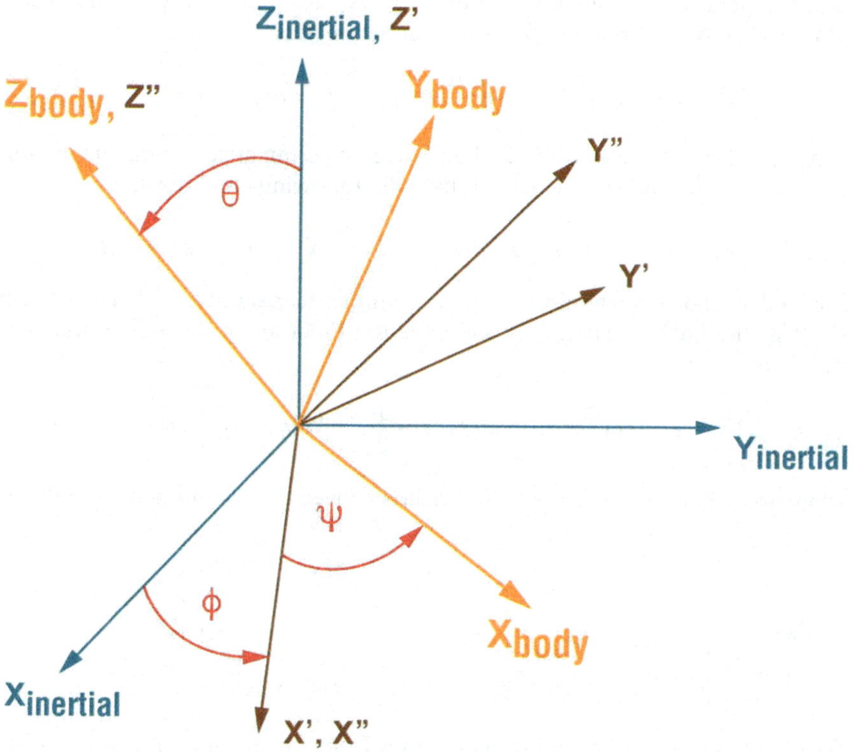

FIG. 4-10. Euler Angle Sequence 3-1-3 (φ,θ,ψ).

Here $\omega = [\omega_x, \omega_y, \omega_z]^T$ is the angular velocity in the body frame. The dynamic equations are:

$$I\frac{\omega}{dt} = \sum_i (r_i \times F_i) - \omega \times (I\omega), \quad (4\text{-}12)$$

where **I** is the constant inertia tensor relative to a fixed point on the rigid body, and $\mathbf{F}_i$ are the forces acting upon the body at a distance r_i from a fixed point on the rigid body. This point is normally the geometric center or the center of mass of the body.

Euler angle representations are very intuitive in the interpretation, but they have one huge disadvantage. The explicit trigonometric functions in the kinematic equations of motion in the denominator lead to singularities. Euler angles therefore most often apply to small attitude variations where we can use small angle approximations.

4.4.2 Quaternions

Alternatively, we can formulate the kinematic equations using the quaternion or Rodriguez parameter formulation [Wertz, 1978; Klumpp, 1976; Whittaker, 1937; Hamilton, 1866]. A quaternion consists of four parameters:

$$[q_1, q_2, q_3, q_4]^T = [\mathbf{q}, q]^T := q_4 + iq_1 + jq_2 + kq_3, \tag{4-13}$$

where q_4 is the real, scalar part, and i,j,k are hyperimaginary numbers building the imaginary part of the quaternion. They satisfy the following conditions:

$$i^2 = j^2 = k^2 = -1 \qquad ij = -ji = k \qquad jk = -kj = i \qquad ki = -ik = j \tag{4-14}$$

We can understand a quaternion as Euler symmetric parameters. Understanding e = $[e_1,e_2,e_3]^T$ as the Euler rotation axis and Φ as the Euler angle, we define the quaternion parameters as:

$$[q_1, q_2, q_3]^T = e^T \cdot \sin\frac{\Phi}{2}, \qquad q_4 = \cos\frac{\Phi}{2}, \qquad q_1^2 + q_2^2 + q_3^2 + q_4^2 = 1. \tag{4-15}$$

This allows us to define the direction cosine matrix between the inertial and the body frame as:

$$\begin{bmatrix} x_{body} \\ y_{body} \\ z_{body} \end{bmatrix} = \begin{bmatrix} q_1^2 - q_2^2 - q_3^2 + q_4^2 & 2(q_1q_2 + q_3q_4) & 2(q_1q_3 - q_2q_4) \\ 2(q_1q_2 - q_3q_4) & -q_1^2 + q_2^2 - q_3^2 + q_4^2 & 2(q_2q_3 + q_1q_4) \\ 2(q_1q_3 + q_2q_4) & 2(q_2q_3 - q_1q_4) & -q_1^2 - q_2^2 + q_3^2 + q_4^2 \end{bmatrix} \begin{bmatrix} x_{inertial} \\ y_{inertial} \\ z_{inertial} \end{bmatrix} \tag{4-16}$$

We can also easily compute the quaternions from any given cosine direction matrix A given the formulation above. This in turn allows us to re-define the kinematic equation from Eq. (4-14) as the following:

$$\begin{bmatrix} \dot{\mathbf{q}} \\ \dot{q}_4 \end{bmatrix} = \frac{1}{2} \begin{bmatrix} 0 & \omega_3 & -\omega_2 & \omega_1 \\ -\omega_3 & 0 & \omega_1 & \omega_2 \\ \omega_2 & -\omega_1 & 0 & \omega_3 \\ -\omega_1 & -\omega_2 & -\omega_3 & 0 \end{bmatrix} \begin{bmatrix} \mathbf{q} \\ q_4 \end{bmatrix} \tag{4-17}$$

4.4.3 Impacts to Rotational Motion

Every force **F** that acts upon the orbital motion can also impact rotational motion. As soon as the forces that act upon the different parts of the object are non-symmetric, they introduce torques and the term $\Sigma_i(\mathbf{r}_i \times \mathbf{F}_i)$ from Eq. (4-15) is not equal to zero.

As conservative forces, gravitational torques influence the attitude motion. Gravitational torques arise from the varying gravity over the (finite) size of the object itself; this means one part of the object is closer to the earth than another part. In the first order, gravity torque for a central potential vanishes for spherical objects. In practice it is relevant for a significant offset of the geometric center of an object and the center of mass.

The most relevant non-conservative forces introducing torques are drag and solar radiation pressure (SRP). We introduced the force expressions for drag and solar radiation pressure in the previous section. Those expressions depend explicitly upon the effective cross-sectional area in the direction of the velocity (drag) or towards the sun (SRP), respectively. Furthermore, on the reflection coefficients of the materials. Torques occur when either the area or the reflection coefficients are not symmetric and uniform over the object. For objects with concavities, self-shadowing effects occur. Self-shadowing introduces torques even when the area and reflection coefficients are otherwise completely uniform [Früh, 2014].

The environmentally induced torques, gravitational torques, SRP, and drag, as well as eddy currents, are coupled with the orbital position of the space object: the distance to Earth, atmospheric density, and distance to the sun depend upon its instantaneous position. In turn, the attitude of the object, and the torque-induced changes thereof, directly influence the translational force upon the object's center of mass and lead to orbital changes. An example is that the overall drag force changes when the effective cross-sectional area in the velocity direction changes.

The full six degree of freedom computation of the fully coupled problem is computationally demanding and often the associated differential equations are numerically stiff. Hence, we often model space objects operationally as uniform spheres—so-called cannon balls—as a first order approximation of the orbital motion, neglecting all attitude effects. The approximation is valid for actual spherical objects or actively stabilized satellites. Orienting their solar panels towards the sun means that the effective area for solar radiation pressure stays constant over time, just as for a uniform cannon ball. In like manner, the same orientation in the velocity direction leads to a constant effective area for drag calculations. For tumbling objects, the cannon ball approximation is not valid as the effective area changes over time and in general does not average out. The problem worsens the lighter the object is relative to the torque determining factor. In case of drag and solar radiation pressure, the area-to-mass ratio is crucial. In terms of gravity gradients, the overall mass relative to the mass non-uniformity is the driving factor.

Figure 4-11 shows the evolution of the angular velocity of a Kapton high area-to-mass ratio object (21.15 kg/m^2), simulated in a near-geostationary orbit [Früh, 2014]. The extremely light piece has uniform reflection properties; however it is modeled as not completely flat but having two edges folded up at an angle of 110 degrees. This leads to gravity gradient torques and solar radiation pressure torques introduced by self-shadowing. The simulation started from complete rest, but soon a significant attitude motion has accumulated. The motion is non-uniform. In turn, the torques influence the orbital motion, leading to a significantly different evolution of the orbital elements compared to a cannon ball object with the same properties.

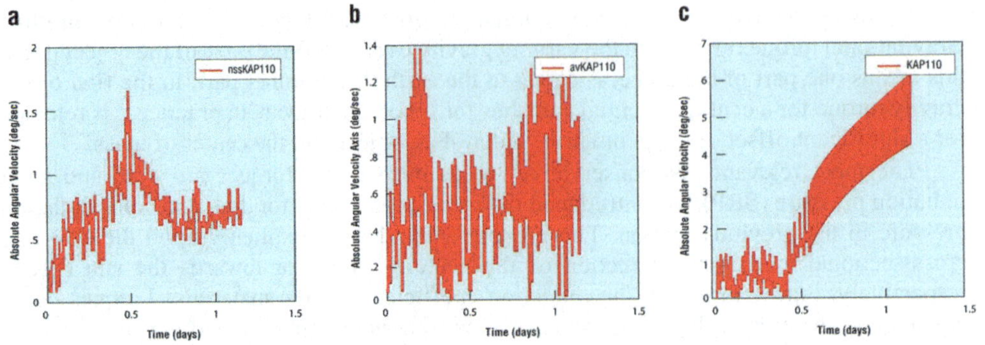

FIG. 4-11. Evolution of the Total Angular Velocity of Object KAP110. These graphs show the averaged avKAP110 simulation as well as the one neglecting self-shadowing nssKAP110 leading to comparable rotation rates, whereas the true simulation leads to a rapid spin up after about half a day.

4.5 Satellite Conjunction Assessment

The satellite event type that is in a sense the most organic result of the basics of satellite natural motion is the satellite conjunction. Left to follow the natural forces that govern and over time alter its orbital behavior, a satellite will encounter close approaches with other satellites; and in rare but real cases such encounters will result in collisions between satellites. On-orbit collisions are among the most grave of space events through their ability to cause both near-term damage by the immediate loss of important satellite missions, and longer-term danger through the pollution of important space corridors with debris fields that can then result in additional collisions. Because conjunction assessment is a simple, direct application of basic principles of satellite motion, and because it is of broad importance to both governmental and commercial space enterprises, it is a natural choice for a mission to treat in some depth as an adjunct to basic astrodynamics principles.

Conjunction assessment (CA) is the practice of determining when two satellites will come close to each other and thus create the possibility of a collision. Such situations are worrisome not just because a collision can cause loss of mission for one or both objects but because resultant debris from the collision creates future hazards to other satellites, potentially rendering unusable entire orbit regimes useful to particular space missions. Conjunction assessment is a preventive set of activities to counter collision hazards to satellites—naturally-occurring possibilities of collision with other orbiting objects. Intentional attempts to maneuver a satellite to collide with another would constitute a satellite threat; and while some of the same detection and assessment approaches might be appropriate in this latter case, the present chapter treats only the mitigation of satellite collision hazards. Once a niche activity pursued only by government agencies, on-orbit CA is now a common owner/operator satellite support task that is one of the central activities of space traffic management, employing sophisticated risk assessment techniques.

CA was originally the bailiwick of military command and control centers to verify that intended launch trajectories would not present dangerous conjunctions with existing space objects (launch CA) and in response to satellite anomalies—the latter to determine whether a satellite that is no longer responsive could have been subject to a collision. Occasionally, a center would do an "all vs all" conjunction assessment run to determine whether any dangerous situations were present in the immediate future, but this was not a regular command center practice.

Part of this CA parsimony was due to the satellite data products available at the time: two-line element sets (TLEs) with no accompanying covariance or other uncertainty data. The lack of orbit uncertainty data meant that the only real parameter available in assessing collision risk was the miss distance between the two satellites at the time of closest approach (TCA). Because TLEs typically have an error of one to two kilometers with this error largely uncharacterized in terms of spatial distribution [Hejduk et al., 2013], it becomes very difficult to ascertain what the collision risk actually is for conjuncting objects with miss distances less than a few kilometers. Because many actual CA miss distances are in this range, using miss distances based on TLEs was not a practical method for determining which conjunctions were truly actionable.

In looking to protect the space shuttle and station from collision dangers, NASA recognized this deficiency and provided funding for the initial catalogue-wide precision orbit maintenance capability. This system was the precursor to the current DoD system that maintains precision state estimate and covariance information on all cataloged objects. With these products, much more sophisticated conjunction risk assessment approaches are possible, as described in the subsequent sections of this chapter; and CA thus became a much more common enterprise for US military and civil space entities. Commercial concerns largely held off until the well-known 2009 Iridium-COSMOS collision, at which point the "big sky" approach to space traffic management—in which the extreme unlikelihood of a collision counseled against active monitoring and mitigation—was no longer tenable. In the wake of this event, DoD began a program of releasing precision catalogue data germane to particular conjunctions, which enabled commercial entities to perform CA much more robustly. Largely due to this information availability, worldwide interest in CA has enjoyed substantial growth, with most space owner/operators performing or arranging for some sort of CA service for their satellites, and astrodynamics technical conferences now including sessions and tracks dedicated to the CA problem.

4.5.1 The Three-Part CA Process

Performing CA to protect a particular satellite follows a three-step process, reference Table 4-2. The process divisions are not immutable and to some degree are the result of the way the practice has evolved, but they do provide a technically informed way to describe the different parts of the CA enterprise.

TABLE 4-2. The Three Part Conjunction Assessment Process.

CA Screenings	Finds satellites that will enter into conjunction with the protected "primary" satellite
Risk Assessment	Determines which conjunctions potentially present a high likelihood of collision and debris production
Risk Mitigation	Determines courses of action for minimizing the likelihood of collision and debris production

CA Screenings

The CA process begins with screening runs, whose purpose is to generate a list of "secondary" satellites that over the screening period of interest will create conjunctions with a protected "primary" satellite. The process, which requires possession of a complete satellite catalogue, typically begins by applying a number of pre-filters to identify satellites that cannot possibly conjunct with the primary (e.g., the apogee height of one object is significantly lower than the perigee height of the other) and therefore reduces the number of secondary objects that we must examine. Such filters include geometric and time prefilters, as well as special additional considerations for nearly coplanar satellites [Hoots, 1984].

Next, a "screening volume" is constructed about the primary object, representing a protected space that, if violated by a secondary object's predicted ephemeris, will cause that secondary to be considered a conjunctor. Screening volumes were at first spherical, but recognition of the non-symmetrical error distribution about orbits and the relative importance of certain error components in the CA process has led to the use of rectangular prism or ellipsoidal error volumes [Narvet et al., 2011]. These error volumes are usually longest in the in-track direction (as this is where the greatest propagation errors reside) and most narrow in the radial direction (as radial differences are most revealing for CA risk assessment).

Ephemerides for the primary and all the secondary objects that survive the pre-filtering process are then produced for the screening interval (a typical duration is seven days into the future), and the screening volume is then moved along the primary ephemeris. Figure 4-12 gives a graphic representation of a conjunctor penetrating the screening volume. There can be multiple such penetrations for a primary-secondary pair during the screening period. For such cases, a subsequent process determines for each penetration the time of closest approach, the primary and secondary object states and covariances at that instant, and other information about the encounter that could be useful in performing subsequent risk assessment. The latter includes force model settings for the orbit determination (OD) that produced both the primary and secondary states, the non-conservative force model parameters (typically ballistic coefficient and solar radiation pressure coefficient), and observation density information for both objects.

These conjunction-related data then go to the satellite owner/operator or CA authority for further analysis; if sent by message, there is a Consultative Committee for Space Data Systems (CCSDS) defined message format called the Conjunction Data Message (CDM; achieved final "Blue Book" status in May 2013). The intent is to issue one such message per discrete conjunction per screening run.

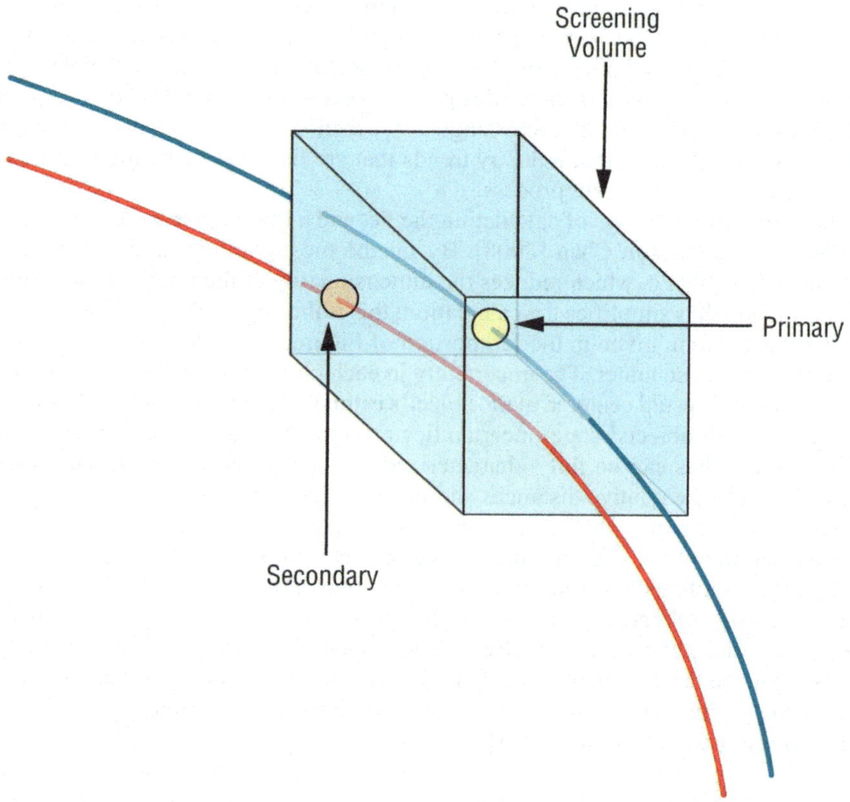

FIG. 4-12. **CA Screening Capture Using Geometric Screening Volume.** A conjunctor (secondary) has entered the screening volume, thus creating an event for which to perform risk assessment.

The present DoD practice at the DoD space correlation center (which is the principal data source for most CA enterprises) is to limit the screening activity to geometric considerations only. This approach gives the risk assessment entity awareness of more possible events. Such awareness can be quite helpful when planning maneuvers, as many of these events that are very low risk with the primary's current ephemeris may become serious should a maneuver alter that ephemeris. However, if one were to wish at this point in the process to reduce the number of alerts by considering collision likelihood, several techniques have been developed to do this quite efficiently [e.g., George and Chan, 2012, and Alfano, 2013].

Risk Assessment

Following the seminal work of Kaplan and Garrick [1981], we generally consider risk to be the consideration of both the likelihood and the consequence of a problematic event.

CA risk assessment at present focuses much more strongly on the likelihood of collision occurrence, so this is the logical place to begin a detailed description. We base this consideration largely (and often entirely) on the calculation of a single parameter called the probability of collision (Pc). Because this parameter is so important to current CA state-of-practice, we explain it here at some length. We then discuss additional risk assessment calculations, as well as current industry trends that are likely to shape the next generation of improvements to the current process.

There are different ways of calculating the Pc, and a thorough treatment of most of the common approaches is in Chan [2008]. By far the most common is the so-called "two-dimensional" or 2-D Pc, which reduces the dimensionality of the problem through a set of assumptions and thus simplifies the calculation; this is the approach that we will detail here. Consider the situation given in the left portion of Figure 4-13, which shows two satellites at their TCA in an encounter. The uncertainty in each object's state estimate is represented by a covariance ellipsoid centered at the object's estimated position (for the present, we will presume that both objects' state uncertainties are represented by a zero-mean, Gaussian distribution and thus can be fully characterized by a covariance matrix). Because one is interested in only the relative distances and uncertainties between the two objects (since it is these that will govern whether a collision will take place), it is acceptable to create a relative uncertainty by combining the two covariances into a single uncertainty statement and place this combined covariance at one of the two ends of the relative position vector. If the primary and secondary objects' position uncertainties are uncorrelated (a reasonable assumption, since they both arise from independent ODs from different observational datasets—but see the comments on 2-D Pc limitations at the end of this chapter), then we can combine the two covariances into a single covariance by simple addition of the two matrices (for a proof, see Chan [2008]).

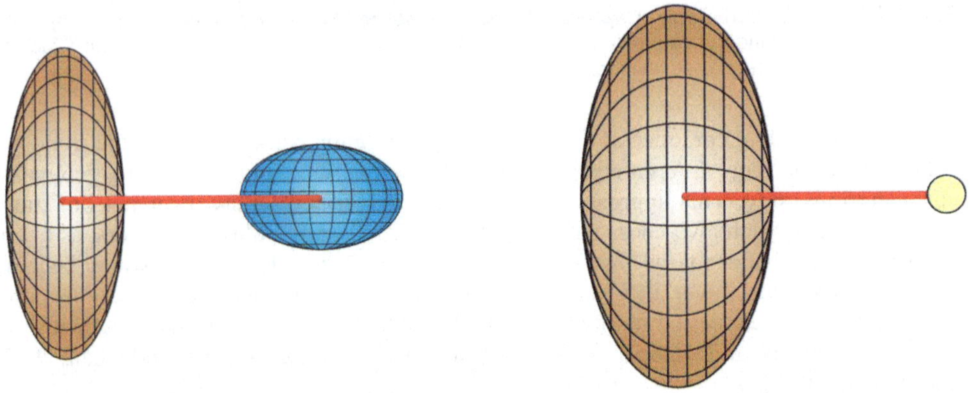

FIG. 4-13. Two Satellites in Conjunction at TCA. The left diagram shows both with their individual uncertainty volumes; the right diagram shows combined uncertainty assigned to secondary object position.

One can proceed analogously with object size. We define the size of an irregular 3-D object by a sphere that circumscribes the object; one can do this for the primary and secondary objects in the conjunction. To determine whether, for a certain miss distance, there is a chance of the two objects' colliding, their circumscribing spheres can be placed one miss distance apart; if the spheres overlap, then there is a collision danger. This situation is in the left portion of Figure 4-14. We can achieve the same effect if one "supersphere," constructed by circumscribing an adjacent placement of the two smaller spheres, is placed at one end of the relative miss vector and compared merely to the point indicating the other end of the miss vector; if the supersphere encompasses the miss vector termination point, then there is a collision danger. The radius of this supersphere is called the "hard-body radius" (HBR). The right part of Figure 4-14 shows this transformation.

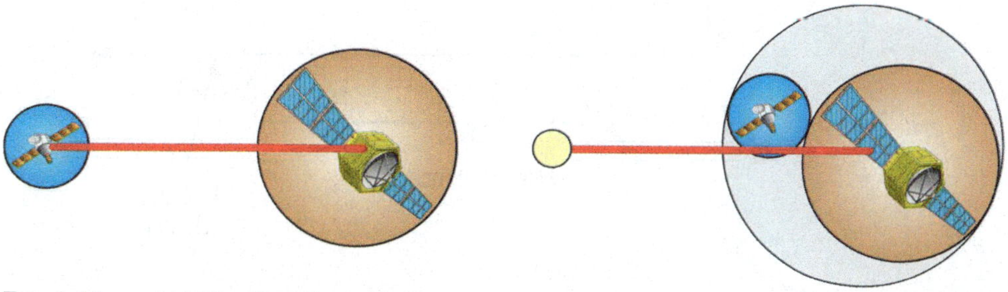

FIG. 4-14. **Satellite Hard-Body Radii.** Left portion shows secondary and primary satellites each within circumscribing spheres; right portion shows circumscribing sphere of combined hard-body radii, assigned to primary satellite position.

The next step is to determine whether for the particular conjunction two simplifying assumptions are possible: that motion be considered essentially rectilinear in the neighborhood of the conjunction and that the combined covariance remain essentially static throughout the encounter (and also possess relatively small velocity uncertainties). Most conjunctions with reasonably large relative velocities (>1 km/sec) will meet these criteria. From these assumptions, one can conclude that if a collision is to occur, it will take place in the plane perpendicular to the relative velocity vector and that the marginal component of the probability perpendicular to the plane will approach unity and can thus be ignored [Alfano, 2005; Chan, 2008]. Since the combined covariance is not changing over this interval, one can project the entire situation into this plane (called the "conjunction plane") and evaluate the collision likelihood in two-dimensional space. A plot of the situation (called, appropriately, a "conjunction plane plot") is in Figure 4-15; one can see the one-sigma combined covariance projected as an ellipse, the miss vector at TCA along the x-axis, and the HBR sphere (projected as a circle) at the end of the miss vector.

The probability of collision (Pc) is thus the amount of uncertainty probability density that falls within the HBR circle, as it is those situations that reflect an actual miss distance small enough to fall within this circle. Since the uncertainty volume (area here) extends to infinity (although with ever-diminishing density), there will always be some probability

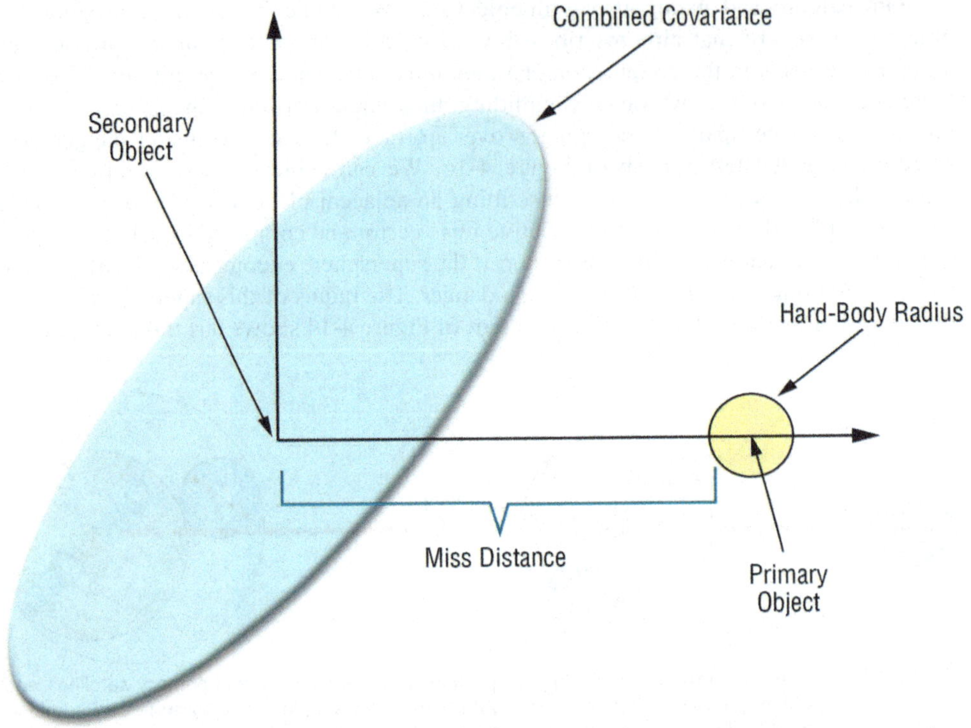

FIG. 4-15. **Conjunction Situation in Conjunction Plane.** In most cases we can use a two-dimensional depiction such as this because certain simplifying assumptions inhere.

density falling within the HBR circle, but the density is often low enough that the Pc is essentially 0 to machine numerical integration limits.

In using the Pc for risk assessment, we must recognize that a conjunction can achieve a low Pc value in two distinct ways. The first and preferable situation is one in which the nominal miss distance is (notably) larger than the HBR and the combined covariance is very small. Because there is little uncertainty in the primary and secondary states, it is highly likely that the actual miss distance is truly larger than the HBR, and the low Pc value thus reflects actual low collision risk. Conjunctions with this attribute are said to reside in the "robust" region because statements of low collision risk are here robustly determined. However, a second, and not infrequent, situation is one in which the nominal miss distance is larger than the HBR but the combined covariance is very large. In this case, because the relative position of the two satellites is so poorly determined, there is a broad range of possibilities for the actual relative miss vector, and therefore it is relatively unlikely that the actual miss will be smaller than the HBR. A low Pc value thus obtains here not because the actual collision risk is known to be low but because we know so little about the satellites' actual positions that we cannot conclude that it is high. Conjunctions with this attribute are said to reside in the "dilution" region because the Pc has been in a sense diluted by increased uncertainty (and not by certainty of a dangerous situation not inhering).

An analogy with car locations in a large parking lot, while not exact, is perhaps helpful here. If two cars are parked in a large parking lot, and we have been told that one car is parked on one side of the lot and the other on the other side, then we can conclude with a high degree of confidence that they are not next to each other. If, however, all we know is that the two cars are each parked somewhere in the lot, then we can also surmise that it is unlikely that they are parked next to each other, but only because in a large parking lot it is unlikely that any two particular cars will happen to be adjacent to each other. In this latter case, the low probability does not mean that the data have shown us the two cars are far apart. It is only a statement of the general unlikelihood of such an alignment if all one knows is that two cars happen to be in the same general area—if better information were available about the two cars' location, we might learn that the two cars are in fact adjacent.

Where a conjunction may stand with regard to these two different strains of a low Pc value (i.e., robust or dilution regions) depends on the ratio of the covariance size to the miss distance, and this ratio tends to evolve over a conjunction's life cycle from discovery to TCA; Figure 4-16 represents this evolution. In this figure, the x-axis is this ratio, although it can also be a proxy for time (in this case proceeding from right to left, to keep the morphology of the figure consistent with its first published version [Alfano, 2005]) in a typical conjunction; and the y-axis is the Pc. When the conjunction first becomes apparent, the covariance at TCA is usually large because we must propagate it several days forward to TCA; the covariance size to miss distance ratio is large, and therefore the Pc is small (due to the relative lack of knowledge about the situation).

As time passes, two things happen: 1) we solicit and receive additional tracking for the secondary object, thus shrinking its epoch covariance; and 2) the propagation time to TCA is shorter, thus even further shrinking the size of the propagated covariance (compared to its size at the event's first discovery). If we presume that the miss distance remains relatively constant, then the Pc at this new point will be higher. Eventually it will reach a peak and then, as the covariance continues to contract with the ever-shortening propagation time to TCA, the Pc will rapidly decrease again (unless the situation is a true collision, in which case the Pc will continue to increase all the way to unity). The Pc decrease at this point is a welcome development in that it represents a durable assessment of a lower risk; the low Pc at the beginning of the event is a correct calculation of the probability, given what we know about the situation, but it is not a guarantee of safety.

Most satellite operators agree that dilution-region events with high Pc values should be seen as high-risk, but low-Pc events in the dilution region present a more ambiguous case: should they be left unmitigated because they cannot be demonstrated to be high-risk; or, because the low-Pc value is a product of poor data, should the event be treated as though the risk were high, as a conservative measure? Essentially this is a question of how the "null hypothesis" for risk assessment should be set: should one presume low risk unless it is demonstrably high, or should one act on the possibility of an unsafe situation unless there is good evidence of low risk? Each satellite operator must answer this question individually. As a point of reference, the NASA Conjunction Assessment Risk Analysis program, which is the largest governmental conjunction assessment enterprise, has opted for the former approach, treating events as high-risk only when the data demonstrate this status. A treatment of all the factors informing the selection of this approach is too large for the present section, but these factors include the inherent risks of performing satellite

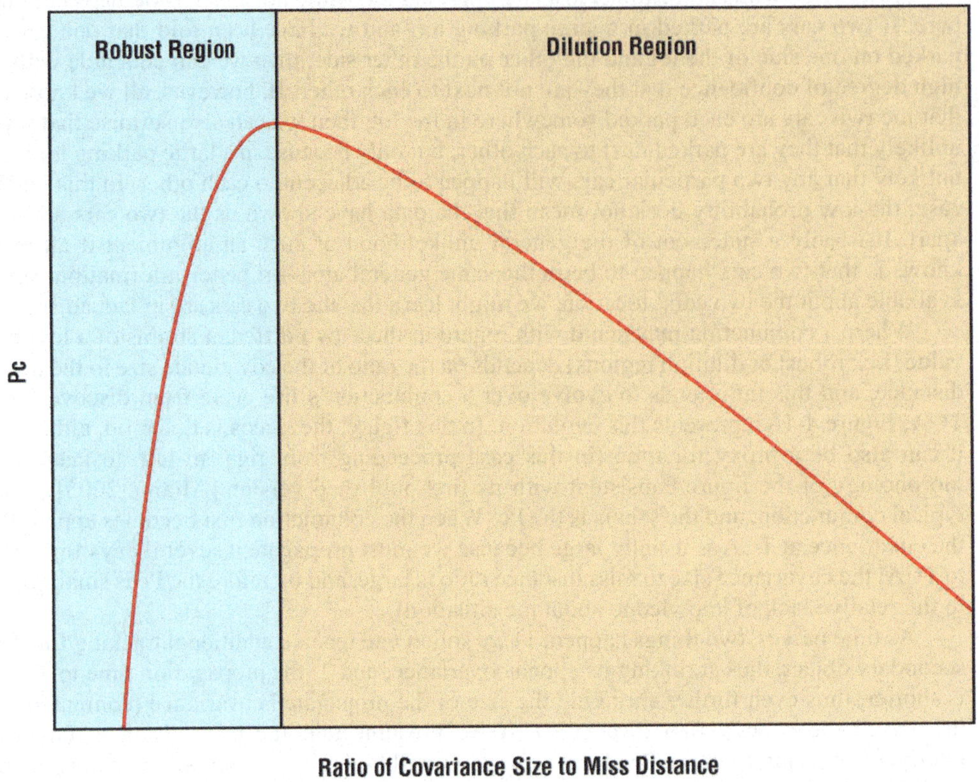

FIG. 4-16. **Conjunction Development Plot Showing Dilution and Robust Regions.** The Pc in the robust region reflects better knowledge of the actual situation than does the one in the dilution region.

conjunction remediation maneuvers, the often substantial interruption of science mission performance during and after such maneuvers, the larger maneuver sizes required when the orbital data are uncertain, and the fact that actual conjunctions themselves are extremely rare. Finally, there is a conceptual similarity between conjunctions with debris objects too small to be tracked, which are estimated to be about 95% of the actual conjunctions endured, and low-Pc dilution region objects: the former are essentially objects with (random) state estimates and infinite covariances, and the latter are objects with an uncertain state estimates and large covariances. In satellite operations we accept the inherent risk of collision with untracked objects, so a small increase to that risk presented by dilution-region low Pc events, which are conceptually similar, is not a substantial change in a satellite operator's risk posture [Hejduk et al., 2019].

While the Pc is the standard method of assessing collision likelihood, the assessment process incorporates important ancillary considerations to help to interpret the Pc value. Principal among these are methods to place uncertainty bars about the calculated Pc or even provide an entire Pc probability density function (PDF), based on the uncertainties in the input data (covariance size and orientation and HBR). Some analysis approaches attempt this to first order by simply varying the input covariance sizes in a reasonable way, calculating the Pc for each one, and using the largest Pc emerging from the exercise as the basis for risk assessment [Laporte, 2014]. If one has access to historical covariance realism data, Monte Carlo analysis drawn from histories of covariance scale factors (factors by which to multiply the covariance in order to make it a more realistic representation of the actual error distribution) can generate a Pc PDF that properly reflects the variation in the main quantifiable source of the Pc calculation's uncertainty [Hejduk and Johnson, 2016].

The error in HBR should also be incorporated [ibid.; Alfano and Oltrogge, 2016]. Figure 4-17 shows example output from such a construct. Here, one would consider the portion of the Pc probability density that exceeds a stated threshold (in this example 10^{-4}, indicated by the red line) in determining whether the risk is great enough to require remediation. In Figure 4-17, about 15% of the Pc probability density exceeds the stated threshold.

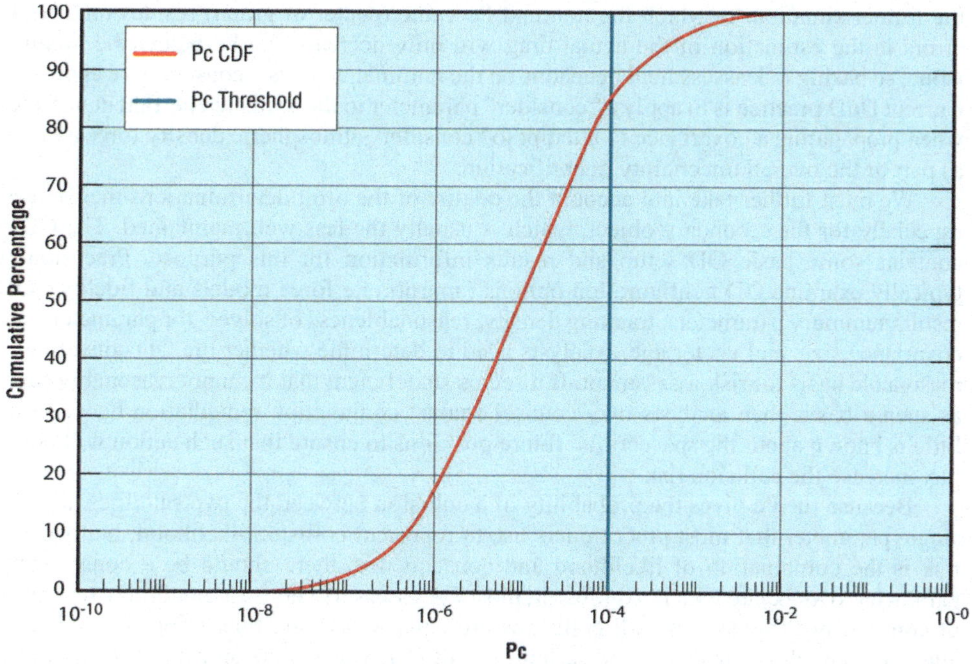

FIG. 4-17. Example Cumulative Distribution Function Plot for Pc Uncertainty Construct. Event severity is determined by the portion of the probability density to the right of the (red) threshold line, about 15% in this example.

One must also consider the uncertainty visited upon the problem through the prediction of future neutral atmospheric density values, which are necessary to propagate the primary and secondary states forward to TCA. Chapter 3 treats this issue at some length. Suffice it to say that these predictions often have large uncertainties during periods of high solar activity or storms, and these uncertainties affect the Pc. Because most of the commonly used atmospheric density models do not supply an estimation uncertainty with each prediction, taking direct cognizance of this source of error is difficult. One first-order approach is to vary the ballistic coefficient of the primary and secondary objects in an ordered, nested way, calculate the Pc for each varied case, and generate a contour plot of the results. If the nominal Pc value sits at or near the highest point on the contour graph, or if the entire plot shows very little variation in Pc, then the particular conjunction is either a "worst case" situation regarding atmospheric density modeling (i.e., any error in the prediction will cause the Pc to decrease) or it's simply not very sensitive to atmospheric density uncertainty. If in either of these cases the nominal Pc value is below a threshold that would necessitate remediation, then an owner/operator can decide not to remediate, reasonably confident that errors in atmospheric density predictions will not increase the actual conjunction risk [Hejduk et al., 2014]. Of course, if the nominal Pc resides in a trough or on a ridge on this contour plot, then the situation could be worse in the future; and no helpful conclusion for risk assessment can be drawn. Figure 4-18 gives an example of the former situation, in which the nominal Pc value (center of graph) resides on a peak; errors in the estimation of the actual drag will only decrease the Pc below the nominal value, so basing risk assessment decisions on the nominal value is a conservative approach. Current DoD practice is to apply a "consider" parameter to the ballistic coefficient variance when propagating a covariance to attempt to "consider" atmospheric density forecast error as part of the overall uncertainty quantification.

We must further take into account the quality of the orbit determinations themselves, especially for the secondary object, which is usually the less well maintained. The CDM contains some basic OD setup and results information for this purpose. Practitioners typically examine OD configuration options (appropriate force models and fidelity), OD results summary parameters, tracking density, reasonableness of solved-for parameters and covariance size, and vector age. Analysts need to determine whether the OD constitutes a reasonable basis for risk assessment. If it seems so deficient that it cannot reasonably serve as such a basis, then analysts may counsel against conjunction remediation because too little is known about the spacecrafts' future positions to ensure that such action will not in fact increase the collision risk.

Because the Pc gives the probability of a collision between the two satellites, it is the single parameter that most practitioners use to represent collision likelihood; but because risk is the combination of likelihood and consequence, there should be a consequence parameter to associate with it. At present, most CA assessors choose a monolithic rendering of consequence, namely that all collisions are equally bad and thus to be avoided. This approach, which originated in the earlier days of CA, was understandable in that context because all objects large enough to be tracked were also large enough to destroy a satellite completely upon impact and to produce a high-energy collision that would generate a large amount of debris.

With the introduction of sensors that track smaller objects, the recognition in the industry of low-relative-momentum conjunctions that would produce much less debris should a collision arise, and a growing awareness that the purpose of CA is to assess risk both to the protected asset and more broadly to its orbital regime, the industry is beginning to take steps to consider consequence as part of risk assessment [Hejduk et al., 2017]. An

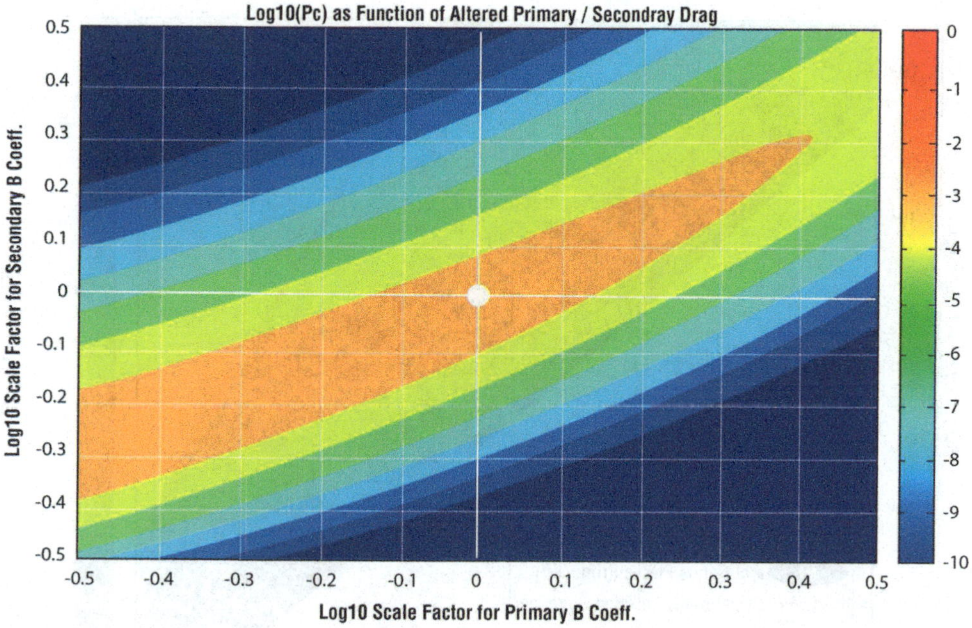

FIG. 4-18. Pc Sensitivity to Atmospheric Density Prediction Error. The nominal Pc (at center) resides on a peak, indicating that any error in atmospheric density (and thus) drag realization should decrease the Pc value.

easy way to do this is to establish different Pc thresholds for considering an event to be serious depending on the consequences. For example, a larger (more permissive) Pc threshold for conjunction remediation might be appropriate for a conjunction with a very small piece of debris in a relatively empty orbital regime than for a conjunction with a large debris object in a heavily-populated orbital regime.

A natural division that suggests itself is the distinction between a catastrophic collision, in which both objects are completely fragmented, and a non-catastrophic collision, in which the smaller object is fragmented but the larger object only cratered [Johnson et al., 2001]. Figure 4-19 is an intensity plot of the number of generated collision debris pieces produced as a function of the two objects' relative velocity and the mass of the secondary object (presuming a primary object mass of 2000 kg). Especially for the smaller secondary object masses, we see a quite modest increase in the number of debris pieces generated until the threshold for a catastrophic collision is reached (magenta line); at that point, the number of generated pieces may jump by orders of magnitude. This consideration is thus a convenient way to separate collisions likely to create only a small amount of debris from situations akin to the 2007 and 2009 large debris generation events.

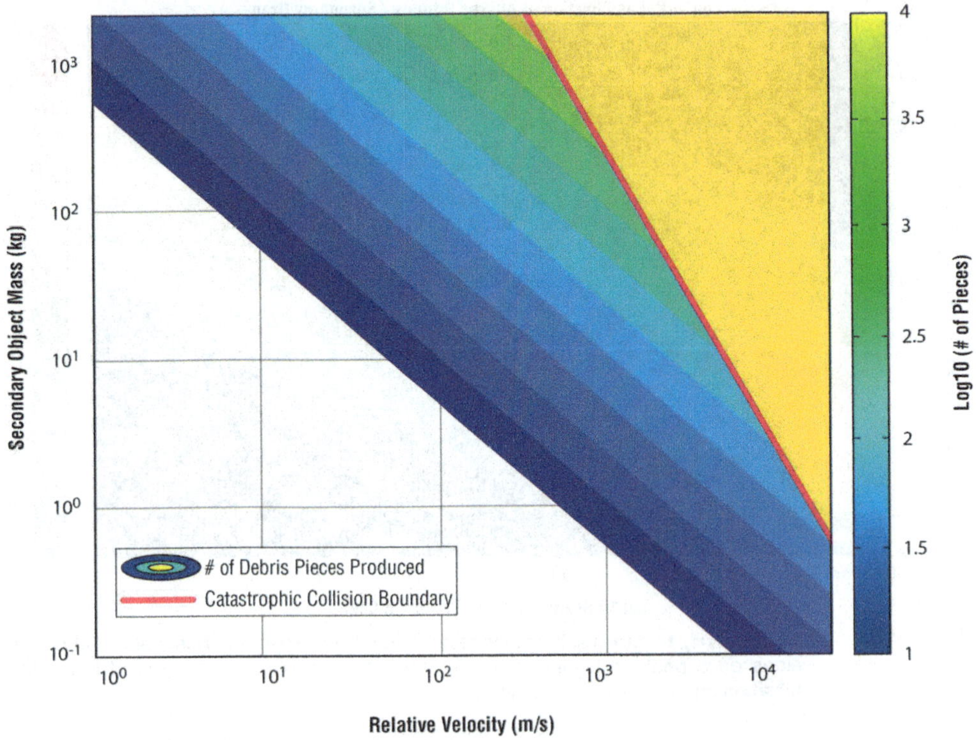

FIG. 4-19. Expected Amount of Collision Debris (Pieces) vs. Secondary Object Mass and Conjunction Relative Velocity. Debris production potential varies widely among different conjunction scenarios, making it a good parameter to help assess expected conjunction severity.

Risk Mitigation

If the conjunction risk is unacceptably high, an exploration of mitigation options begins. Because normally only one of the two space objects is an active payload, we usually limit mitigation options to that object, although there are proposed techniques for altering the orbit of the secondary (e.g., by laser illumination) to remediate conjunctions [Stupl et al., 2015]. Mitigation options are spacecraft-dependent and include the following:

- The inability to do anything at all for mitigation
- Being able to change the spacecraft's attitude to minimize its cross-sectional area along the relative velocity vector and thus reduce the collision likelihood
- Changing the spacecraft's attitude to alter the drag profile and thus the trajectory
- Slow-motion electric propulsion orbit modifications
- Impulsive chemical burns in the direction of choice. Because most conjunction remediations are presently of this last type, we discuss it here at greater length.

The basic idea is simple: the primary satellite will adjust its orbit to decrease the likelihood of collision with the secondary object. The details of the selection of the orbital maneuver, however, add several additional considerations that make the overall procedure somewhat complex. First, any change to the primary object's nominal orbit to decrease the risk of the current conjunction may increase the conjunction risk with another object. For this reason even low-probability conjunctions are identified by conjunction screenings and monitored until they reach their TCAs—these conjunctions, along with the principal conjunction, must be evaluated against any proposed orbit change and their associated Pc values recalculated and examined. Second, the mitigative effect of a maneuver is a function both of the maneuver size and the time of maneuver execution, and this sets up a problematic opposition: an earlier maneuver can usually achieve the needed mitigative effect with a smaller fuel expenditure, but waiting as long as possible to maneuver increases the likelihood of the receipt of additional tracking data and a fresh Pc calculation, which may reduce the conjunction likelihood sufficiently to obviate the need for a maneuver at all.

Generally, owners/operators prefer in-track mitigation burns because they preserve key orbital elements and often double as a drag make-up maneuver; so we tend to explore this type of maneuver first. Through a brute-force method of introducing impulsive delta-V changes of varying sizes at varying points in the orbit over a period of interest, one can generate a "Maneuver Trade-Space" plot, an example of which is in Figure 4-20. The burn size is on the x-axis, the burn time on the y-axis, and the color shows the base-10 logarithm of the Pc value after the maneuver. One can see in this example that a maneuver of non-trivial size will be necessary to bring the Pc down to $\sim 10^{-07}$, which is often a minimum value for a reasonable mitigation; and the maneuver will have to take place at one of the times associated with the darker groups of ovals on the plot. We note also how smaller maneuvers earlier in the event achieve the same mitigative effect for which larger maneuvers are required later.

A similar analysis is possible for any of the known non-principal conjunctions individually, and the results can be combined into a single presentation like Figure 4-20. If the Pc values for each of the conjunctions are uncorrelated (which they should be, since they are independent events), then we can use DeMorgan's Law of Complements to calculate a cumulative probability for each grid point that combines the Pc values for each of the conjunctions [Frigm et al., 2015].

4.5.2 Conjunction Event Processing Timelines

While individual CA practitioners have their own proclivities for the timelines associated with event processing, some general practices are common. CA screenings normally are run at least once per day and often multiple times per day, with usually at least a seven-day propagation (look-ahead) period; so the great majority of conjunctions show up about a week before TCA. "Late-notice" conjunctions, in which the conjunction may not be detected until only a day or two before TCA, do occur but are relatively rare; and they often arise from external influences, such as space weather events. The Pc calculations far from TCA have rather little predictive force, given the expected orbit refinement and shrinking of uncertainty volumes (as the propagation interval decreases), but they do identify potential CA events, thus allowing for review of the objects' OD solutions and, if

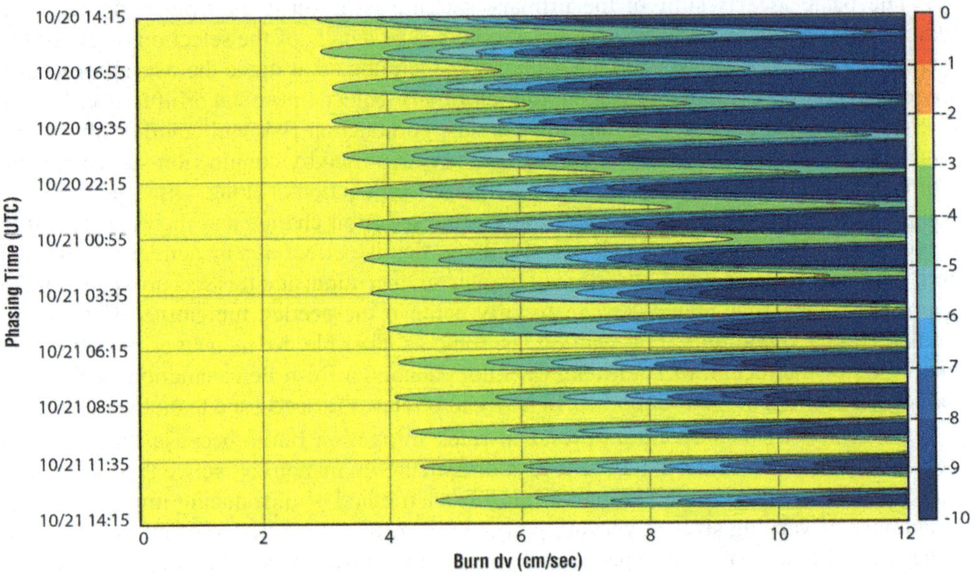

FIG. 4-20. **Maneuver Trade-Space Plot.** This plot illustrates resultant Pc (actually base 10 log of Pc) as a function of burn time and intensity.

it is thought that tracking increases will be helpful, requests for additional tracking data are made. Usually about two to three days before TCA a tentative decision to remediate is taken and, if it is affirmative, maneuver planning begins (anticipatory activities may have been taking place well before this time in order to have ready a viable maneuver plan if desired). A final predicted maneuver ephemeris must be generated and screened before the maneuver can be uploaded to the spacecraft, and this requires time as well. Modifications to the maneuver plan can continue, with the freeze point dictated by the particular spacecraft and its operating policies. However, maneuvers can usually be waived off very close to the final maneuver time; and this is a failsafe that is frequently executed if fresh tracking data indicate that the collision risk has significantly decreased [Newman et al., 2014].

4.5.3 Additional Considerations

While the basics of CA operations are straightforward, remediations are sufficiently risky and disruptive, and collisions sufficiently catastrophic, that risk assessment must proceed with a high degree of sophistication. This spawns a number of additional issues, the most significant of which are listed in Table 4-3 and discussed below.

2-D Pc Limitations

In an earlier section, we outlined the set of assumptions lying behind the 2-D Pc calculation; and while these assumptions are valid for most conjunctions, for certain classes of encounters they are not appropriate. Low-velocity conjunctions, for which the relative

TABLE 4-3. Additional Considerations in Conjunction Assessment.

2-D Pc Limitations	Monte Carlo occasionally required for Pc calculation
	Decorrelation of primary and secondary covariances improves calculation for high-drag orbits
Maximum Pc Constructs	Maximum Pc calculations are not true probabilities but only worst-case assessments
Non-Gaussian Covariances	Options exist for calculating Pc with non-Gaussian covariances, but high-Pc events generally not adversely affected

velocity is below ~100 m/s, show deviations, often significant, between the 2-D calculation and a Monte Carlo rendering [Frigm and Rohrbaugh, 2008; Alfano, 2009]. Encounters in which the combined covariance is distended in one of the dimensions can manifest rapid changes in the covariance orientation, violating the assumption of an invariant covariance during the encounter and in some cases resulting in actual Pc values orders of magnitude higher than the 2-D calculation [Hall et al., 2017]. In such cases, one can make use of Monte Carlo calculation techniques. Generally, applying perturbations from the covariances propagated to TCA, as opposed to the use of the epoch covariances and all of the necessary numerical propagations, is adequate to produce an operationally serviceable Pc estimate [Hall et al., 2018].

A parallel presumption of the 2-D Pc is that the primary and secondary object covariances are uncorrelated and thus can be combined into a joint covariance through simple addition. Because different tracking data are used to produce each object's orbit determinations, this is *prima facie* a reasonable assumption. However, as mentioned in the earlier discussion of space weather effects, propagated covariances are enhanced, through the application of a consider parameter, to reflect the expected atmospheric density forecast error. Because this error will for most situations be common to the primary and secondary objects (since they typically occupy similar orbit altitudes), it is a correlated error that should be removed from the joint covariance. Approximate and exact techniques have been developed to identify and remove this correlated error [Casali et al., 2018], which has an operationally relevant effect only when the satellite drag is relatively high. However, at relatively low altitudes, such as 400 km, Pc differences of several orders of magnitude sometimes obtain; so the effect is significant in certain situations.

Maximum Pc Constructs

If the covariance of one or both of the conjunction objects is unknown, there are techniques to determine the maximum Pc ("PcMax") that the event would produce if the missing covariance or covariances could assume any possible size, aspect ratio, and orientation [Alfano, 2008; Frisbee, 2015]. Because this value represents a worst-case situation, some practitioners have attempted to use it as an event filtering mechanism or even a risk assessment parameter. It is important to note that the PcMax is not an expression of collision likelihood; rather it is a statement of the maximum likelihood possible if a certain set of conditions were present. To make it into a likelihood statement one would have to introduce a second parameter that gives the likelihood that this set of conditions will in fact inhere (the developers of this construct recognize this fact and have never advocated for the PcMax as an independent risk assessment parameter). The construct gains more legitimacy as a pre-filtering parameter, but again its utility rests on the assumption that the nominal miss distance will not change as the event develops and that for the screening volume size selected it is actually a meaningful discriminator. For the typical set of CA risk assessment parameters used currently against the present precision space catalogue, the construct may not be that useful [Hejduk, 2016].

Non-Gaussian Covariances

If observations are plentiful, unbiased, homoscedastic, and uncorrelated, the resultant OD errors should be Gaussian in each component and properly represented by the OD covariance. If the data do not meet these criteria, or even if they do if a covariance is propagated for an excessive amount of time, then the actual uncertainty volume may take on non-Gaussian features and thus not be properly characterized by the formed covariance. In such a case, the Pc calculated by traditional methods using these deficient covariances will be incorrect. Such scenarios have been synthesized and analyzed [Horwood, 2016], although a competing study found that for higher-Pc events the effect was unlikely to be operationally significant [Ghrist and Plakalovic, 2012]. If concern persists about whether the uncertainty volume at TCA is Gaussian, an alternative is to perform "brute force Monte Carlo" Pc calculations using perturbations of the two satellites' epoch states, as most CA researchers agree that OD error volumes at epoch will nearly always be Gaussian or very nearly so. Moreover, one can perform their analyses via Gaussian Mixture Models [DeMars et al., 2014].

4.6 Brief Summary on Causes of Perceived Motion

At the beginning of this chapter, we covered astrodynamics, namely the science that studies motion of objects in space. We will preclude covering this again, but suffice it to state that one of the possible causes of perceived motion is actual motion.

In an ideal universe, actual motion would be the only cause of perceived motion. However, other realities preclude that from being the only possible cause. Observe Figure 4-21 below. In this figure what is shown is the difference between the actual measured range to a space object and presumably our predicted range (i.e. the measurement residuals) to the same space object, as a function of the Day-of-year (DOY). A salient feature of this plot is a discontinuity around DOY 242. One hypothesis for a cause in this discontinuity is that the space object changed its orbit and thus our predictions are offset. Did the space object actually move to a different location? However, another possible explanation for this evidence is that our sensor has experienced a bias instead. This is to illustrate that our measurement systems are imperfect and corrupted with noise and biases. A bias could be erroneously interpreted as a change in space object location. Therefore, we must take great care in making conclusions without assessing all possible hypotheses that could satisfy our evidence.

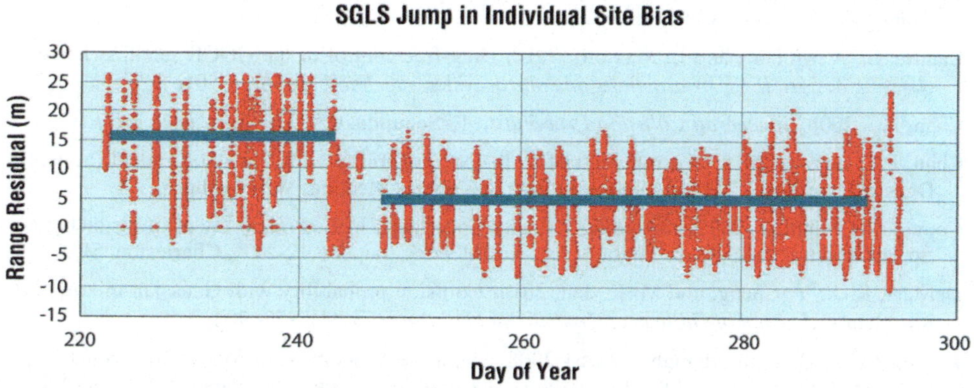

FIG. 4-21. **Space-ground Link System Jump in Individual Site Bias.** This figure illustrates an example of the difference between the measured range and predicted range from the ground to a space object.

Another possible cause of perceived motion can result from errors in our models of the underlying physics of motion as well as the specific methods we use to infer space object motion. Limitations in both of these can also result in reaching an erroneous conclusion of space object motion. If we have significantly non-representative motion models, our predictions will diverge from the truth. Moreover, if our inference models fail to account for real and significant sources of error, our estimates will also be biased from the truth. A robust framework for Space Domain Awareness would employ some version of Multiple Hypothesis Testing (MHT) and evaluate the consistency of the results against available evidence. A meaningful hypothesis must be able to explain past observations and evidence and predict future observations as well, to quantifiable levels of accuracy and precision.

References

Alfano, S. 2005. "Relating Position Uncertainty to Maximum Conjunction Probability." *Journal of the Astronautical Sciences*, Vol. 53, No. 2 (April–June 2005), pp. 193–205.

Alfano, S. February 2009. "Satellite Conjunction Monte Carlo Analysis." 2009 AAS Space Flight Mechanics Meeting (Paper #09-223), Savannah, GA.

Alfano, S. February 2013. "Determining a Probability-Based Distance Threshold for Conjunction Screening." AAS Space Flight Mechanics Conference (Paper #13-352), Kauai, HI.

Alfano, S. and D. Oltrogge. September 2016. "Probability of Collision: Valuation, Variability, Visualization, and Validity." 2016 AIAA/AAS Astrodynamics Specialist Conference, Long Beach, CA.

Battin, R. H. 1987. *An introduction to the mathematics and methods of astrodynamics*, first ed., 796 pp., AIAA, New York.

Bowman, B. R. et al. 2008. The JB2006 empirical thermospheric density model, *Journal of Atmospheric and Solar-Terrestrial Physics*, *70*(5), 774–793.

Calef, Brandoch et al. 2006. "Photometric Signature Inversion." *Unconventional Imaging II ed. Gamiz, V., Proceedings of SPIE*, 6307.

Canuto, E., A. Molano, and L. Massotti. 2010. Drag-free control of the GOCE satellite: noise and observer design, *IEEE transactions on control systems technology*, *18*(2), 501–509.

Chan, F.C. 2008. *Spacecraft Collision Probability*. El Segundo, CA: The Aerospace Press.

Chan, F.C. January 2015. "Formulation of Collision Probability with Time-Dependent Probability Density Functions." 2015 AAS Space Flight Mechanics Meeting, Williamsburg, VA.

Coppola, V.T. January 2012. "Including Velocity Uncertainty in the Probability of Collision between Space Objects." AAS Space Flight Mechanics Meeting (paper # 12-247), Charleston, SC.

DeMars, K. J., Y. Cheng, and M. K. Jah. 2014. Collision probability with Gaussian mixture orbit uncertainty. *Journal of Guidance, Control, and Dynamics*, *37*(3), 979–984.

Frigm, R.C. and D. Rohrbaugh, August 2008. "Relative Velocity as a Metric for Probability of Collision Calculations." 2008 International Astronomical Congress (Paper # IAC-08-A6.2.5), Glasgow, Scotland.

Frigm, R.C. et al. September 2015. "Total Probability of Collision as a Metric for Finite Conjunction Assessment and Collision Risk Management." 2015 AMOS Technical Conference, Kihei, HI.

Frisbee, J.H. August 2015. "An Upper-Bound on High-Speed Satellite Collision Probability when Only One Object has Position Uncertainty Information." AAS/AIAA Astrodynamics Specialist Conference (Paper # 15-717), Vail, CO.

Früh, Carolin and Moriba K. Jah. 2014. "Coupled orbit-attitude motion of high area-to-mass ratio (HAMR) objects including efficient self-shadowing." *Acta Astronautica*, 95: 227–241.

George, E.R. and F.K. Chan. September 2012. "Covariance-Based Pre-Filters and Screening Criteria for Conjunction Analysis. 2012 AMOS Technical Conference, Wailea, HI.

Ghrist, R.W. and D. Plakalovic, August 2012. "Impact of Non-Gaussian Error Volumes on Conjunction Analysis Risk Assessment." 2012 AAS/AIAA Astrodynamics Specialists Conference, Minneapolis, MN.

Goldstein, Herbert. 1950. *Classical Mechanics*. Cambridge: Addison-Wesley Publishing Company, Inc.

Haines, B. et al. 2004. One-Centimeter Orbit Determination for Jason-1: New GPS-Based Strategies, *Journal of Marine Geodesy*, *27*(1-2), 299–318(220).

Hall, D.T., M.D. Hejduk, and L.C. Johnson. January 2017. "Time Dependence of Collision Probabilities during Satellite Conjunctions." 2017 AAS Space Flight Mechanics Meeting (Paper # AAS 17-271), San Antonio, TX.

Hall, D.T. et al. August 2018. "High-Fidelity Collision Probabilities Estimated using Brute Force Monte Carlo Simulations." 2018 AAS/AIAA Astrodynamics Specialist Conference (Paper #18-244), Snowbird, UT.

Hamilton, Sir William Rowan. 1866. *Elements of Quaternions*. London: Longmans, Green and Co.

Hejduk, M.D. et al. September 2015. "Predicting Space Weather Effects on Close Approach Events." 2015 AMOS Technical Conference, Kihei, HI.

Hejduk, M.D. and L.C. Johnson. February 2016."Approaches to Evaluating Probability of Collision Uncertainty." 2016 AAS Space Flight Mechanics Meeting (Paper # 16-241), Napa, CA.

Hejduk, M.D. September 2016. "Covariance Manipulation for Conjunction Assessment." 2016 AAS/AIAA Astrodynamics Specialist Conference, Long Beach, CA.

Hejduk, M.D. et al. June 2017. "Consideration of Collision 'Consequence' in Satellite Conjunction Assessment and Risk Analysis." 26th International Symposium on Space Flight Dynamics, Matsuyama, Japan.

Hoots, F.R., L.L. Crawford, and R.L. Roehrich, June 1984. "An Analytic Method to Determine Future Close Approaches between Satellites." *Celestial Mechanics* Vol. 33, No. 2, pp. 143–158.

Hornwood, J.T. et al. September 2016. "KRATOS: Kollision Risk Assessment Tool in Orbital Element Spaces." 2016 AMOS Technical Conference, Kihei, HI.

Kaplan, S. and B. Garrick. 1981. "On the Quantitative Definition of Risk." *Risk Analysis*, Vol. 1, No. 1, pp. 11–27.

Klumpp, Allan R. 1976. "Singularity-free Extraction of a Quaternion from a Direction Cosine Matrix."*AIAA Journal of Spacecraft and Rockets*, 13, no. 12: 754–755.

Laporte, F. September 2014. "JAC Software: Solving Conjunction Assessment Issues." 2014 AMOS Technical Conference, Wailea, HI.

Linares, Richard et al. 2014. "Space Object Shape Characterization and Tracking Using Light Curve and Angles Data." *AIAA Journal of Guidance, Control, and Dynamics*, 37, no. 1: 13–25.

McCarthy, D. D. 1996. IERS Technical Note 21 *Rep.*, International Earth Rotation Service, Paris.

Montenbruck, O., and B. Gill. 2000. *Satellite Orbits: Models, methods and applications*, Springer-Verlag, Berlin.

Narvet, S. et al. August 2011. "Assessment of Uncertainty-Based Screening Volumes for NASA Robotic LEO and GEO Conjunction Risk Assessment. 2011 AAS Astrodynamics Specialist Conference (Paper #11-432), Girdwood, AK.

Newman, L.K. et al. September 2014. "Evolution and Implementation of the NASA Robotic Conjunction Assessment Risk Analysis Concept of Operations." 2014 AMOS Technical Conference, Kihei, HI.

Petit, G., and B. Luzum. 2010. IERS conventions *Rep.*, DTIC Document.

Sauceda, Felipe et al. 26 October 2001. *Space Station Reference Coordinate Systems*, Revision F.

Souchay, J., and M. Feissel-Vernier. 2005. The International Celestial Reference System and Frame*Rep.*, IERS Technical Note 34, BKG Frankfurt.

Stupl, J. et al. September 2015. "LightForce Photon-Pressure Collision Avoidance: Updated Efficiency Analysis Utilizing a Highly Parallel Simulation Approach." 2015 AMOS Technical Conference, Kihei, HI.

Tapley, B., G. Born, and R. Schultz. 2004. *Statistical Orbit Determination*, Elsevier Academic Press.

Tapley, B. D. et al. 2004. The gravity recovery and climate experiment: Mission overview and early results, *Geophys Res Lett*, *31*(9).

Vallado, David and Wayne McClain. 2001. *Fundamentals of Astrodynamics and Applications*. El Segundo, CA: Microcosm Press.

Wertz, James R. 1978. *Spacecraft Attitude Determination and Control*. Dordrecht: D. Reidel Publishing Company.

Whittaker, E. T. 1937. *A Treatise on the Analytical Dynamics of Particles and Rigid Bodies*. Cambridge: Cambridge University Press.

Ziebart, M. et al. 2007. Cracking the GPS-SLR orbit anomaly, paper presented at ION GNSS2007, Fort Worth, Texas, 2007/09/01/.

Chapter 5

Sources of Information (Hard and Soft Inputs)

Dr. Jim Shell, *Novarum Tech*
John Garnham, *University College London*
Dr. Francis Chun, Dr. Roger Tippets, and Dr. Rex Kiziah
United States Air Force Academy, Department of Physics

5.1 Introduction (Noise, Incompleteness, Bias, Corruption)

In this chapter we discuss various sources of information necessary to conduct space domain awareness (SDA) operations. A significant amount of data are available related to space-related organizations, space activities and space objects, and this plethora of data must be organized, filtered, and combined in the proper way to obtain the information necessary to support SDA operations.

The most important ingredient in space domain awareness is evidence. To know something, we must measure it; to understand something, we must predict it. Both require observations. As long as we have a hypothesis that explains all past observations and can predict future ones—to within a quantifiable measure of precision—no one can prove that the hypothesis is not true. However, no observer tells the complete truth, because all observers (physics- and human-based) are corrupted by noise or bias, and are of finite resolution, precision, and accuracy.

Most people attempt to force the data to answer a specific question (i.e., estimate a specific state), yet no one has ubiquitous observations. In other words, the set of data available to the analyst is almost certainly incomplete. Yet we almost always make decisions based upon incomplete data, and we make ad hoc (and often simplifying) assumptions to arrive at such decisions. Even when the answer is obviously "insufficient information to decide," many analysts make a decision anyway without truly understanding the consequences of deciding outside of the available information or inferable knowledge.

Robust and meaningful space domain awareness requires the analyst to embrace the complexity of the problem at hand and seek to infer knowledge from the available data, and to assume nothing external to what the data imply. We call this the "Moriba Jah Refrigerator Approach" to analysis, which is similar to how a refrigerator cools an object. Cooling is not achieved by actually cooling the object but rather by eliminating or removing heat from it. Therefore, for space domain awareness, we should try to remove ambiguity from the system instead of attempting to estimate a sole and specifically posed state. In other words, the analyst must seek to be prejudice-free in the inference process and hypothesize as much as possible, only using the available data to discard a hypothesis that cannot explain them. All surviving hypotheses have some likelihood of being true.

The analyst should make decisions based on all the possibilities that survive the scrutiny of evidence. Not doing so exposes the end user to unquantified or erroneous risk. We imagine that in the beginning, the analyst has a body of hypotheses, much like a block of marble of arbitrary shape. As data are gathered and analyzed, the analyst must subject each hypothesis to predicting what has been observed. If the hypothesis fails to do so within acceptable precision, then the analyst should discard it. This is like chipping away pieces of marble, with the goal of revealing the statue within the block itself. Additionally, it would be appropriate for the analyst to use data alone to form hypotheses, not subject *a priori* knowledge (bias) to the analysis. There may be a role for this in such methods as machine learning and artificial intelligence, but we should invoke it judiciously and not in the absence of known laws of physics.

In this text, "hard inputs" refers to information from physics-based sensors such as radars and telescopes, and "soft inputs" refers to information from human observations or interpretations. In general, hard inputs are numerically valued functions, and soft inputs are semantically valued. But this is a generality and not universally true. What follows are finer descriptions of these two sources of information.

5.2 Physics-based (Hard Inputs)

5.2.1 Electro-Optical (EO) Systems

Background

Observations of reflected sunlight from artificial Earth-orbiting objects began with the onset of the space age. The Soviet Union's launch of Sputnik in 1957 prompted the mobilization of international teams of citizen observers under the Smithsonian Astrophysical

Observatory's "Moonwatch" program. By the use of simple telescopes, human-based geometric optical "fences" were established, enabling observations of Sputnik and the associated rocket body. By mid-1958, twelve film-based telescopes were deployed internationally to provide the West with satellite observations [Hayes et al., 1963].

It was not until the 1980s that electro-optical (EO) systems came into use, when electronic detectors such as silicon intensified target tubes with analog readouts supplanted film-based detection [Otten et al., 1980]. The "modern" era of optical systems began with low-noise, silicon-based charge-coupled device (CCD) sensors in the 1990s. In 1997, electro-optical sensing left the ground on a space-based platform in low Earth orbit [Stokes et al., 1998]; geosynchronous orbit-based sensing made its appearance in 2014 [GSSAP Fact Sheet, 2015]. We elaborate on the distinctive attributes of ground-based versus space-based optical sensing below.

By the mid-2000s, improved CCD performance and decreasing costs coupled with computational capabilities rapidly accelerated optical system deployment. Substantial performance from low-cost systems is now the norm for space surveillance by commercial [Sibert et al., 2016] and international scientific groups [Molotov et al., 2008]. Some of the most capable optical sensing networks for high-orbiting objects now reside outside the domain of nation-states.

In this section, we review the attributes of optical systems, and then specific SDA missions performed by optical systems. An examination of the "imaging chain" will provide an understanding of the optics, detectors, and processing techniques used to extract SDA information.

The term "optical" spans the electromagnetic spectrum from ultraviolet through the thermal infrared, or approximately 0.2 to 14 micrometers wavelength. Optical systems detect optical energy from space objects via either *active* or *passive* means. Active optical sensing supplies the illuminating energy source, such as a laser. Passive sensing exploits naturally present energy, typically reflected solar energy, but sometimes also thermal emissions. Passive sensing of reflected solar energy over the visible to near-infrared portion of the spectrum is by far the most common use of optical systems.

Optical systems have unique advantages and disadvantages compared to other SDA sensing modalities. Passive ground-based optical sensors are among the least expensive, but daylight and weather impose obvious limitations, as Figure 5-1 shows. In addition to required night time operating, the space objects must be illuminated by the sun. The concurrent solar illumination of a space object accompanied by a ground telescope having night time visibility is *terminator conditions*. For higher orbiting objects, such as HEO and GEO orbits, terminator conditions are frequent, as shadowing by Earth is relatively rare. However, LEO objects are often in umbra limiting visibility to telescopes, leading to less frequent terminator conditions.

Passive space-based optical sensors are free from the limitations of bright daytime skies. However, viewing geometries must still avoid pointing these sensitive sensors too close to the sun. GEO-based optical systems diversify these geometric limitations because the viewing geometry of an object relative to the sun in GEO complements the view from ground- and LEO-based systems.

Passive optical systems use sunlight for illumination, with the reflected solar energy from a space object inversely proportional to the square of the range to that object (R^{-2}).

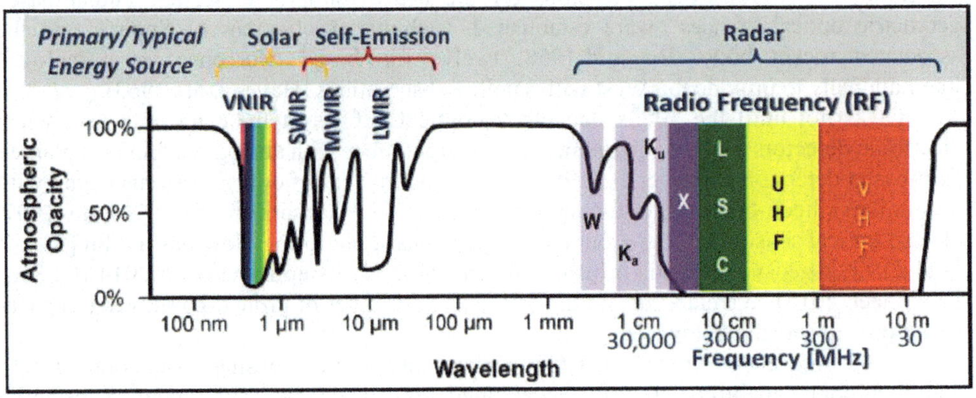

FIG. 5-1. Key Attributes of the Electromagnetic Spectrum for SDA Optical Sensing. Most optical sensors for SDA use silicon detectors, which respond from 0.4 to 1.0 μm, or in the visible to near infrared (VNIR) portion of the spectrum where solar-reflected energy dominates.

Radar systems, providing their own illuminating energy source, are subject to R^{-4} energy dissipation. These properties result in passive optical sensors being preferable for higher orbiting objects, while radars are more appropriate for lower altitude objects. These properties also mean that optical systems have significant advantages in searching broad areas for the purpose of new object discovery or post satellite maneuver recovery.

In terms of measurement information, passive optical systems lack the range and range-rate of radar systems but provide superior angular resolution. Active optical systems such as satellite laser ranging provide positional measurements superior to that of radars, but the object sensing capacity is significantly lower.

Optical systems, described in Table 5-1, are suitable for two distinct SDA missions:

1) *astrometric* or line-of-sight measurements of space objects and

2) *characterization* of space objects through either non-resolved radiometric techniques or resolved imaging.

Non-resolved radiometric measurements are often incidental to astrometric measurements, while resolved imaging requires a distinctly different optical system.

Astrometric measurements of objects by optical systems provide an angular position (line-of-sight unit vector) of an object at a point in time. This is the most common application of optical systems and warrants the most extensive discussion. Astrometric measurements serve to update the orbital state of an object, a process we call *orbit determination*. Chapter 4 addresses this in more detail. Astrometric measurements also help to identify satellite maneuvers* and to update the new orbital state of the satellite after the maneuver. Finally, the "search and discovery" mission detects previously unknown space objects as well as

* We define a satellite orbital maneuver as any self-generated event or process resulting in a non-isentropic change to the kinematic state of a human-made, resident space object.

TABLE 5-1. Strengths and Weaknesses of Various Optical Systems.

	Optical System	Advantages	Disadvantages
Astrometric / Radiometric	Ground-based	Low cost, high sensitivity from larger apertures, high search rates	Subject to weather and night operating constraints
	Space-based (in LEO)	Continuous operation without weather constraints, synoptic view of GEO region	Higher cost, sensitivity challenging
	Space-based (in GEO)	Continuous operation without weather or daytime constraints, complementary solar geometry to ground- and LEO-based systems	Multiple systems required for synoptic view of the GEO region Higher cost
	Ground-based laser ranging	High-precision positional measurements	Low sensing volume, objects without retroreflectors challenging
Resolved Imaging	Ground-based	Greater flexibility and usually lower cost compared to space-based solutions.	Opportunities may be subject to terminator conditions, insufficient resolution for deep space objects
	Space-based	Diffraction limit of optical system easier to achieve due to lack of atmospheric distortion.	Limited aperture sizes compared to ground-based systems. Effective over limited ranges

existing satellites that may have maneuvered, supplying data to initiate uncorrelated track processing to establish an initial orbit for the object.

In making astrometric measurements, we use either a *task-based* or *search-based* paradigm. The task-based approach seeks measurements of specific objects via directed observation based on the forward propagated orbital position at the time of the measurement. Search-based tasking provides coverage by collecting multiple fields of view over specified solid-angle regions. This approach is particularly suitable for objects with slowly varying positions relative to the earth-fixed coordinate system, such as GEO and near-GEO objects. Search-based tasking is an efficient means of new object discovery and also effectively provides positional updates of existing objects. Active optical systems, namely laser ranging systems, may also yield astrometric measurements. We address these further in the laser ranging section.

The final section of 5.2.1 discusses this technique. Section 5.2.2 covers the radiometric characterization of objects via optical measurements in detail, so we provide only a brief mention here. For historical reasons, we often use the term *photometric* due to its origins with human visual response. Measurements of the non-resolved optical signature of a space object enable insight into such attributes as its physical size, stability, material properties, and surface orientations. We generally look for measurements over varied spectral, temporal, and geometrically diverse illumination conditions. For this purpose we may leverage radiometric data incidental to measurements taken for the astrometric mission, but often employ dedicated sensing using extended duration observations or sampling rates required for particular characterization measurements. Although relatively immature, polarimetric measurements hold the promise of deriving additional information.

Finally, resolved imaging offers directly human-interpretable data on an object's physical appearance. For ground-based sensing, atmospheric effects inherently limit the resolution of these systems, requiring techniques to overcome the atmospheric limitations. The resolved imaging section gives a more detailed discussion.

Gathering Signal—Telescopes and Optics

This text does not delve into the details of optical system design, but focuses on the key design parameters of optical systems for space surveillance performance. The *aperture size* is the foremost design parameter for an optical system and is most commonly expressed in terms of the telescope diameter, d (m). The aperture determines how much light may be gathered and made available for detection, and it also establishes a limit on the angular resolving power of the optical system. This is critical for spatially imaging objects, and we address it further in the polarimetry section of 5.2.1. The second design parameter of an optical system is the *focal length, f* (m). Focal length determines the effective angular extent of a single detector or "pixel" on the focal plane. It also establishes the field of view (FOV) when coupled with the detector physical size. Figure 5-2 shows a basic overview of a generic optical system with these design elements. We represent the optics with a simple lens, though in reality they comprise multiple reflective and refractive components.

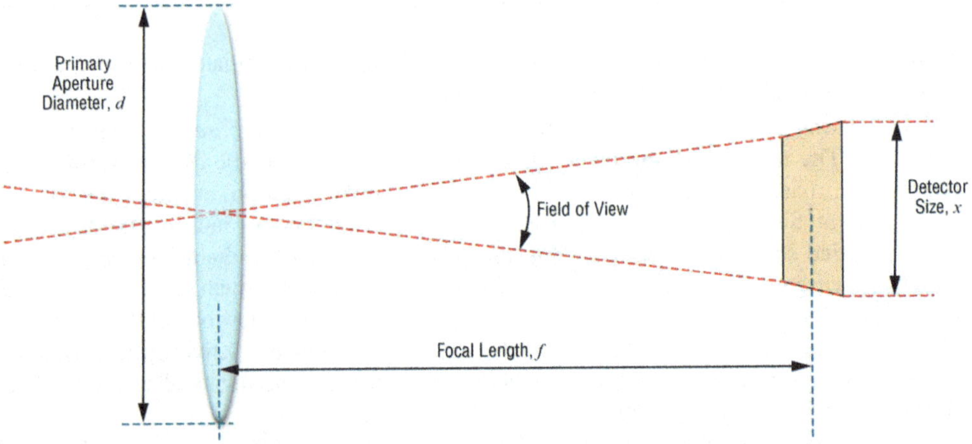

FIG. 5-2. **Components of a Basic Optical System.** A generalized optical system with the key design parameters relevant to space domain awareness.

A space surveillance telescope gathers *signal photons*, or those originating from space objects, as shown in Figure 5-3. Measurement of this signal is the objective, independent of the specific SDA missions the optical system supports. In addition to signal photons, the optical system also detects *noise photons*, or those from undesired sources. Noise sources compete with the signal photons and degrade the desired measurement. We discuss and quantify each type.

What are typical "signatures" of space objects? The astronomical heritage has a common visual magnitude scale for space objects. The signature of an object in visual magnitudes, mv_{ob}, is given by [McCue et al., 1971]:

$$mv_{ob} = -26.73 - 2.5\log\left[\frac{A_{ob} \cdot \rho \cdot F(\varphi)}{R^2}\right] \tag{5-1}$$

where A_{ob} is the area of the object, ρ is the reflectance, R is the range, and $F(\varphi)$ is the solar phase angle function for various object shapes. Space object optical signatures are highly complex due to the varied geometries and reflectance properties of multiple surfaces, but we may make a simple estimate by approximating the signature of a diffusely reflecting sphere with diameter d_{ob}. The solar phase function for a diffuse sphere is

$$F(\varphi) = \frac{2}{3\pi^2}[\sin\varphi + (\pi - \varphi)\cos\varphi] \tag{5-2}$$

Expressing the area of the space object as

$$A_{ob} = \frac{\pi d_{ob}^2}{4} \tag{5-3}$$

results in a diffusely reflecting sphere signature of

$$mv_{ob} = -26.73 - 2.5\log\left[\frac{d_{ob}^2\rho}{6\pi R^2}[\sin\varphi + (\pi - \varphi)\cos\varphi]\right] \tag{5-4}$$

The visual magnitude scale is not a radiometric engineering unit for quantifying system performance. Visual magnitude is equivalent to an irradiance, or power per unit area. Common units for irradiance are watts per square meter, but many sensors detect individual photons, so we use photons per second per square meter (ph/(s · m^2)). Without further development, we assert that a visual magnitude of zero is equivalent to approximately 5.6×10^{10} (ph/(s · m^2)) when considering the normalized response of a typical silicon detector.* This results in a visual magnitude to photon irradiance (E_{ob}) conversion of

* These photon-based approximations are the result of the spectral content of the sun along with the spectral response of a typical silicon-based sensor.

$$E_{ob} = 5.6 \times 10^{10} \cdot 10^{-\left(\frac{mv_{ob}}{2.5}\right)} \quad (5\text{-}5)$$

We have so far neglected the effects of the atmosphere on the signal photons for ground-based telescopes. The atmosphere both absorbs and scatters energy from the object, decreasing that available to the telescope. The telescope site altitude and elevation pointing angle are the primary factors dictating the atmospheric energy losses. We represent these losses by an atmospheric transmittance parameter, τ_{at}, with usual ranges of τ_{at} spanning 0.6 to 0.8 over the visible solar spectrum.

A telescope of aperture area A (m^2) may then collect the incident space object irradiance. In terms of the more commonly used telescope diameter d, the signal photon rate collected at the telescope aperture P_{sap} is given by

$$P_{sap} = E_{ob} \cdot \tau_{at} \cdot \pi \frac{d^2}{4} \quad (5\text{-}6)$$

With the signal photons now having reached the telescope aperture, we must consider losses en route to the detector. Most telescope designs incorporate secondary optical components that obscure some of the effective aperture area. We quantify this reduction by an obscuration value, τ_{ob}, representing the unobstructed fraction of the aperture area. As with the other transmittance variables, this unitless value ranges from 0 to 1, with typical values from 0.5 to 0.8. In addition to light blocked by internal telescope components, some light is lost at each optical interface, either through imperfect reflecting surfaces or transmittance losses through refractive elements; we represent this by τ_{tr}. Typical transmittance losses result in τ_{tr} ranging from 0.7 to 0.9. Table 5-2 contains definitions and ranges of values for optical signal power variables.

Collecting all the terms and accounting for the atmospheric and telescope transmittance penalties, the final equation for the signal photon rate (ph/s) available at the detector is:

$$P_s = E_{ob} \cdot \tau_{at} \cdot \pi \frac{d^2}{4} \cdot \tau_{ob} \cdot \tau_{tr} \quad (5\text{-}7)$$

We have not discussed how the optical system focuses this energy, only that the telescope has collected this amount from the space object. For astrometric measurements, we can effectively treat space objects as point sources. The size of the focus spot, or *point spread function* (PSF) is primarily driven by the atmospheric-induced blurring termed "seeing." Typical seeing conditions result in the best angular resolution of 1-3 arc seconds. Additional discussion on resolution is in the resolved imaging section of 5.2.1.

Example #1: Signal Photon Rate P_s

Consider a one-meter object in geosynchronous orbit. Approximately how much of the space object signal is available for detection by a 40-cm ground-based telescope?

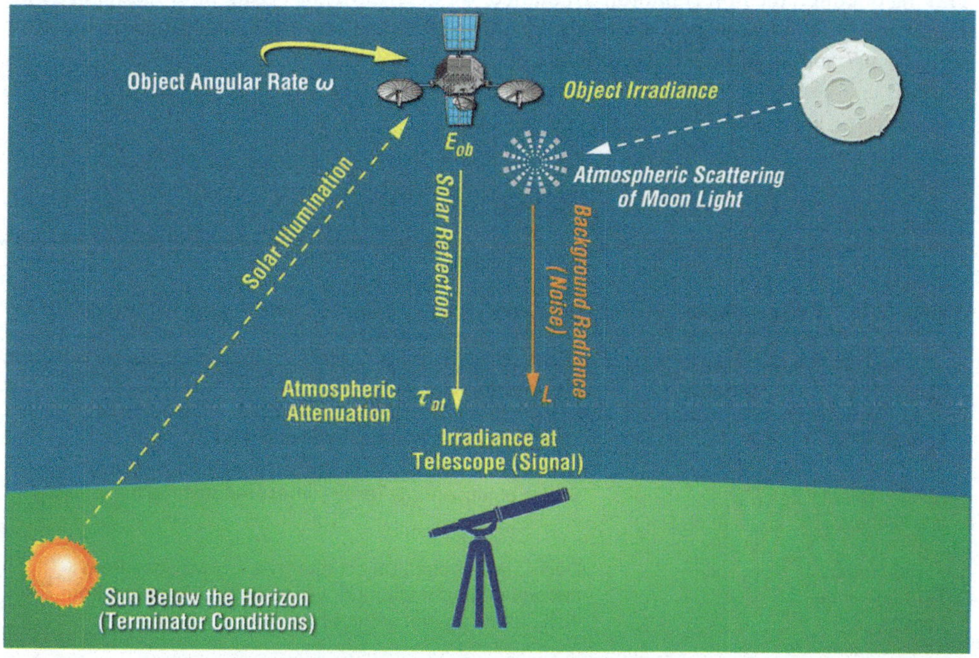

FIG. 5-3. Collection of Signal Power from a Space Object. Photon sources made available to the optical system include signal photons from the space object but also noise photons resulting from background radiance such as moonlight scattered in the atmosphere.

First we determine the signature of the object. A diffuse 1-m diameter sphere with reflectance of 0.2, at range of 40,000 km and sun illumination at a solar phase angle of 45° ($\pi/4$ radians) results in a visual magnitude of 15.3 mv. We obtain this result using Equation (5-4), shown below with the relevant quantities substituted for the variables:

$$mv_{ob} = -26.73 - 2.5\log\left[\frac{1^2 \cdot 0.2}{6\pi(40E6)^2}\left[\sin\frac{\pi}{4} + \left(\pi - \frac{\pi}{4}\right)\cos\frac{\pi}{4}\right]\right]$$

We now convert this visual magnitude into the object photon irradiance according to Equation (5-5):

$$E_{ob} = 5.6 \times 10^{10} \cdot 10^{-\left(\frac{15.3}{2.5}\right)}$$

$$E_{ob} = 43{,}364\ (\mathrm{ph}/(\mathrm{s}\cdot\mathrm{m}^2))$$

Having obtained the object photon irradiance, we determine the signal photon rate by using Equation (5-7). Assuming an atmospheric transmittance of $\tau_{at} = 0.8$, an optical obscuration of $\tau_{ob} = 0.7$ and an optical transmittance of $\tau_{tr} = 0.9$, we obtain

$$P_s = 43,364 \cdot 0.8 \cdot \pi\frac{0.4^2}{4} \cdot 0.7 \cdot 0.9$$

$$P_s \approx 2700 \text{ (ph/s)}$$

TABLE 5-2. Optical Signal Power Variables. These variables enable an accurate estimate of signal photons, P_S.

Item	Variable	Comments
Aperture Size	d	The telescope's primary optic diameter.
Space Object Irradiance	E_{ob}	Space object energy/(area · time) presented to the telescope.
Atmospheric Transmittance Losses	τ_{at}	Decrease in space object energy transmittance from absorption and scattering. Losses normally range from 20-40%, resulting in a value of 0.6 to 0.8
Telescope Obscurations	τ_{ob}	Secondary mirrors or components in the telescope that block a portion of the primary telescope aperture. Typical values range from 0.5 to 0.8.
Telescope Transmittance	τ_{tr}	Losses incurred at each optical surface in the telescope due to reflection and transmission losses for each element. Values usually range from 0.7 to 0.9.

Having quantified the signal photons originating from the space object and gathered by the telescope, we must now account for the *noise photons*. Noise photons include any photons not originating from the space object but transmitted through the optical system and made available for detection. The performance and low-noise characteristics of modern detectors often result in noise photons being the dominant noise source, rather than electronic noise introduced from the detector.

Direct noise photons are those that are present in the same field of view of the telescope as the space object. For ground-based telescopes, atmospheric sources of direct noise photons usually dominate and include the daylight sky, the twilight sky, moonlight scattering, light pollution, and a phenomenon termed *air glow* resulting from the recombination of ionized constituents. These sources generally cover a large spatial extent without significant variation across the field of view. The telescope makes no distinction among these sources and unfortunately collects them just as it does signal photons. Direct noise photons also originate from exoatmospheric sources, the impacts of which extend to space-based optical systems. These sources include zodiacal light, or solar scattering from particles predominantly in the solar ecliptic plane, and the diffuse galactic background in the galactic plane.

What are typical values for the background* radiance? As with astronomical point sources, the visual magnitude scale is commonly used and is expressed as visual magnitudes per square arc second. A good astronomical site on a dark, moonless night will have a background radiance of about 21 mv/asec^2. Just as we convert the visual magnitude of the object into radiometric units, we convert the background visual magnitude mv_b into radiometric units of background radiance, L (ph/(s · m^2 · sr)), according to:

$$L = 5.6 \times 10^{10} \cdot 10^{-\left(\frac{mv_b}{2.5}\right)} \cdot \left(\frac{180}{\pi}\right)^2 \cdot 3600^2 \tag{5-8}$$

$$= 2.4 \times 10^{21} \cdot 10^{-\left(\frac{mv_b}{2.5}\right)}$$

Light pollution is encroaching upon much of the world, with negative impacts on backyard observers and space surveillance sites equally [Falchi et al., 2016].

This background radiance received by the telescope must now be translated into the direct noise photon irradiance, E_{nd} (ph/(s · m^2)), incident on the detector array. This is accomplished according to

$$E_{nd} = \frac{L \cdot \pi \cdot \tau_{ob} \cdot \tau_{tr}}{1 + 4(f/d)^2} \tag{5-9}$$

where L is the background radiance, and f and d are the focal length and primary aperture diameter. The noise photons are also subject to the same telescope obscuration and transmittance losses as the signal photons. Photographers will recognize f/d as the "f number" or f/#, and that the signal from an extended radiant source (i.e., subjects for photography) is significantly enhanced for lower f/# values per Equation (5-9).

Unlike the signal photons, which effectively originate from a point source, the extended background radiance results in a distributed signal across the detector array. Given a square photodetector with a side of dimension μ, the noise photon power P_{nd}, in units of photons/second, is therefore the product of the photon noise irradiance (ph/(s · m^2)) with the photodetector area (m^2) or

$$P_{nd} = E_{nd} \cdot \mu^2 \tag{5-10}$$

which now has the same units as the photon signal power (Equation (5-7)) or photons per second.

---

* Although the term "background" is used, it is recognized that sky radiance is in fact technically in the foreground. However, this distinction has no impact in quantifying the sensing effects.

Example #2: Direct Noise Photon Irradiance E_{nd}

What is the photon noise irradiance on the focal plane from the 40-cm ground-based telescope from the previous example? The sky background is 20 mv/asec^2, and the telescope has a focal length of 120 cm. The individual detectors are 10 μm on a side.

First, we convert the visual magnitude-based background into radiometric units according to Equation (5-8) or

$$L = 2.4 \times 10^{21} \cdot 10^{-\left(\frac{20}{2.5}\right)}$$

resulting in $L = 2.4E13$ ph/(s · m^2 · sr).

We find the photon noise irradiance by Equation (5-9):

$$E_{nd} = \frac{2.4E13 \cdot \pi \cdot 0.7 \cdot 0.9}{1 + 4(1.2/0.4)^2}$$

$$E_{nd} = 1.28E12 \text{ ph/(s} \cdot \text{m}^2\text{)}$$

Finally, the photon noise power according to Equation (5-10) is

$$P_{nd} = 1.28E12 \cdot (10E{-}6)^2$$

$$P_{nd} = 128 \text{ ph/s}$$

Indirect noise photons are from stray light sources outside the field of view but scattered into the optical system due to insufficient baffling or suppression. Photographers are familiar with this effect when attempting to photograph a subject with a bright light immediately outside the frame of the image and the resulting ghosting of the image. This effect becomes particularly significant for space-based optical systems, which may observe objects at high solar phase angles, or when they are angularly close to the sun. Such observation geometries may also be highly desirable, as this determines the maximum observation gap time of that object from any given space-based optical system. Consider the situation for a one-meter object at GEO with the sun 45° away from the object, or a solar phase angle of 135°. The one-meter object in this situation is approximately 18.3 mv, compared to the sun at –26.73 mv. This corresponds to a brightness ratio of the sun relative to the object of ~10^{20}, meaning the optical system must provide 10^{20} suppression of the stray solar light to enable equal signals from the space object and the sun.

We may determine indirect photon noise by how well the optical system baffling suppresses sources outside the field of view, quantified by the *point source transmittance* (PST). The PST provides the fraction of the source energy scattered into the optical axis as a function of the angular extent away from the optical axis. Typical PST values for a source 45° off axis range from 10^{-12} to 10^{-8}.

Detecting Signals—Sensors

With the telescope having supplied signal and noise photons to the focal plane, they now must be detected and measured. There is a wide variety of photodetectors, but the principal detectors for SDA are photoelectric detectors, in which incident photons produce electrons that we can then measure. Silicon-based detectors are the ubiquitous workhorses due to their outstanding performance, near theoretical performance limits, and low cost. Furthermore, their spectral response is well matched to the sun and hence to reflected energy from space objects. Charge-coupled devices (CCDs) have been the most usual detector type, but complementary metal-oxide semiconductor (CMOS) devices are becoming increasingly common. We address three detector characteristics: physical size, detection efficiency, and noise.

The physical size of the detector, or the length x along a side, determines the field of view (FOV) for a given optical system. There are several classes of standardized sizes, and increasingly the improved performance of consumer-grade photography detectors makes them useful for space surveillance applications. A full-frame digital SLR detector is 36 mm × 24 mm. Referring back to Figure 5-2, we see that the FOV is geometrically determined according to

$$FOV = 2\tan^{-1}\left[\frac{x}{2f}\right] \tag{5-11}$$

Just as the overall size of the detector determines the field of view, the size μ of the individual photodetectors determines the *instantaneous field of view* (IFOV), or the angular extent of a single detector. The term *plate scale* is common and is synonymous with IFOV. The IFOV is geometrically determined by

$$IFOV = \frac{\mu}{f} \tag{5-12}$$

where we note that the small angles associated with an IFOV allow a small angle approximation that makes trigonometric terms unnecessary. The IFOV is a key design parameter and should match the mission of the optical system. For astrometric measurements, a relatively large IFOV may be appropriate and will provide sufficient accuracy. A larger IFOV enables improved sensitivity, as well as a greater dynamic range for the single photodetector. For resolved imaging, the IFOV should critically sample the size of the point spread function (PSF), or focus spot. This results in an IFOV of one-half of the PSF width.

We assume that the detector array size is the limiting factor in determining the FOV. This is often the case, but the optical system must also be able to provide adequate performance with the increasing angular field. Optical aberrations often significantly increase with increasing field of view, and if a larger physical detector is selected to provide an increased FOV, the optical system performance must also be examined to see if the aberrations are acceptable at larger field angles.

Example #3, Field of View (FOV) and Instantaneous Field of View (IFOV)

What is the field of view of the 40-cm telescope in the previous example? From earlier we saw that the focal length f is 120 cm. In this instance, we use a 25-mm CCD so $x = 25$ mm. Using Equation (5-11), we find that

$$FOV = 2\tan^{-1}\left[\frac{25}{2 \cdot 1200}\right]$$

which is 0.021 rad or 1.19 deg.

Similarly, given an individual detector size of 10 um, we find the angular extent of a single detector or IFOV by Equation (5-12):

$$IFOV = \frac{10 \times 10^{-6}}{1.2}$$

or 8.3 μrad. Conversion from radians into the more commonly used arc seconds results in 1.72 asec.

Returning to the signal and noise photons incident on the detector, we must now convert these photons into electrons, which are made available in the detector readout for conversion into digital counts. The conversion efficiency of individual photons to electrons depends on the *quantum efficiency QE* of the detector. As with other values, it is spectrally dependent, but for simplicity we use a single weighted-average value across the solar spectrum. Typical values of QE range from 0.3 to 0.8, depending on the type and quality of the detector. We determine the electron production rate in the detector by taking the product of the signal and noise photon rates with the quantum efficiency. Collecting the individual terms from Table 5-2, the resulting signal electron rate (e_{b_rate}) is

$$e_{s\_rate} = P_s \cdot QE = E_{ob} \cdot \tau_{at} \cdot \pi\frac{d^2}{4} \cdot \tau_{ob} \cdot \tau_{tr} \cdot QE \quad (5\text{-}13)$$

Similarly, we find the background noise electron rate (e_{b_rate}) by using E_{nd} from Equation (5-9) and P_{nd} from Equation (5-10):

$$e_{b_{rate}} = P_{nd} \cdot QE = \frac{QE \cdot L \cdot \pi \cdot \tau_{ob} \cdot \tau_{tr} \cdot \mu^2}{1 + 4(f/d)^2} \quad (5\text{-}14)$$

We ascertain the total signal photoelectrons (e_s) and background noise photoelectrons (e_b) from the product of the generation rates with the exposure or *integration time* (t). The resulting total signal and background noise electrons are therefore:

$$e_s = e_{s\_rate} \cdot t \text{ and} \quad (5\text{-}15)$$

$$e_b = e_{b\_rate} \cdot t \tag{5-16}$$

We recall that the background direct noise photons extend spatially over the detector, but so far have not discussed the spatial extent of the signal photons. The best-case scenario for detection is one in which a single photosite has all the space object energy or signal photons incident on it. In such a case, the focus spot is less than the photosite size (i.e., PSF $< \mu$). This is an idealized assumption, but serves us well as an estimate of the optimal sensitivity performance. However, we must also consider the following: in general, the space object is moving relative to the telescope, resulting in streaking and distribution of the signal over multiple photosites, thus decreasing the per photosite energy.

We have to consider several competing factors in selecting an optimal integration time. One of the primary considerations is the apparent angular rate of earth-orbiting objects, which is much higher than that of astronomical sources. The object angular rate reduces the useful integration time and hence achievable sensitivity. For a geostationary object, where there is minimal motion relative to Earth, one could conclude that very long integration times are possible, with a resulting SNR improvement. However, the accumulation of star streaks moving at 15 asec/s in the image frame soon becomes counterproductive.

A simple illustration readily highlights "streaking losses" that result from longer than optimal integration times. Assume the background noise photons produce an average of 100 digital counts per pixel per second. In reality this signal would have a natural variance following Poisson statistics, but treating the signals as a constant is sufficient for illustrating the concept of streaking losses. The space object produces a signal of 500 digital counts per second and is moving at an angular rate of one pixel per second. In this case, a one-second integration time produces an image having 600 counts in a pixel where the object is present, with adjacent pixels having the background noise of 100 counts (Figure 5-4). Now, allow the integration time to be extended to three seconds; the result is a background signature which is three times higher, or 300 counts. However, there is no increase to the space object digital count per pixel because after one second it has moved to the next detector. The result of the three-second integration provides 800 counts per pixel where the space object signal is present, compared to the background of 300 digital counts—or a lower signal-to-noise ratio than that from the one-second integration time.

The telescope tracking mode is a primary consideration for determining optimal integration times. Often space surveillance telescopes use *sidereal tracking*, or tracking with the stars such that the stars remain fixed as point sources and space objects streak through the field. *Rate-track* is the opposite: when the approximate position and angular rate of an object are known, moving the tracker at the same rate as the object permits longer integration times to improve sensitivity, resulting in the object appearing as a stationary point and the stars streaking. Finally, there is a "*fixed mount*" mode whereby the telescope remains stationary. Fixed-mount tracking is particularly useful for tracking GEO-region objects, which have low angular rates relative to Earth. They may also serve as an "optical fence" construct, suitable for tracking LEO objects transiting the field of view.

Angular rates of space objects span a large range. A fast-moving LEO object observed from the ground may exceed 1 deg/s (3600 asec/s). GEO-region objects have much lower rates, with a 15-degree inclined GEO object exhibiting a maximum angular rate of ~ four asec/s relative to a fixed Earth reference. We can readily determine an optimal integration

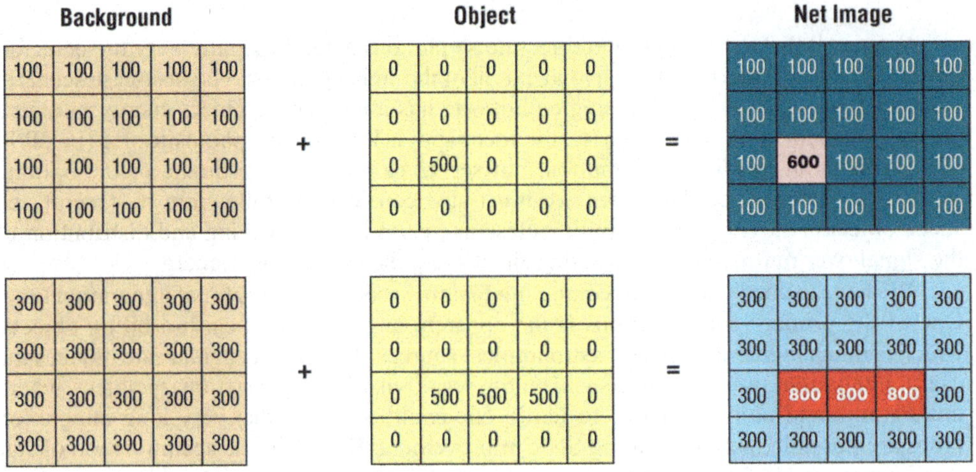

FIG. 5-4. **Example of "Streaking Losses."** This simple example illustrates the effect of optimal integration time on signal-to-noise. The top shows an optimal integration time, where the space object transits one pixel width. At bottom, an integration time three times as long results in the object energy being distributed over three detectors, while the background counts accumulate for all detectors.

time by calculating the time an object transits a single photosite width. Given an apparent angular rate of ω, the transit time across a single pixel, or optimal integration time, is

$$t_{opt} = \frac{\mu}{f \cdot \omega} = \frac{IFOV}{\omega} \tag{5-17}$$

where the previously defined variables for pitch size μ and focal length f apply.

We now consider detector-generated noise, which is an inherent property of all detectors. The two basic categories of detector noise are *dark noise* and *read noise.* Dark noise is undesired thermally induced electrons, which occurs even without external input. Its units are electrons/second (e^-/s) and it increases with integration time. Read noise consists of spurious electrons produced by the electronic circuitry of the detector. Space-based sensors also experience noise produced by gamma rays, which may cause spurious signals. Given the performance of many modern detectors and the typical integration times of interest, the dark noise is practically negligible, so we may treat the read noise as the total detector noise. We often term this situation *background-limited* or *shot-limited performance*. For this reason, we represent the total detector noise as a time-independent value e_d.

At this point we present the overall signal-to-noise ratio (SNR). The previous calculations have resulted in a specific value for signal and noise photons, but in reality the photon arrival rate follows a Poisson distribution, such that the error in the total value is proportional to the square root of the total. This statistical behavior results in the signal also having a noise component. The resulting SNR equation, in terms of photo-generated electrons, is therefore

$$SNR = \frac{e_s}{\sqrt{e_s + e_b + e_d^2}} \tag{5-18}$$

Photoelectrons produced from signal and noise photons in the detector are ultimately converted to an integer digital value in a resulting image file. The available output range of values is a function of the bit depth. Most sensors range in output from 12 to 16 bits, resulting in a numerical range varying from zero up to 4096 (12 bit) or up to 65,536 (16 bit). Normally, it takes several electrons to produce a sufficient signal to account for a single digital count. The relationship between the photogenerated electrons and the digital counts is a function of the detector gain. Although it is ultimately the digital counts which matter in the downstream imaging processing sequence, the discussion continues in the context of photogenerated electrons, which also enables a discussion of signal and noise.

Example #4, GEO Object SNR

Building upon the previous examples, what SNR is achievable from a 40-cm telescope while in sidereal track for a 1-m GEO-region object under ideal conditions?

Here we build upon Examples #1-3 and provide an end-to-end calculation. In Example #1, we determined that our 40-cm telescope provides a signal photon rate of approximately 2700 ph/s for a 1-m object in GEO. In Example #2, the direct photon noise rate was determined to be 128 ph/s. In Example #3, we found that our 10-μm detector size resulted in an IFOV of 1.72 asec. So we start with

$$P_s = 2700 \text{ ph/s}$$

$$P_{nd} = 128 \text{ ph/s}$$

$$\text{IFOV} = 1.72 \text{ asec}$$

Our detector has a quantum efficiency of 0.7 ($QE = 0.7$) and detector noise of 20 e. For this scenario we will operate the telescope in a sidereal (star tracking) mode, which for a GEO satellite, results in an angular rate of 15 asec/sec.

We first determine the signal photon and noise photon electron production rates from Equation (5-13) and Equation (5-14):

$$e_{s\_rate} = P_s \cdot QE = 2700 \cdot 0.7$$

$$e_{b_{rate}} = P_{nd} \cdot QE = 128 \cdot 0.7$$

resulting in $e_{s\_rate} = 1890$ e/s and $e_{b_{rate}} = 90$ e/s.

In order to obtain the signal and noise electrons, we must first determine the optimal integration time to avoid streaking losses. From Equation (5-17) we find

$$t_{opt} = \frac{IFOV}{\omega} = \frac{1.72}{15}$$

or $t_{opt} = 0.114$ sec.

Using Equations (5-15) and (5-16), we find that the signal and noise electrons are

$$e_s = e_{s\_rate} \cdot t = 1890 * 0.114 = 215 \text{ e and}$$

$$e_b = e_{b\_rate} \cdot t = 90 * 0.114 = 10 \text{ e}$$

Finally, we turn to the SNR equation (Equation (5-19)) to find that

$$SNR = \frac{e_s}{\sqrt{e_s + e_b + e_d^2}} = \frac{215}{\sqrt{215 + 10 + 20^2}}$$

Resulting in an $SNR = 8.6$.

Since the rule of thumb for single frame detection is an $SNR > 6$, for this scenario we have enough signal for a viable detection.

Image Processing

Image processing is required to extract the information needed for space domain awareness, as we see in Figure 5-5. We start with a raw image composed of integer digital count values resulting from the signal photons, noise photons, and detector noise. The output of the image processing is an accurate position of the object at the time of the detection, and the photometric value or total energy from the object. We present only the basics of typical image processing steps.

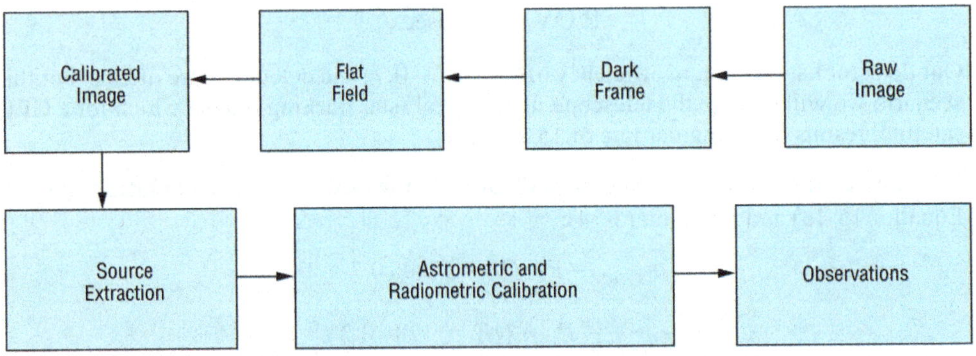

FIG. 5-5. Image Processing Steps. A sequence of basic image processing steps converts a raw image frame into information content and the data are reported out as an observation.

Two conditioning steps are required. We use dark frames and flat field images to calibrate the image. Dark frame images remove the bias of dark noise in the image. Flat field images correct for non-uniform illumination effects across the detector.

First, *dark frame* images are taken to quantify the pixel values in the absence of radiometric sources, that is, blocked from any light incident on the detector array. These images may be subtracted from the actual on-sky images to minimize the contributions from dark current or dark noise (see above). It is important to acquire dark frames under the same conditions as those of the on-sky images, primarily by keeping the integration time and sensor temperature the same as the actual on-sky images. Dark frames may also be used to correct for natural non-uniformities in pixels and minimize the effects of "hot pixels" that have anomalously high output.

Second, all optical systems result in a fall-off of the irradiance incident on the detector away from the optical axis or with an increasing field of view. This is particularly true for wider field-of-view optical systems that use vignetting as part of their design to minimize optical aberrations at high field angles. To correct for this nonuniformity, a *flat field* image may be acquired via imaging an extended, uniform radiometric source across the field of view. Flat field images may be obtained via an optical integration sphere, or in a less uniform case, an area of sky having a uniform background radiance. The flat field image is then used to "flatten" the as-received irradiance across the detector array by increasing the digital count value as a function of the distance away from the optical axis. The flat field calibration will remain viable so long as the optical prescription and sensor used remain fixed.

After having applied the above dark frame and flat field corrections, we now have a calibrated image with digital counts corresponding to the radiometric conditions under which it was acquired. To first order, the image is composed of energy from the background radiance (background noise), space objects (signal), stellar sources, and sensor noise (predominantly read noise). Depending on the tracking mode, the star sources may be fixed points (sidereal tracking) with the space objects streaking, or the space objects may be fixed points (rate tracking) with the stars streaking, or finally a combination of streaking stars and space objects (such as that from fixed mount tracking). Individual pixels with higher digital counts due to stars and space objects are identified via an image processing algorithm that identifies energy sources, or a "source extraction" algorithm. We wish to obtain as much sensitivity as possible from the image to identify dim space objects while minimizing spurious detections or false alarms. A minimum signal to noise ratio of six is the canonical standard for identification of signals in a single image frame. Recently, open source software has matured and can provide highly capable source extraction processing (for instance, SExtractor https://www.astromatic.net/software/sextractor).

After energy sources are identified, the position and digital counts must be calibrated to a truth source to obtain the position in the stellar sphere along with the radiometric value. Star catalogs are such references that contain both precision positions and brightness values. Stellar sources from the source extraction step are matched to stellar positions in star catalogs. Through a coordinate transformation, each pixel may then be mapped to a corresponding position in the stellar sphere usually expressed as a right ascension (RA) and declination (DEC) value. In a similar manner, the digital count amplitude of each source's signal is calibrated to the radiometric truth provided in the star catalog. The position and brightness processing are respectively termed astrometric and photometric processing.

Similar to source extraction, public domain software such as that at astrometry.net has enabled significant capabilities for astrometric and photometric frame reduction. Development of astrometric reduction software has been sponsored by the US National Science Foundation and NASA in order to provide a common processing pipeline for numerous astronomical programs as well as telescopes searching for near Earth objects (e.g., asteroids) that may pose an Earth collision hazard. As previously discussed, visual magnitude is commonly used as the standard brightness unit. The image processing chain culminates in the output of "observations" or a set of textual data containing the first order information content from observing the space object.

Observations consist of 1) the time, 2) object location (expressed as two angles, RA & DEC), 3) the object brightness, and 4) data sufficient to identify the sensor location. As with any measurement, the inclusion of the measurement uncertainty is also beneficial. Sequential observations of the same object may be collected into a *track*. The natural motion of a space object results in a smooth, piecewise continuous change in the position as a function of time. Individual sensors can exploit this property to assert that individual observations taken in sequence all correspond to the same object. A track is preferred over individual observations as it provides further confidence against a false detection. Furthermore, the information from multiple observations in a track may be used to more effectively associate the detections to an object catalog as well as initiate processing for the formation of new orbits which do not correspond to a known object in the catalog (uncorrelated track processing).

Finally, computational advances coupled with high rate sensors (such as scientific CMOS detectors) are enabling sensitivity improvements beyond that possible with single frame detections. For instance, we can visualize that if we have a series of faint detections across multiple frames with a motion consistent with a space object that these individual high noise detections are indeed viable detections.

Thermal Infrared

Sensing space objects in the thermal infrared offers unique benefits as well as challenges. Thermal radiation from space objects provides a signal independent from the requirement for solar illumination. Space object signatures in the thermal infrared are governed by the object surface temperatures and emissivity area product projected to the sensor. For this reason, thermal radiation measurements are highly valuable for space object characterization, as multiple band techniques enable an accurate estimation of temperature and hence projected area. Multi-spectral infrared observations of GEO-region debris objects are beneficial for establishing the objects' origin [Skinner et al., 2015]. Concurrent thermal and visible light signature measurements are complementary and further inform object properties. Space object thermal infrared signatures remain an active research area, with more capable systems enabling new data for investigation [Lederer, 2016].

Ground-based thermal infrared telescopes face significant challenges. Atmospheric water content significantly degrades thermal signatures, and ground-based thermal sensing usually requires unique, high-altitude locations. Space-based thermal sensing overcomes these limitations, and satellites such as NASA's Wide-field Infrared Survey Explorer

(WISE) have provided significant thermal infrared data on geosynchronous region space objects [Murray et al., 2015; Lee et al., 2016].

Resolved Imaging

Imaging of space objects or the ability to discern spatial details of an object is obviously useful. Imaging provides data in a form directly interpretable by humans and can allow direct differentiation of object types (e.g., satellite, rocket body, debris). It also supports satellite anomaly resolution such as identification of solar array deployment failures. For example, in 2016 the US Geosynchronous Space Situational Awareness Program (GSSAP) satellite was used to investigate an anomaly with a new satellite [Neighborhood Watch, 2016]. Optical systems designed for imaging have vastly different properties than those used for astrometric and photometric measurements. Long focal lengths and detectors that permit critical sampling of the focus spot or point spread function (PSF) are typical.

The aperture size of the system determines the minimum angular resolution of an optical system. Resolution is not a simple, binary value, but rather a complex function. However, we can obtain a reasonable approximation by the Rayleigh resolution criterion. The Rayleigh criterion approximates the minimum angular resolution of a system under the assumption that the optical system is diffraction-limited, or in other words, the optical elements provide a perfect focus without aberration. The Rayleigh criterion, θ_R, is given by

$$\theta_R = \frac{1.22 \cdot \lambda}{d} \tag{5-19}$$

where d is the telescope aperture, and λ is the wavelength of the light. For a typical silicon-based detector, we may use a wavelength of 600 nm.

Unfortunately, the atmosphere degrades the image resolution, unless of course the optical imaging system is space-based. The atmosphere acts as a distorting medium for imaging space objects, and we have to compensate for these distortions in order to acquire images that approach the telescope's diffraction limit. Thermal gradients in the atmosphere refract the light and have a blurring effect on imaging. The "seeing conditions" refer to the level of atmospheric blurring, and vary according to local conditions at the telescope site. A typical site has atmospheric seeing conditions on the order of two arc seconds, so no matter how large the telescope aperture, imaging a point source such as a star results in a best focus of two arc seconds. If we use two arc seconds as the resolution in Equation (5-19), we see that the equivalent optical system diameter is only ~7.5 cm—so much for using a large aperture telescope!

There are two ways to overcome atmospheric distortion. Speckle imaging is a class of techniques exploiting multiple short exposures. The most direct and simplistic approach is the use of *lucky imaging*, in which a series of short exposures less than the time scale of the atmospheric fluctuations is acquired, exposures on the order of 10 ms or less. Such short exposure times necessitate bright objects to obtain sufficient signal. Occasionally during these short durations, the atmospheric conditions are significantly better, enabling an image much improved over the average seeing conditions. From the series of short duration, high repetition images, the sharpest images may then be selected and spatially registered with one another and stacked to further improve the quality.

For larger telescopes, however, we use *adaptive optics* to correct the blurring from atmospheric effects. Adaptive optics enable the dynamic compensation of wavefront errors imparted by the atmosphere. They do this by monitoring a reference source in the field of view and using a wavefront sensor that feeds a dynamic control loop driving actuators on an optical element, which compensates for the distortions. The reference source may be from the space object itself, or from a laser guide star. Laser guide stars provide a signal for the telescope using Rayleigh backscatter or more effectively from a sodium laser. A Rayleigh scattering laser provides the most dominant signal from scattering in the lower atmosphere and thus is not as effective as a sodium laser, which ionizes a naturally present layer of sodium concentrated at 90 km altitude (Figure 5-7 and Figure 5-6). This higher altitude reference source enables the sampling of the entire atmospheric optical path and hence yields improved correction. Modern adaptive optics techniques enable ground-based telescope imaging resolution performance of almost the diffraction limit of the aperture. Numerous astronomical telescopes now employ this technique.

FIG. 5-6. **Adaptive Optics.** A sodium guide star laser at the 3.5-meter telescope at the Air Force Research Laboratory's Starfire Optical Range [Pellerin, 2016].

Example #5, Telescope Imaging Resolution

What is the theoretical imaging resolution limit of a 3.6-meter telescope for imaging objects in low Earth orbit (LEO)?

We use the Rayleigh resolution Equation (5-19), assuming adaptive optics provides full restoration of atmospheric blurring. Thus,

FIG. 5-7. **Image Produced by Adaptive Optics.** A ground-based telescope image of the International Space Station by Air Force Research Laboratory's 3.6-meter telescope at the AMOS site.

$$\theta_R = \frac{1.22 \cdot 600 \cdot 10^{-9}}{3.6}$$

Or $\theta_R \cong 0.2\,\mu\text{rad}$. For a LEO in a 400-km altitude orbit, we use a range of about 500 km to the telescope. The linear extent of 0.2 μrad at a range of 500 km is the product of these values, or approximately 0.10 m.

Laser Ranging

The laser and the space age had a nearly concurrent birth, with the first laser operating three years after the launch of Sputnik. Four years later, in 1964, the Beacon B satellite was launched, and the first laser-ranging measurement of an orbiting object was obtained. Laser ranging measurements of the moon took place in 1969 after Apollo 11 placed a reflector on the surface. Today, a robust network of more than 40 satellite laser ranging (SLR) systems operates in 30 countries and provide sub-centimeter level measurements of satellites hosting retroreflectors [ILRS, 2016]. The primary historical mission for SLR systems has been in support of *geodesy* or the precision measurement of the earth's surface, tectonic movements, etc. Precision orbits of satellites provide a truth reference position by which to derive SLR station positions. Specific satellites hosting retroreflectors have been historically useful in support of this mission.

Advances in lasers and associated detectors have now enabled *non-cooperative* laser ranging of space objects, or measuring the laser reflections from objects that do not host retroreflectors. Retroreflectors effectively return the cross-section of the intercepted laser energy back to the source consistent with the divergence of the laser, or a specular return over a broad range of orientations. The absence of retroreflectors results in higher irradiance requirements on the space object, with the signal generally treated as a diffuse reflection of the object.

While modern SLR systems provide high-precision measurements on non-cooperative objects, their throughput is still very limited compared to phased array radars and passive optical sensing. The high precision, low-volume measurements from SLR systems are suitable for refining the orbital states of objects for which a conjunction is predicted, and could be an effective contributor for more rapidly eliminating "false positive" conjunction events. This will increase significantly with improved tracking of LEO debris, with the inauguration of the United States "Space Fence" phased array radar system, placed in operation in 2020.

Laser illumination of space objects also introduces the potential to damage operational satellites. For this reason, the US Department of Defense (DoD) has published policy managing the risk for DoD laser systems [DoD, 2016]. In fact, high-power laser systems may be a way to actively remove debris objects. Laser-induced ablation of the surface of debris objects has shown some promise of imparting sufficient force to lower the perigee of objects, thus accelerating their natural orbital decay rate [Phipps et al., 2012].

New Opportunities for Optical Systems

We conclude the optical section by discussing two important applications of interest: daytime optical imaging and closely spaced object detection.

The potential of extending viewing opportunities into daytime hours for ground-based telescopes is of high interest as it enables improved custody or revisit of geosynchronous objects. Daylight imaging of LEO objects was demonstrated as early as 1982 [Rork et al., 1982]. Improvements to sensors operating in the short wave infrared (SWIR) have enabled significant new capabilities in this key spectral band, enhancing daytime performance. In the SWIR, the daytime sky is significantly darker than that of visible light, yet there is still sufficient solar energy in the band to illuminate space objects (Figure 5-1) [Jim et al., 2012].

Another application of increasing concern is the ability to discriminate closely spaced objects. Increased interest in rendezvous and proximity (RPO) missions for potential service and debris removal of GEO objects has highlighted the need and associated sensing challenges for monitoring such activities [Scott and Ellery, 2015]. In such scenarios, a small satellite may be close to a larger and much brighter satellite. This situation requires enhanced resolution to angularly discriminate the two objects, but also improved dynamic range such that the smaller, dimmer space object may be detected concurrent with the brighter one. Acceleration of RPO missions at GEO will place a premium on imaging techniques, enabling closely spaced object discrimination.

Polarimetry for SDA

Historical Context

Recognition that polarization is a property of light dates to the 17th century. Sir Isaac Newton developed the corpuscular theory of the propagation of light while Christiaan Huygens developed the wave theory [Cigal, 2002]. Huygens derived the law of reflection and refraction and explained birefringence phenomena using his wave theory [Huygens, 1690]. Malus discovered how polarization phenomena occur by reflection [Malus, 1809]. In 1852 Sir George Stokes published a milestone paper providing a complete theoretical foundation for polarization [Stokes, 1852]. In 1892 Poincaré showed that the polarization ellipse could be represented on a sphere. This allowed visualizing and solving of complex interactions of polarization [Poincaré, 1892]. The Mueller calculus simplifies calculations describing the effects of polarizing devices on light using the Stokes vector [Mueller, 1947]. The Stokes vector (S_0, S_1, S_2, S_3) is a set of values that describe the polarization of light. Here, S_0 represents the intensity of the light; S_1 represents the preference of the light between vertical (0°) and horizontal (90°) linear polarization; S_2 represents the preference of the light between 45° and –45° linear polarization; and S_3 represents the preference of the light between left-handed and right-handed circular polarization.

The Jones calculus became an alternative way of representing polarization states as electric fields [Freniere et al., 1999], [Jones, 1941], [Jones, 1942]. In the late 1940s, S. Chandrasekhar developed groundbreaking work on radiative transfer and needed to include polarization effects [Collett, 1993]. He found Stokes' work, and the modern era of astronomical polarization emerged [Chandrasekhar, 1946]. While astronomical polarization theory is well developed, the application side of the subject is only now coming into its own. Historically, the principal reason for this discrepancy was the lack of instruments with sufficient measurement sensitivity. Since the late 1960s, astronomy has pushed the development of polarization instrumentation with increased sensitivity.

Background

We can represent light as a transverse electromagnetic wave with mutually perpendicular electric (*E*) and magnetic (*B*) fields, which are in turn orthogonal to the direction of propagation. Electromagnetic waves exhibit vibrations transverse to the propagation direction. The vibration pattern is generally random and when summed over a period exhibits no preference as to transverse direction. Polarized light is light whose vibration pattern does exhibit such a preference. Such light is one of nature's fundamental concepts. A monochromatic polarized beam cannot be subdivided into simpler components. Study of this preference provides insight into the nature of the origin of the light source and any intervening matter these waves have interacted with on their journey.

We classify light in one of four polarization states: linear polarization, circular polarization, elliptical polarization, or unpolarized. Linear polarization usually pertains to a situation in which the *E* field's direction is constant over some time. If there are two monochromatic linearly polarized plane electromagnetic waves, the direction of polarization of the resultant E vector is determined by the phase difference between the two waves. If two waves are in phase (e.g., the positions of both waves with their respective periods are identical) or 180° out of phase, the resultant wave is linearly polarized. If, however, the two waves have equal amplitudes, their E fields orthogonal to each other, and are not in phase or 180° out of phase, then the tip of the resultant E field traces out a circle in a counterclockwise direction as seen by a viewer observing the oncoming wave. This circular polarization is either left- or right-handed depending on the direction of the rotation of the circle. If the two waves are of unequal amplitudes, then the resultant E field traces out an ellipse, leading to elliptical polarization. Finally, the lack of preference for any one polarization direction or phase for a group of photons results in unpolarized light.

Satellite Polarimetry

Although polarization measurements of astronomical objects have been made, there are relatively few studies of polarization of manmade satellites in orbit around Earth. The earliest polarimetric observations of manmade satellites were by Stead [Stead, 1967], and measured the percent polarization of non-resolved satellites. This work took place in the early days of photoelectric photometry when observational sophistication (as compared to modern-day capability) was limited.

More recently, Speicher [Speicher, 2015], [Speicher et al., 2014] observed several geosynchronous communication satellites and simultaneously measured the horizontal and vertical linear polarization in their unresolved optical signature. One key finding was that the amount of linear polarization generally decreased with decreasing phase angle, consistent with Stead's results. There were also repeatable polarization signatures for individual satellites from night to night, and the average satellite polarization also appeared to be grouped by the manufacturer (Figure 5-8) and the size of their solar panels (Figure 5-9). These results indicate that satellite polarimetry can aid in discriminating and characterizing manmade space objects, thus increasing space domain awareness in support of safe, stable, and sustainable space activities.

5.2.2 Radio Frequency (RF)

A General Overview of Radars–the Basic Physics

Radar is a powerful sensing tool for SDA. It's an acronym that stands for Radio Detection and Ranging. Its origins date back to the beginnings of radio communications at the turn of the last century, but it was primarily World War II that rapidly changed radar from a research interest to a fully operational wartime system. In the post-war years, it found use in many civilian applications in addition to being a very important military system around the world. The continual advancements in radar make it likely that it will continue to increase in importance. One example is the recent development of autonomous cars, which make extensive use of highly sophisticated short-range radars.

For SDA applications, we can use radar to detect an orbiting object and measure its range. Most radars are also capable of measuring the time-varying change in range (also

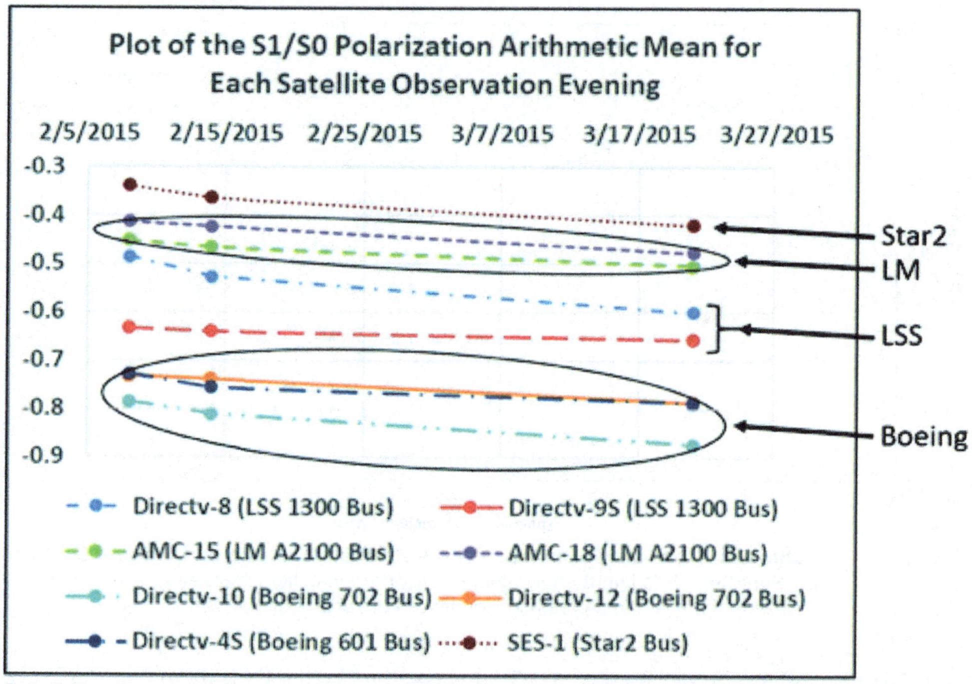

FIG. 5-8. **Summary of the Average Ratio of S_1 to S_0 for Eight GEO Satellites.** This figure shows that the average satellite polarization appears to be grouped by the manufacturer, with Boeing spacecraft exhibiting higher average polarization compared to Loral Space Systems (LSS), Lockheed-Martin (LM), and Star2 (Orbital Sciences Corporation).

called range-rate) to high accuracy by exploiting the Doppler effect. Radars, unlike optical surveillance systems, usually have comparatively poor angular resolution. This makes the combination of optical and radar systems potentially very powerful as each excels in ways that the other does not. Optical systems can't measure range or range-rate, which radars excel in, while radars have poor angular resolution, which optical systems are excellent at. Additionally, because radars use radio frequency (RF) signals, they are useful in all weather (just like broadcast radio or television) and also in both daytime and nighttime.

Radars work by transmitting a signal and then listening for the return echo from any objects that the transmitted signal has interacted with or "illuminates" [Lacomme et al., 2001]. The farther the object, the longer the delay between signal transmission and reception. By measuring this delay, we can estimate the range to the object. The amplitude of the return signal is a function of both the distance to the reflecting object and the combination of its size and reflectivity. In radar, this combination of the size and reflectivity of an object is called the radar cross-section (RCS). In general, the larger an object's RCS, the easier it is to detect. This fundamental aspect of radar is the basis for the development of "stealth" technology since the 1970s.

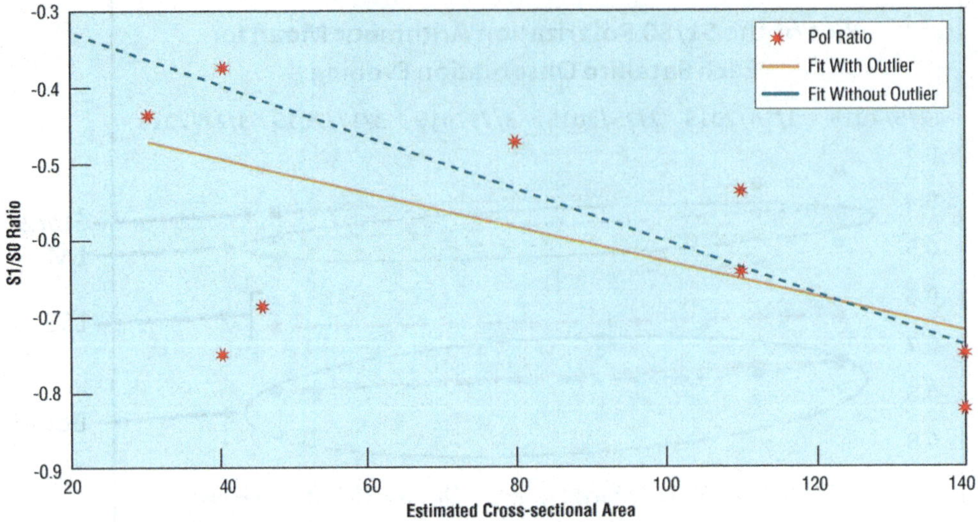

FIG. 5-9. **Estimated Cross-sectional Area Versus Polarization, With and Without Outlier in Fit Function.** This figure shows that the average satellite polarization appears linearly related to the estimated cross-sectional area of various spacecraft.

The focus of stealth technology is to find ways to make the RCS of an object as small as possible while still maintaining the system's functionality. Decreasing the RCS makes the reflected signal smaller. This decreases the range where the radar received signal is large enough to be detectable. We discuss details of relevant SDA-specific RCS issues with orbiting objects, particularly space debris, later in this chapter. Some types of orbital objects have a very small RCS due to small physical size or being made of materials that do not reflect well. Sometimes the space objects tumble, resulting in rapidly changing RCS values, which complicates detection.

The radar range equation quantifies the general sensitivity of radar. It quantifies the system signal-to-noise ratio for a given radar design and the range to an object with a specified RCS. Figure 5-10 shows the high-level relationship.

It is instructive to follow the signal through the various objects in Figure 5-10. It begins with the transmitter that generates the signal and amplifies it to the designed power level. The transmit antenna then broadcasts the signal. The details of the antenna and the transmitted signal result in a specific transmit antenna pattern. The tighter the beam, or narrower the angular spread, the less the signal energy density decreases with range. In the limit of a single patch antenna above a ground plane, we can approximate the pattern as a half-sphere. At the other extreme, a very large antenna can have a very small beam width, sometimes referred to as a pencil beam. The two types of transmit antennas have very different energy density at the same range due to the different areas over which the beam energy is spread. This is antenna directivity or antenna gain.

Next, the signal reflects off the object. The reflected signal power is the product of the signal energy density and the object RCS (which has units of area). This energy is then

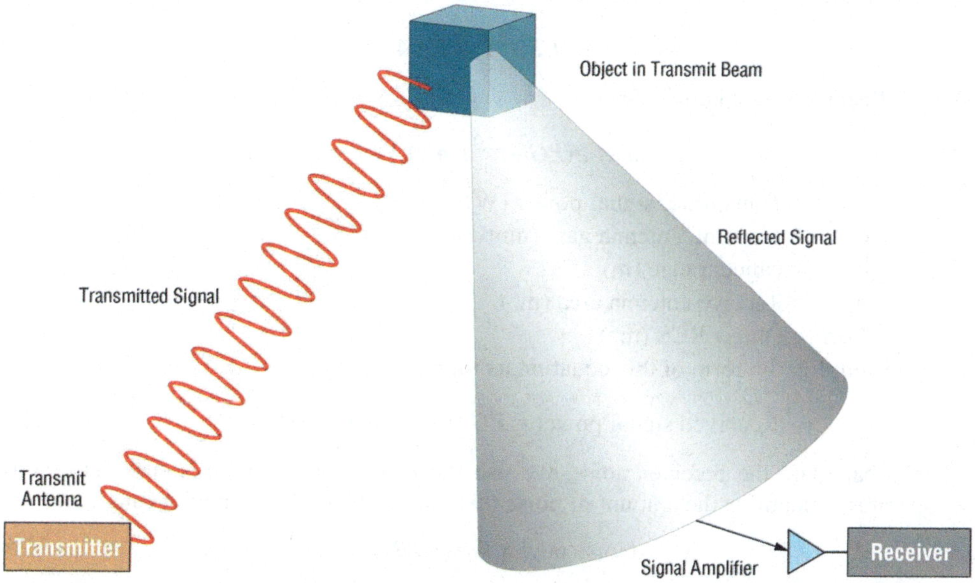

FIG. 5-10. **Typical Radar Architecture Model in the Radar Range Equation.** The transmitter sends a signal towards an object, which reflects the energy that illuminates it in many directions. Some of the reflected energy is received by the receive antenna and sent to the signal receiver hardware. Energy is lost at each stage of this process. The radar range equation models this process of energy loss from transmission to reception.

reflected towards the receive antenna (which may or may not be the same as the transmit antenna). The reflected signal energy then spreads out over distance—resulting in a loss of reflected signal energy density—and is finally captured by the receive antenna. The area of the receive antenna determines how much reflected signal energy it captures. This signal is then amplified, along with additional noise. If we ignore all other noise sources except the noise added by the receiver, we can calculate the radar signal-to-noise ratio for the radar design specifics and the object's range and RCS.

The radar signal-to-noise ratio (SNR) is a key performance measure of the radar. This value predicts how well the radar will detect an idealized notion of the object. SNR is the "sufficient statistic for detection." This means that all the information needed to detect an object is in the signal's SNR; hence, the higher the SNR, the more detectable the object is.

We now combine these concepts into a single equation called the radar range equation:

1) Signal power density at object range

$$(\mathrm{W/m^2}) = PG/4\pi r^2 \tag{5-20}$$

2) Reflected power of object

$$\mathrm{W} = RCS(PG/4\pi r^2) \tag{5-21}$$

3) Power density at receive antenna

$$W/m^2) = RCS(PG/4\pi r^2)/4\pi r^2 \quad (5\text{-}22)$$

4) Received signal power W =

$$A(RCS(PG/4\pi r^2)/\ 4\pi r^2) \quad (5\text{-}23)$$

P = Transmitter signal power (W)
G = Transmit antenna gain (unitless)
r = Object range (m)
A = Receive antenna area (m^2)
RCS = Object RCS (m^2)

We can simplify the form of this equation as follows:

$$\text{Received signal power} = PGA(RCS)/(4\pi r^2)^2 \text{ (W)} \quad (5\text{-}24)$$

To calculate the receiver noise, we use the receiver noise temperature. The noise temperature quantifies the amount of noise that the receiver adds using the equation:

$$\text{Noise power W} = \mathrm{k}TB \quad (5\text{-}25)$$

k = Boltzmann's constant ($1.38*10^{-23}$ J/K)
T = noise temperature (K)
B = signal bandwidth (Hz)

Thus, we calculate the radar SNR as:

$$SNR = \text{Signal power/noise power} = PGA(RCS)/\ [\mathrm{k}TB(4\pi r^2)^2] \quad (5\text{-}26)$$

Most radars use a form of processing called a matched filter. A matched filter increases signal power by the signal bandwidth times signal duration ($B\tau$). This results in a modified SNR equation:

$$SNR = \text{Signal power/Noise power} = PGA\,\tau(RCS)/\ [\mathrm{k}T(4\pi r^2)^2] \quad (5\text{-}27)$$

Finally, we can increase the system SNR by transmitting multiple pulses (n), each τ seconds long.

$$SNR = \text{Signal power/Noise power} = PGA(\mathrm{n}\tau)(RCS)/\ [\mathrm{k}T(4\pi r^2)^2] \quad (5\text{-}28)$$

Sometimes this equation replaces $n\tau$ with the total transmission duration t, and also replaces the power P with the average power transmitted over the entire time t, including the time when the transmitter is not transmitting, when the receiver is receiving. This pulse transmission operation is how monostatic radars work (discussed in more detail later—transmitter and receiver share the same antenna). For some bistatic radars [Cherniakov, 2008] (also discussed in more detail later—transmitter and receiver have separate antennas), the transmitter transmits continuously, so the average and peak power are the same, and $t = n\tau$.

We can improve the system SNR by increasing transmit power (P), transmit antenna gain (G), receiver antenna size (A), or transmission time ($n\tau$), or decreasing receiver noise temperature (T). From the standpoint of radar applied to SDA, the SNR drops off as $1/r^4$. Given the long ranges to space objects, especially geosynchronous orbit objects at ~36,000 km, getting sufficient SNR can be a serious challenge.

RF Systems Attributes

Radar is one example of an RF system. Other types of RF systems include radio communications, radio navigation (e.g., GPS), and broadcast television. Active RF systems broadcast and receive RF energy. Passive ones only receive RF energy. An example of a passive RF system is radio astronomy systems like the large Arecibo antenna in Puerto Rico [Arecibo Observatory home page], and the Very Large Array [Very Large Array home page] in New Mexico. Most RF systems are active systems. While a GPS receiver (user segment) only receives RF energy, the GPS satellites (space segment) transmit the RF navigation signal, and therefore the entire GPS system of satellites, ground stations, and receivers is considered an active RF system.

For SDA applications, active RF systems are most interesting, and provide an observable signal (the RF signal). If an object is actively transmitting an RF signal, that implies the object is functional and not a dead satellite or debris. We use multiple attributes of a transmitted signal to classify its type and provide insight into its intended function. The commonly used RF signal attributes are:

1) Signal power
2) Signal frequency
3) Signal bandwidth
4) Signal modulation type
   a. Amplitude modulation
   b. Frequency modulation
   c. Phase modulation
5) Pulsed vs. continuous signals

Communications systems are typically continuously modulated signals. Modern communication signals are either phase modulated or a combination of amplitude and phase modulation. Older communication systems may be frequency modulated. As a historical example, the earliest terrestrial radio signals were amplitude modulated (AM), later frequency modulated (FM) radio came into use, and most recently satellite radio uses combined amplitude and frequency modulation (QAM for quadrature amplitude modulation). Satellite television also uses QAM modulation. Frequency modulation is still common in satellite command and telemetry up/downlinks.

Radar signals are normally pulsed. The modulation type is most commonly frequency, but sometimes phase modulation is used. The most common radar signal is the linear FM chirp. The waveform starts at a single tone that rises or falls rapidly, which if in the audible region (between 50 – 20k Hz) sounds like a chirp, although radar signals are at RF frequencies normally many orders of magnitude higher than audible. The range resolution of radar is a function of the bandwidth; the higher the transmitted bandwidth, the finer the range resolution. The equation for a radar range resolution is:

$$res = c/(2BW) \text{ (resolution)} \tag{5-29}$$

where c is the speed of light, and BW = transmitted bandwidth.

All RF system receivers "tune" the receiver to the frequencies being transmitted. As previously discussed, the receiver noise is a function of the bandwidth (kTB), so to minimize receiver noise, we want to minimize the received bandwidth to only what's necessary. This also rejects any unwanted signals that are outside the required system bandwidth. We call this *receiver selectivity*, which describes how well a receiver rejects unwanted frequencies. To avoid undesired interference between different RF systems, they will generally operate on different frequencies (or channels).

When signals have overlapping bandwidth, they may mutually interfere. This is why frequency allocation for RF systems is important. In a military conflict, intentional jamming signals are broadcast at high power in the bandwidth of a signal we wish to interfere with. The different techniques to both jam and avoid jamming are collectively called electronic attack (EA)/electronic protection (EP), and jointly electronic warfare (EW) [Adamy, 2001]. There are many books on both EA and EP techniques and approaches. To avoid unintentional interference between RF systems, we have DoD, national, and international frequency allocation governing bodies that govern all RF transmitters (military, civilian, commercial). This is broadly called spectrum management [Spectrum, 2000].

RF System Missions and Classes

Passive RF systems are exclusively sensing systems that use the natural or manmade RF emissions of an object to sense it in some way (e.g., radiometer). We also use reflected RF energy from a natural or artificial RF source not under the control of the receiver to gain information about the objects that reflect the RF energy (e.g., passive bistatic radar). Active systems intentionally emit RF energy that either directly encodes information receivable (communications) or the reflected signal from an object can allow us to gain information about that object (active radar). From an SDA perspective, the RF systems of interest are the RF transmitters of an active system, which broadcast an RF signal that can be detected (either directly or reflected) and processed to extract information about the signal source. Passive RF systems can receive these signals.

The system missions for active RF systems are sensing and communications. These two broad missions have two main classes of either a point-to-point system focusing energy in a very specific direction or a broad area system that radiates RF energy in many directions simultaneously. An example of point-to-point communications is common high-speed data downlinks from imaging satellites to ground stations. The higher the communications data rate, the higher the transmit power needed. Using a larger antenna, however, increases the power density at the receive antenna, minimizing the RF signal power we need to transmit. The RF energy is concentrated at the desired receiver location. An example of broad area communications is a satellite telemetry downlink. The telemetry is broadcast in all directions, because in the case of a satellite anomaly the satellite may lose the ability to point a high-gain antenna properly. The loss of telemetry would make anomaly resolution and recovery much more difficult. The transmitted telemetry is usually low data rate and therefore the required transmit power is reasonable for this application.

The implication of the two different classes of active RF systems is that for the high-power transmitter case, the likelihood of an SDA RF sensor being in the antenna high-power main beam is low. Therefore, the signal to be sensed is most likely in the antenna side lobe, which will often be more than 100 times smaller. The broadcast system signal will not be in a side lobe, but will generally be lower power. Thus, an SDA sensor that attempts to exploit an active RF transmitter needs to have significant sensitivity. Understanding the transmit antenna pattern as well as the relative viewing geometry between the transmitter and receiver is necessary to properly design and use such an SDA sensor.

The SDA radar system avoids these complications, as both the transmitter and receiver are controlled. The range losses for a radar are proportional to $1/\text{range}^4$, while a passive receiver sensing a transmitted communications signal only has a $1/\text{range}^2$ signal loss. Thus radar is much more negatively impacted by long range, making use of radar for SDA at MEO and GEO/HEO much more difficult.

Gathering Signal

The reception of an RF signal uses an antenna to convert the RF electromagnetic energy to an electrical signal that can be transmitted through electrical conductors. The types of electrical conductors that pass RF signal energy are specialized and differ as a function of the frequency they are designed to operate at. At low frequencies, simple electrical wiring is fine, but as the frequency rises, coaxial cables are more common. At frequencies above 10 GHz, the coaxial cable losses become significant due to energy absorption in the cables, and RF waveguides become necessary, since they have much lower signal losses. The typical RF waveguide is a rectangular hollow piece of metal. The cross-section of a waveguide varies with the intended frequency of operation—the higher the frequency, the smaller the wavelength and the smaller the waveguide cross-section.

The receive antenna needs enough area to capture sufficient signal power for the receiver electronics to detect. The size of the antenna also changes the angular directivity of the received signal. Signals in the receive antenna main beam are passed on without significant loss; however, signals not in the main beam can be significantly suppressed by the antenna's angular selectivity. Thus they are passed with significantly lower energy levels. The signal gain of the antenna is much less outside the main beam than inside it. Signal suppression of 20–30 dB is common outside the main beam. This provides the ability to focus a receiver to a signal at a known location and minimize unwanted signals at different angular locations.

Common types of antennas are reflector and phased arrays. Reflector antennas must be physically pointed in the desired direction, while phased arrays can be electronically pointed without physically moving the antenna. Figure 5-11 shows two common types of reflector antennas. Both types of reflectors concentrate the received energy into the feed horn, and from there it goes to the receiver either by coax cable or waveguide, depending on frequency. The Cassegrain design is usually the more compact of the two, at the expense of the sub-reflector blocking more of the incoming signal.

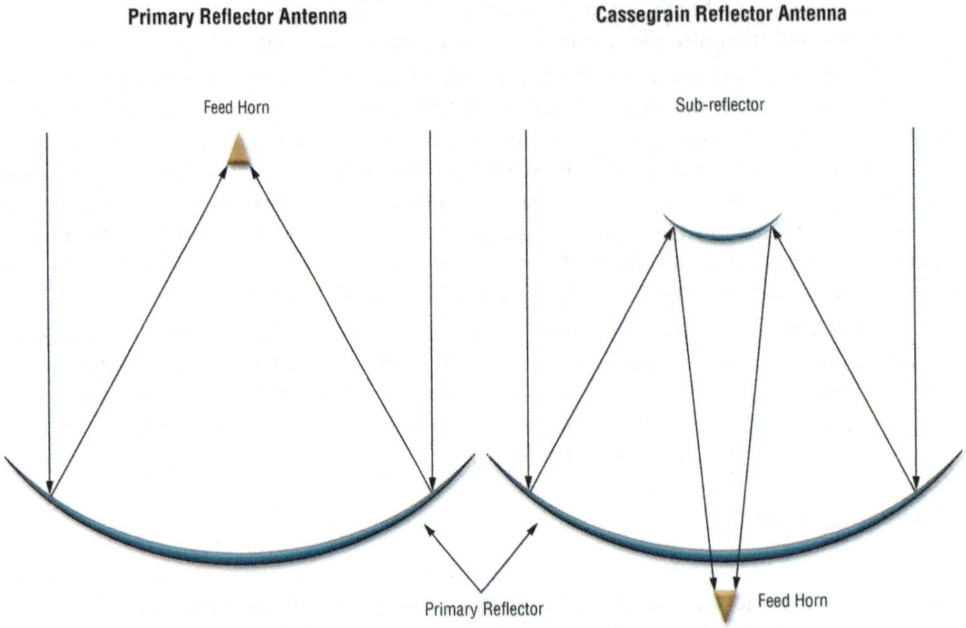

FIG. 5-11. **Two Typical Reflector Designs.** The two main reflector types trade central blockage (Cassegrain with the large subreflector) for a more compact design. An added advantage is that the Tx power and receive low noise amplifier can be mounted behind the main antenna. This minimizes mass suspended out over the antenna, or decreases the distance and losses between the antenna and the location of Tx or Rx amplifiers if not mounted at the primary feed.

The phased array antenna is more expensive but has the advantage of not needing to be physically pointed. This enables near-instantaneous re-pointing of the receive beam by electronic control. The phased array (Figure 5-12) consists of an array of small individual antennas (elements). The signal from each element is fed into a signal-combining network called a manifold of either coax cables or waveguides, depending on the operational frequency. These signals are then combined into a single high-power signal. The electronic steering is accomplished by modifying the signal phase of each signal (thus the "phase" in phased array) in a pre-determined manner that results in a change of which signal directions are passed through after summing, and which signal directions are suppressed [Van Trees, 2001b].

Detecting Signal—Sensors

The signal received by a sensor generally has three components: the desired signal, unwanted signals, and receiver noise. The focus of detection is to detect the wanted signal (or signal of interest) in the presence of these other signals (often called clutter or interference). Previously we discussed the types of RF signals that we want to detect to support the SDA mission (radar, communications). Unwanted signals may be other radars

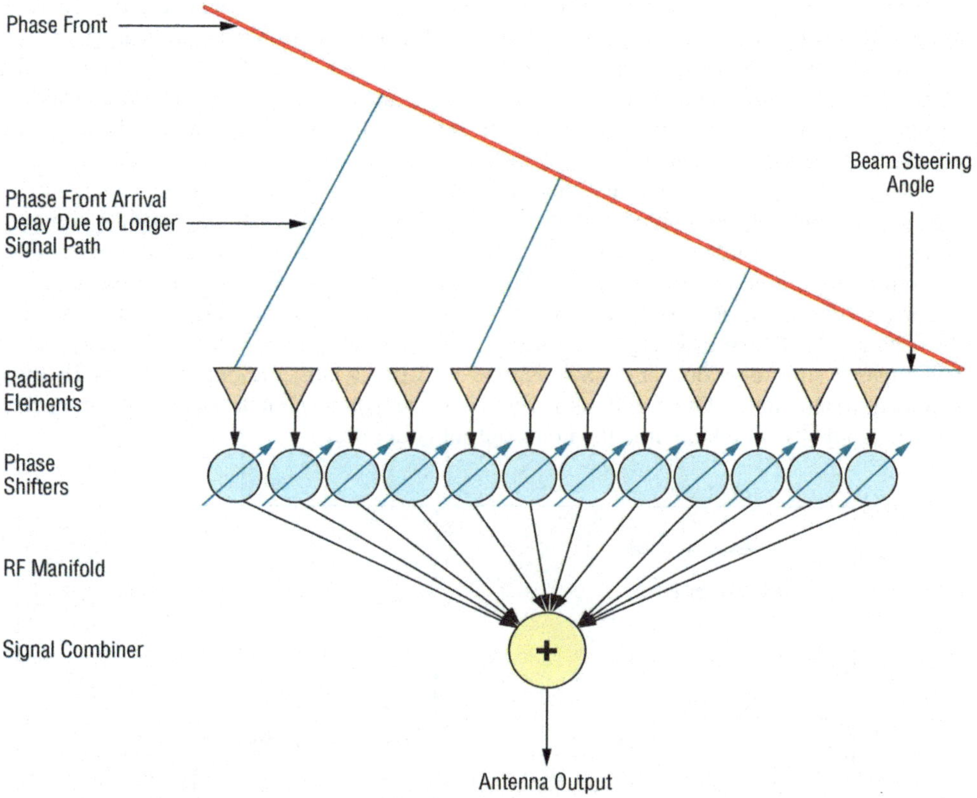

FIG. 5-12. Simple Phased Array Antenna Architecture. For this antenna design, the phase shifters are adjusted so that the extra phase rotation of the incident signal at the desired steering angle is "subtracted" so the output phase of each element is 0-degrees for a signal at the desired angle, resulting in full coherent power addition.

or communications systems (in space or on Earth), natural RF emissions (like the Sun or Jupiter), and possibly intentional jamming signals. We can separate the unwanted signals from the wanted signals in a few different ways. This is *signal filtering*.

The first form of signal filtering is to suppress all signals outside of the bandwidth of the signal of interest. This is *band-pass filtering*. If the unwanted signal either completely or partially shares the same bandwidth, this will not totally suppress the unwanted signal.

A second common filtering technique is *spatial or angular filtering*. Only signals from a specific angle of arrival at our antenna are allowed through and the rest are suppressed. This uses the antenna beam pattern as an angular filter. The closer the unwanted signal is to the desired signal, the narrower or tighter the beam needs to be to suppress the unwanted signal.

The majority of receiver noise in modern receivers is due to noise from signal amplifiers. The noise in the amplifier comes from the motion of electrons in the receiver due to temperature, resulting in random signals that add to the amplified signal. We call this *additive thermal noise,* and it's independent of the input signal. The thermal energy results in random motion of the electrons in the wiring in addition to the motion due to the input signal.

Signal detection typically uses a *matched filter* (MF). The MF is "matched" to the desired signal to be detected. The MF can be an analog or digital circuit. Most modern systems use a digital MF, although some still use an analog MF, especially radars that use a linear FM "chirp" waveform.

We can implement the MF in either the time or frequency domain. In the time domain the received signal is convolved with the time-reversed complex conjugate of the signal to be detected. This results in the MF output being a function of the total signal energy over the entire temporal length of the MF. Undesired signals or receiver noise that are uncorrelated with the desired signal do not experience the same output signal gain, but only the square root of the total energy. An example of an MF output for a linear FM chirp signal is a squared sinc function $(\sin(x)/x)^2$. Figure 5-13 gives both the output of an MF for a matched signal input (linear FM chirp) and an equal power Gaussian noise input. The matched output is in blue, while the noise output is in orange.

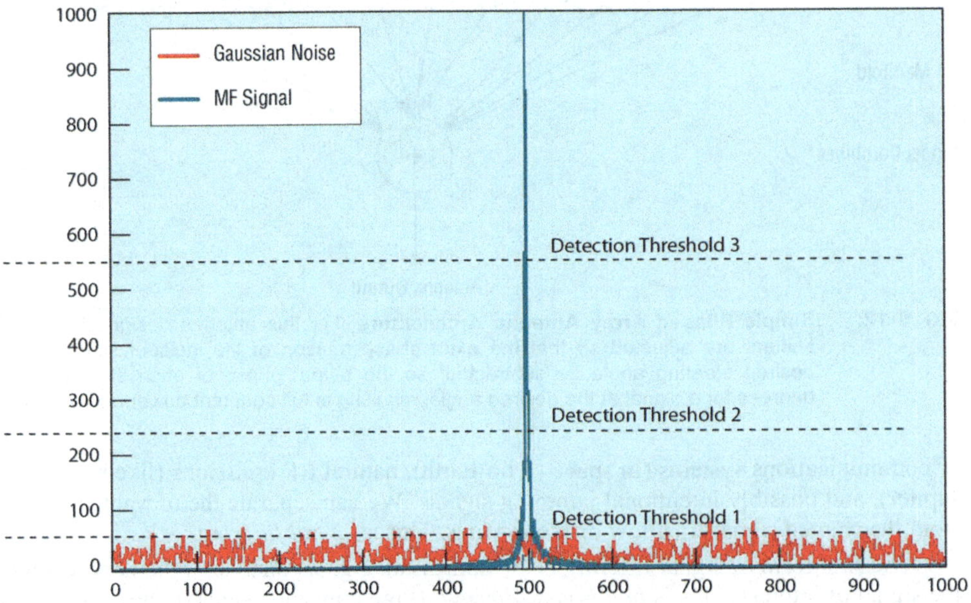

FIG. 5-13. Comparison of MF Output for Matched Linear Frequency Modulation (LFM) and Gaussian Noise. The signal clearly experiences gain when the Rx signal is temporally aligned with the MF signal, while noise is qualitatively (and statistically) unchanged. This is how an undetectable signal buried in noise can be detected with the MF. Effectively all of the signal power is "concentrated" while the noise is not. Note that as detection thresholds decrease a lower power signal can be detected, but eventually if the threshold becomes too low, noise will be "detected." A higher detection threshold will be more tolerant of noise.

The average level of the noise is approximately 33, while the MF signal peak is 1000. If we were looking at a signal that had the same signal power as noise, the signal would be clearly detectable, and so would the time the signal was fully temporally aligned with the MF signal. The time of temporal alignment (starts at the same time as the MF signal), is denoted by the large peak, and that delay corresponds to a specific object range.

We see by inspection that there is a signal in the noise in this case. If we want to automate the detection process, we set a threshold value [Van Trees, 2001a]. If the MF output is greater than the threshold, then we have "detected" the signal. Clearly, if we set the threshold too high, some small signals will not be detected, and if we set it too low, we may occasionally falsely detect only noise. We prefer to set the threshold as low as possible but no lower. A common way to do this is to decide on the maximum allowable probability of falsely detecting noise as signal. Given noise power, we know the mean noise level, as well as the noise standard deviation. If we want a maximum false alarm of 0.1%, we need the threshold to be slightly more than 3σ (three standard deviations) above the mean noise value. For Gaussian noise where voltage values follow a normal probability distribution, the noise power goes as voltage squared, and this probability distribution is called the chi squared distribution. For the chi squared distribution 1σ is the square root of twice the mean (μ), *so to have a maximum false alarm probability of 0.1% the detection threshold (δ) is approximately*:

$$\delta \geq \mu + 3\sigma = \mu + 3(2\mu)^{1/2} \tag{5-30}$$

For lower false alarm probabilities, a larger threshold is necessary (4σ or more).

Four detection outcomes are possible: signal present and detected, signal present but not detected, no signal present with false detection, and no signal present and no detection. The first and the last are correct, and the second and third cases are errors. We see both from Figure 5-13 and from the threshold equation that as the noise power increases, the desired signal must also increase in power for the same false alarm rate. If we do not change the threshold as noise power increases, we can detect the same signal power but with a higher false alarm rate.

SDA Radar Operations

One of the primary applications of radar to SDA is surveillance. The Oxford Dictionary defines surveillance as *Close observation, especially of a suspected spy or criminal.* In military applications it describes the act of observing over a large area to detect previously unknown objects of interest. Once a surveillance radar has detected an object of interest, it is often handed off to a tracking radar to keep track of, so that its location is always known, and we can reasonably predict where it is going. Fortunately, space objects follow orbital mechanics and travel in predictable orbits unless forces act on them to modify their path. If these forces are unknown or unmodeled, then the orbital predictions become less accurate over time. Space objects are much more predictable than aircraft that can turn in any direction at any time as well as change altitude and speed. The volume of space to perform surveillance is extremely large, however, compared to Earth's surface and atmosphere. Figure 5-14 gives an example of an SDA surveillance radar operated by the US Air Force Space Command.

FIG. 5-14. The AN/FPS-85 Phased Array Space Surveillance Radar. The radar is the only US military SDA dedicated phased array radar.

"The 20th SPCS C-6 operates and maintains the AN/FPS-85 Phased Array Radar, the only phased array radar dedicated to tracking more than 16,000 near-Earth and deep-space objects. The AN/FPS-85 Phased Array Space Surveillance Radar provides actionable space situational awareness for USSTRATCOM's space control mission area. It is one of 29 sensors that constitute the global Space Surveillance Network, or SSN, and is the only phased array radar dedicated to space surveillance. It collects more than 16 million observations of satellites per year, accounting for 30 percent of the SSN's total workload. The unique aspect of this radar is the phased array antenna technology. Unlike mechanical radar, which must be physically aimed at an object in space to track and observe it, the phased array antenna is steered electronically by controlling the timing, or phase, of incoming and outgoing signals. This increases capabilities and decreases response time, which allows near-simultaneous tracking of multiple targets throughout the system's area of coverage" [20th Space Control Squadron, 2009].

The primary purpose of surveillance radar is to detect an object and determine its location and (sometimes) radial velocity. Radial velocity is the component of the object's velocity in the direction along the vector between the radar and the object. The radar transmits a known signal or waveform. This signal reflects off of objects in space, with some of that signal energy returning to the radar. The radar then processes the received signal with a matched filter, matched to the transmitted waveform. An object is detected if its signal after matched filter processing is above the detection threshold. The range to the object is also determined at detection, as the match filtered signal will exceed the threshold when that corresponds to the object range. The delay from transmission to detection is the round-trip time from transmitter to receiver with the time radar signal traveling at the speed of light.

The azimuth and elevation angles to the object are determined by where the radar is pointing the beam. Generally, the radar's angular accuracy is much less than its range accuracy. The two angles and range determine the three-dimensional location of the object relative to the radar. We calculate the object's radial velocity by the Doppler shift of the received signal. If the object is moving towards the radar, the return signal is shifted higher in frequency, while if it is moving away, the return signal is shifted lower in frequency.

The choice of radar frequency can be a complicated engineering trade analysis. Several major relationships between frequency and radar performance commonly result in surveillance radars operating at lower frequencies than tracking radars. There is a minimum frequency limit set by the smallest physical dimensions of an object we wish to detect. When the object gets significantly smaller than the wavelength of the radar signal, it reflects much less energy. Generally, we do not want the object to be much smaller than 1/2 the wavelength, so to see 10-cm objects (like a 1u CubeSat), then the longest wavelength we would want to use is 20 cm. This is an example of Mie scattering first observed in optics. This wavelength corresponds to a frequency calculated as follows:

$$f = c/\lambda = (3 * 10^{8}\text{m/s})/0.2\text{m} = 1.5\text{ GHz}, \tag{5-31}$$

where f is the frequency, c is the speed of light, and λ is the wavelength. If we want to detect smaller objects the minimum frequency is greater, and for larger objects the minimum frequency is less.

Previously we described the major radar and object parameters that predict the radar SNR. The higher the SNR (more powerful signal relative to the noise) the better the detection performance. We can only control the radar specifics, and those are transmit power, transmit antenna gain, transmit time, and receiver area. Lower frequencies are easier to efficiently amplify, so it takes less electrical power for lower frequencies to generate the same transmit power. Antenna gain is a function of antenna area and wavelength, so for an antenna of a given size, we get higher gain at higher frequency. But higher-frequency antennas are more expensive to build.

Additionally, higher gain means a smaller angular beam is transmitted, illuminating a smaller area of space. So the beam has to move rapidly to different locations. Moving the beam more quickly means our transmission time τ must be smaller, which then decreases our SNR. Therefore, larger transmit gain doesn't help surveillance radar SNR that must operate over a large area. This leaves receiver antenna size to optimize. Because it is cheaper to build a large low-frequency antenna than a high-frequency antenna of the same size, most surveillance radars operate at low frequencies, commonly 1–2 GHz.

In order to track an object, we must first detect it. Next we must ascertain the object's location, and preferably its radial velocity relative to the radar. Over time we have to group multiple detections and correlate them to specific objects to be tracked, and then combine them to create a historical model of the object's path through space. This model can then predict where the object will be in the future. This constitutes an object track. Figure 5-15 illustrates an example. Shown are four different times that radar detections are plotted in range and azimuth (elevation angle removed for clarity). The range rate color tells if the target is moving away from the radar (red) or toward the radar (blue). The figure shows the proper grouping of detected objects. We note that in Figure 5-15, some detections at some times are not grouped as a tracked object. These could be false detections or a very small object that is only detected sporadically, making it difficult to decide which detections we should group as a single object. The calculated range rate allows determination of the track direction (toward or away from the radar). Object 3 is moving towards the radar while objects 1 and 2 are moving away. The automation of proper grouping of detections to specific tracked objects, or the start of a new track for a new object, is an area of significant detail beyond the scope of this book.

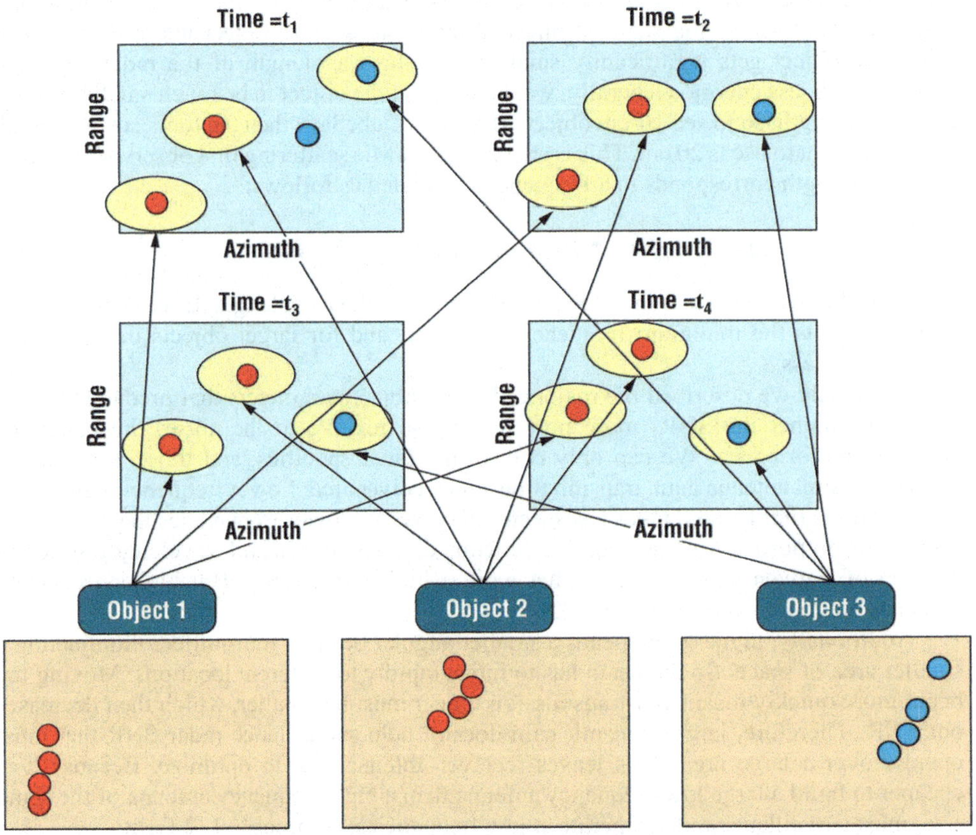

FIG. 5-15. Example of How Radar Detections Over Time are Correlated into Object Tracks. The tracker uses range, angle, and Doppler information to correlate individual detections, and then estimate heading and velocity for each detected object. Generally, false detections are uniformly spread over angle, range, and Doppler and so do not correlate over time and can be discarded. If the number of false detections becomes too high, then the false correlations become more likely. This is one practical reason why the probability of false detection can't be "too high," limiting how low a detection threshold can be set.

The required data input to the tracker for each detected object is, its range, azimuth, elevation, and radial velocity. These are all parameters that radar can estimate, although range and range-rate are generally much more accurate than azimuth and elevation angles. The range estimation is determined by the delay between the transmitted signal and the received signal. The range is the time difference between transmit and receive multiplied by the speed of light.

We calculate the range rate through the use of the signal Doppler. The radar sends many pulses very rapidly. The rate of signal re-transmission is called the pulse repetition frequency (PRF). We usually measure PRFs in kilohertz. At this very rapid rate objects are "almost" stationary. The range bin is the range resolution of the radar. The range resolution of radar is

$$\text{Range resolution} = c/2BW \tag{5-32}$$

However, the small motion per radar pulse (δr) results in a change of return phase. For motion that is a fraction of a radar wavelength (λ) (for 1.5 GHz this is 20 cm) the resulting phase shift (in radians) is:

$$\text{Phase shift} = (2\pi)\delta r/\lambda \tag{5-33}$$

The change in signal phase over time is a measure of the signal Doppler shift. This is commonly calculated by using a fast Fourier transform of the measurements of phase over time (multiple signal pulse transmissions), where each phase measurement is at the PRF rate. The combination of the range and Doppler information creates a range/Doppler map. Doppler is converted to range-rate (v) as:

$$v = cf_d/(2f_t) \tag{5-34}$$

where c is the speed of light, f_d is the Doppler frequency, and f_t is the transmitted frequency.

We determine the azimuth and elevation angles of the object by using the transmit and receive antenna patterns. The first and simplest (and least accurate) method of estimating angular position is using the direction of the transmit and receive beams. The angular estimate is simply the pointing direction of the combined transmit and receive beams (normally they both point in the same direction). The angular accuracy is the beamwidth. We calculate the half-power beamwidth (HPBW) (Figure 5-16) as:

$$HPBW = \lambda/D \text{ (radians)} \tag{5-35}$$

where D is the antenna diameter for circular antennas (length or width for rectangular antennas) and λ is the radar wavelength.

The angular beamwidth gets narrower at higher frequencies and with larger antennas. A second, more advanced (and complicated), technique is monopulse [Rhodes, 1980]. This technique enables much better estimations of angular accuracy at the expense of more complex antenna hardware, more complex processing, and the assumption that at any given target range, only one target is in the beam. The technique uses multiple simultaneous receive beams pointing in slightly different directions (this can be separately pointed beams, or a sum and difference beam, of two separate receive horns typically immediately next to each other), as Figure 5-17 shows. An object will have different amplitude and phase

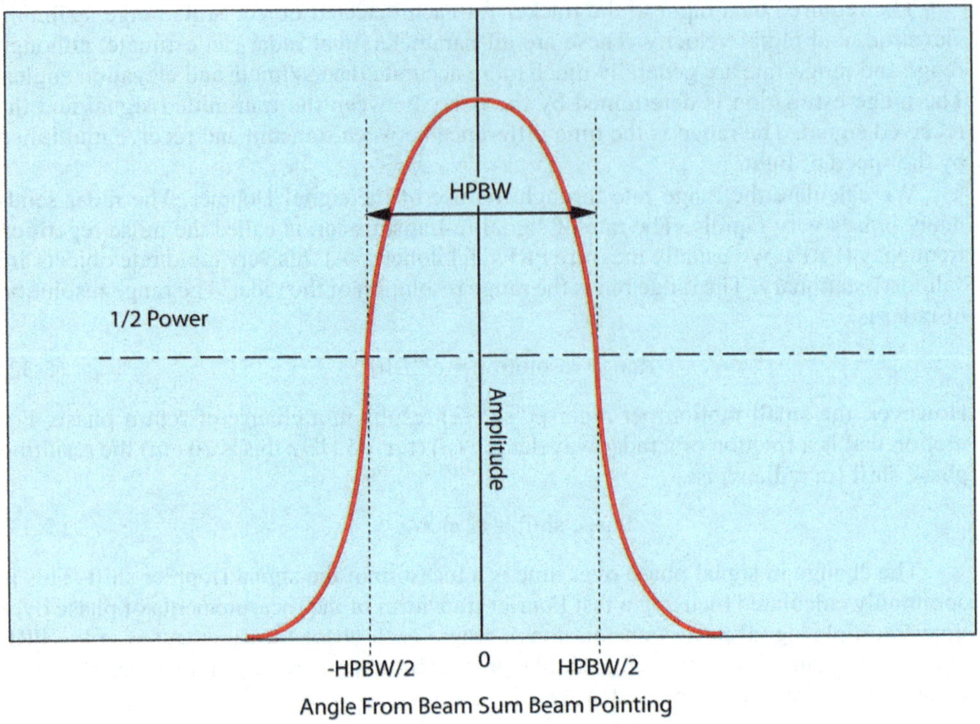

FIG. 5-16. Half-power Beamwidth. The half-power beamwidth is exactly what the name implies; it is the angular width of the beam from 1/2 power on one side, to 1/2 power on the opposite side. This is the fraction of the beamwidth that contains the majority of the total power. The plot shows clearly how the power drops off rapidly outside the 1/2 power beamwidth.

response for each beam as a function of its location, and this helps determine the location more precisely in the main beam. Two locations on either side of the sum beam will have the same amplitude values and ratio. We can resolve this ambiguity by knowing that the phase difference on one side of sum beam center is positive and that on the other side is negative. There is actually much more information in the phase difference as a function of angle, however.

Two types of monopulse are available: amplitude monopulse and phase monopulse. They can function by themselves or jointly. Most US systems use only the phase monopulse. We calculate the angle estimation accuracy of monopulse in the following manner [Mosca, 1969], [McGinn, 1966], [Ewell, 1972]:

$$\sigma = \theta/(k_m[SNR]^{1/2}) \tag{5-36}$$

where σ is the angle estimation error due to noise, θ is antenna 3dB beamwidth, k_m is the difference channel error slope, and SNR is the radar signal-to-noise ratio.

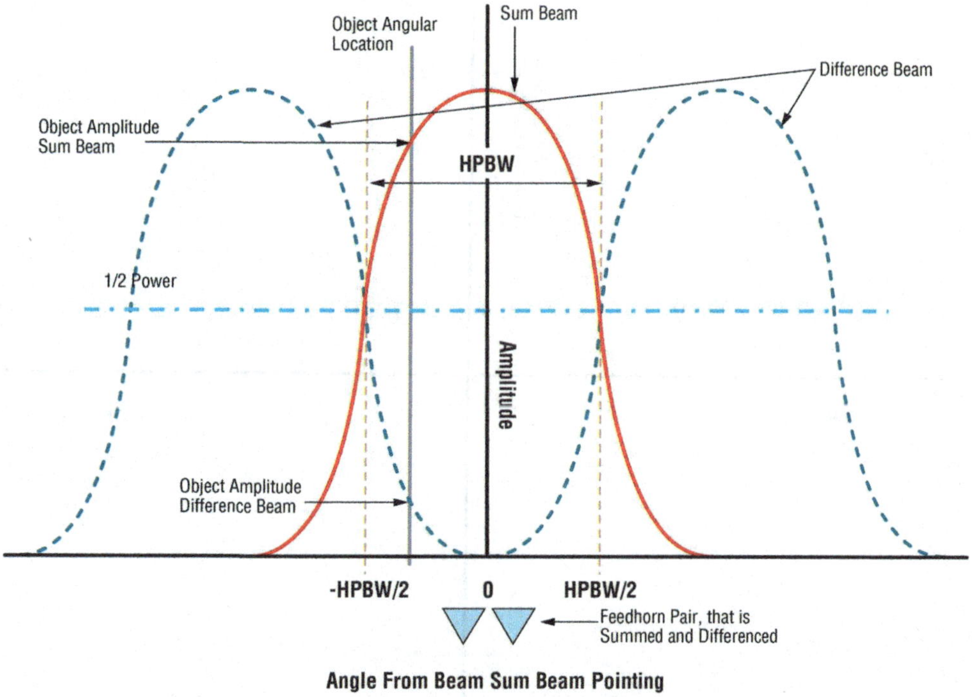

FIG. 5-17. Illustration of the Sum and Difference Beam in Monopulse Processing. The monopulse technique enables angular resolution finer than the beamwidth for point objects. The two feeds are summed and subtracted to create the sum and difference beams. The amplitude and phase differences between the two beams are used to determine a fine angle. This figure illustrates the different signal powers for a specific signal angle of arrival. Signal noise limits the signal power estimation accuracy, which limits angular resolution.

The difference channel error slope is the slope of the line in Figure 5-18. The system phase estimation error has an inverse relationship to the square root of the SNR, and the resulting angle estimation error is a function of the slope of the differential phase that results due to target angle. The actual value of k_m depends on the specifics of the monopulse system design. A conservative value for k_m is one.

In calculating the range, precise timing is important, while for angular monopulse, consistent amplitude and phase between the sum and difference beams are critical. We call the relative amplitude and phase accuracy *channel balance*. Real-world hardware (HW) amplitude and phase response typically change with temperature and time, and different HW have slightly different responses. We have to calibrate the system to correct for these differences.

The general method of calibration is to either inject a known signal into the hardware path or to look at a very well known object. In both cases we can determine what the correct system output should be. We compare the actual system output to what we expect, and the difference is the calibration error. We can correct this error by modifying "tunable"

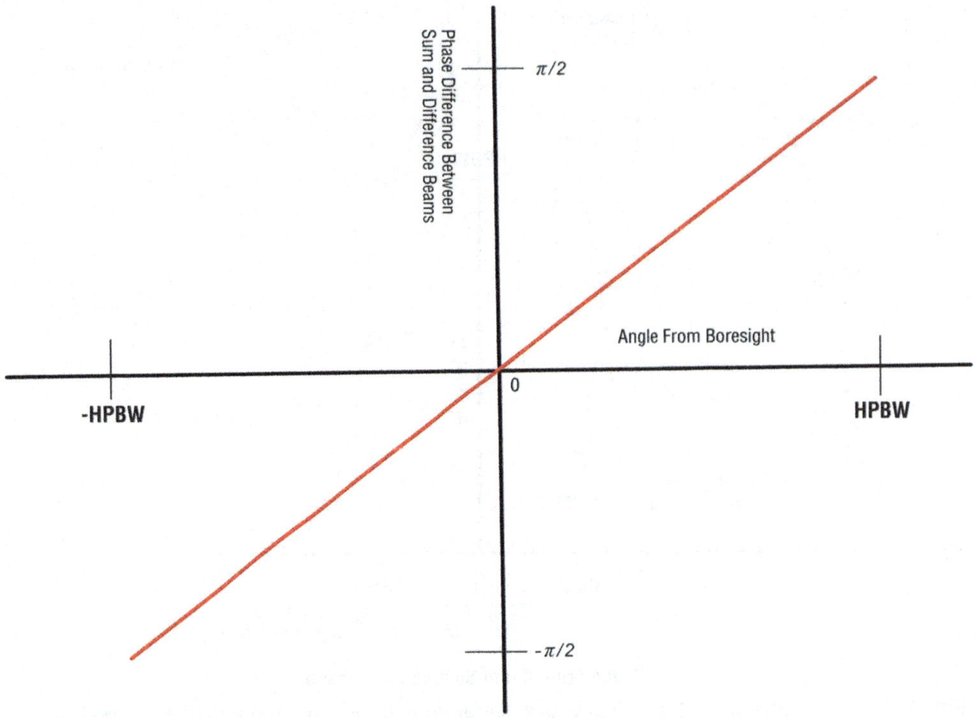

FIG. 5-18. Differential Phase between the Sum and Difference Beams as a Function of Object Angular Location Relative to Sum Beam Center. The slope of the line is k_m, the difference channel error slope.

hardware or performed in-signal processing. Modern systems normally apply calibration correction in the digital processing chain.

The advantage of the direct signal injection is that it can be done quickly—sometimes in seconds—at any time, even while the radar is operational. This allows us to perform the calibration frequently. The disadvantage is that it only calibrates the receive half of the radar. Additionally, unless there is a signal generator away from the radar, the signal will need to be injected into the hardware after the antenna. All hardware before the injection point is uncalibrated. The advantage of using a well-known object is that it tests the entire transmit and receive signal path. The disadvantage, particularly for an SDA radar, is having a calibration satellite fly over the radar only occasionally, so the radar scheduling must adjust its operation based on the calibration satellite orbit. The satellite should have a well-known radar cross-section (RCS) as well as a well-known orbit. The advent of GPS receivers on satellites allows them to autonomously determine their orbit. In some cases, calibration satellites echo back the transmission signal after amplifying and frequency shifting the signal. This provides a much higher power receive signal and therefore improved SNR. The improved SNR improves the calibration accuracy.

We can use radars to generate images. The best-known technique is called synthetic aperture radar (SAR). In an SAR the radar moves and sweeps out a large virtual aperture that provides excellent resolution in the direction of motion, while the radar's high bandwidth waveform provides the precise range resolution. The simplest SAR version is called strip-mode SAR, where the radar looks perpendicular to the velocity vector. This results in the range resolution being orthogonal to the cross-range. Figure 5-19 shows the relative geometry for strip-mode SAR. An important point is that the range resolution is in the beam pointing direction, which is not parallel to Earth's surface, while the cross-range resolution is parallel to the ground when the satellite or aircraft remains at a constant altitude. The combination of the range and cross-range resolution is a 2D plane defined by the radar velocity vector and the radar beam pointing direction. The calculated range resolution in this slant plane must be projected onto the ground for us to calculate the effective range resolution on the ground.

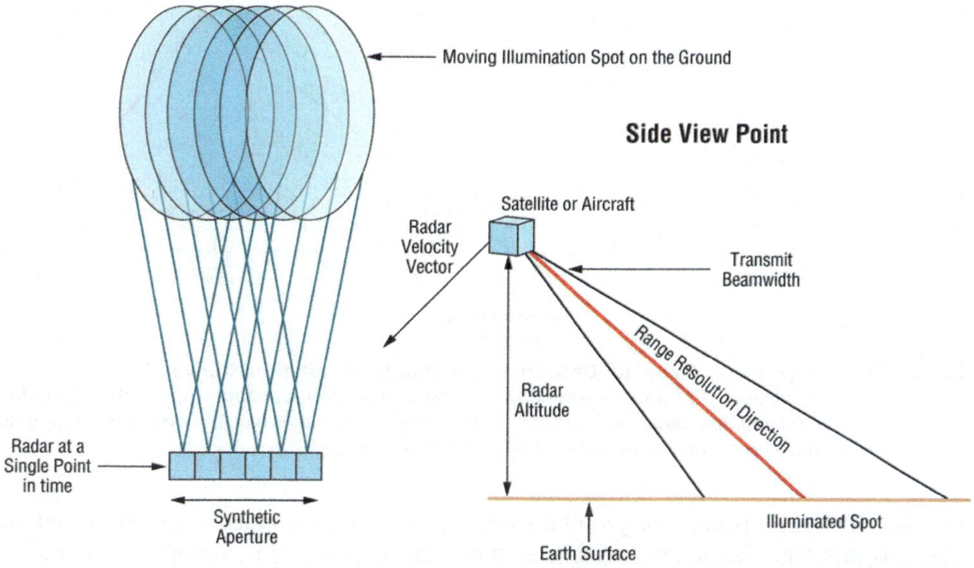

FIG. 5-19. **Strip SAR Relative Geometry and Resolution Directions.** The synthetic aperture radar technique synthesizes a large effective aperture in the velocity direction, by exploiting the radar coherent sensing at each radar transmit pulse and across multiple pulses. The angular resolution in the velocity direction is the Rayleigh resolution of λ/D, where D is the synthetic aperture length. The range resolution from the transmit bandwidth is perpendicular to the angular resolution, providing a 2-dimensional image.

We can calculate the cross-range resolution in the slant plane by the length of the synthetic aperture. A counter-intuitive implication is that for better cross-range resolution we want a physical antenna that is shorter in the cross-range (radar velocity) dimension. This results in a wider beam, which then results in a larger effective synthetic aperture. We can calculate the cross-range beamwidth as with any conventional antenna as wavelength divided by antenna length. The range resolution is the speed of light divided by twice the signal bandwidth (Equation (5-32)). The range resolution on the ground is a function of the angle the slant plane makes with the ground, as Figure 5-20 shows.

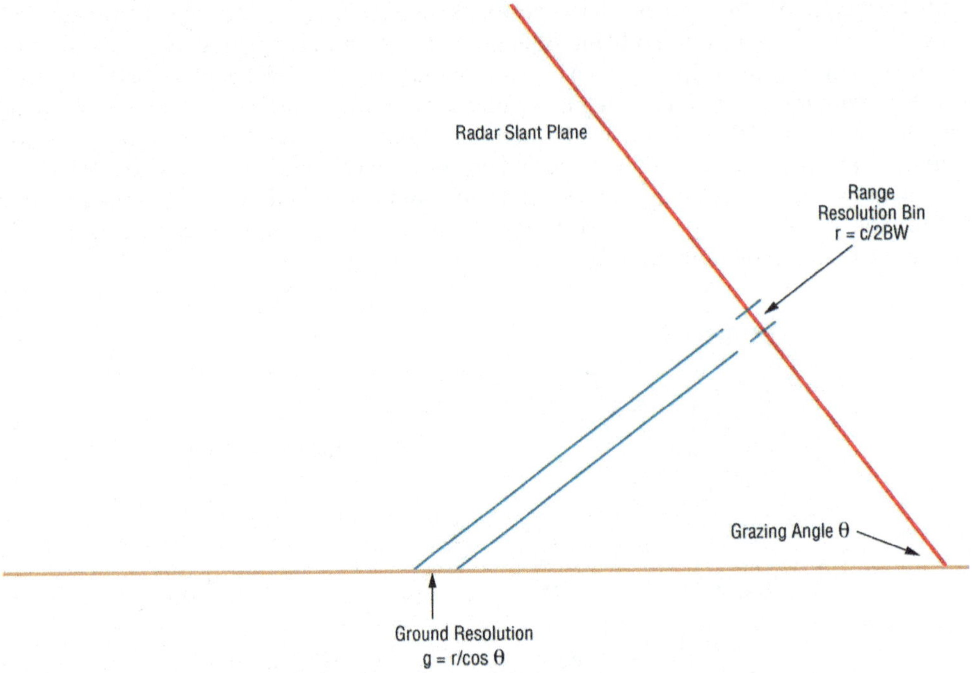

FIG. 5-20. SAR Slant Plane to Ground Plane Range Resolution Conversion. The range resolution of a radar looking down at the ground is a function of geometry. It's in the direction the radar is "looking", which then must be projected onto the ground to determine the effective resolution in the ground plane.

For SDA SAR is not a very useful technique, because the radar is stationary and the object is moving. The general principles, however, do apply, due to the relative motion between the object and the radar, combined with a second effect caused by the nominal rotation of the object relative to the radar. This is called inverse SAR (ISAR), originally developed by the Navy to resolve the SAR image blurring of ships due to wave motion during the time of a synthetic aperture. ISAR is a much more complex signal processing problem. For ISAR to work, we develop a model of the object's rotation based on the radar data, and any other available information, and then use the estimated object rotation to create the ISAR image. This class of mathematical techniques is *blind deconvolution*, where the sensor data is used to both develop a model of the object behavior and generate the image. Figure 5-21 gives an example of an ISAR image of a satellite created by the MIT/Lincoln Laboratory Haystack radar [Czerwimski et al., 2014].

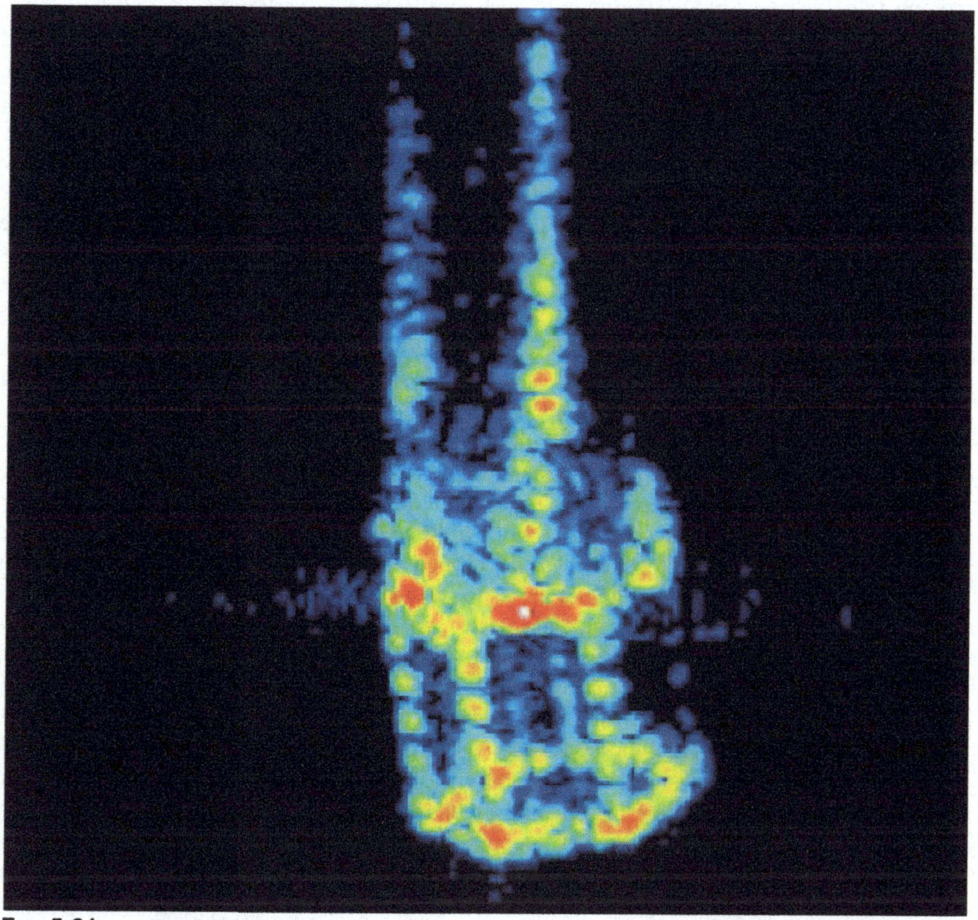

FIG. 5-21. Satellite ISAR Image. This is an example of a satellite ISAR image [Czerwimski et al., 2014].

One of the radar design constraints is that the radar must have a very stable local oscillator to synthesize the transmit waveform and to demodulate the received signal. During the time of the synthetic aperture (either SAR or ISAR), the local oscillator phase must be very stable, or it will create an apparent Doppler (range-rate) that does not reflect the actual relative motion of the object with the radar. In the case of SDA, we may have very slowly rotating objects, requiring long synthetic apertures. Long-term stable local oscillators are difficult to maintain and expensive.

The range to an object has a very large effect on system SNR. SNR drops as range to the 4th power. The ranges to LEO objects (typically 1200 km and less altitude), including the effects of slant range, are up to 3000 km, ranges to MEO objects (typically ~15,000 km altitude) on the order of 25,000 km, and objects in GEO (~35,000 km altitude) may be up to

60,000 km away from the radar. Clearly, radar optimized for LEO objects will have very poor sensitivity for GEO objects. A GEO-focused radar needs a very large aperture and transmit power. Because GEO objects are nearly stationary relative to a point on the earth, the radar does not need to move its transmit and receive beam quickly to track the object. But in LEO, objects move much more quickly and are only in view for 5–10 minutes. This requires an ability to quickly move the radar beam to keep a LEO object in the beam or move quickly between objects to detect and track them before they are no longer in view. The need for an agile beam for LEO objects favors phased array antennas that do not have to move the entire antenna to point the beam. The USAF SDA radar is a phased array radar for this reason, even though it was built in the 1960s when this approach was enormously expensive.

A different way to implement a LEO radar is to create a large "fan-beam" across the sky and detect all objects that pass through it. This mitigates the need to move the beam, but means the transmit antenna gain is low. Since transmit antenna gain is important for a radar SNR, this approach needs much more power, possibly combined with a larger receive antenna. The now deactivated (September 2013) AN/FPS-133 Air Force Space Surveillance System used a series of three fan beam transmitters and six multiple receivers across the southern US to make a detection fence that spanned the 48 states of CONUS. This system operated at very low frequency (~200 MHz), which limited its sensitivity to physically small objects. In the '60s, when this system was developed, the radar was designed to detect larger satellites, and artificial debris was not an issue. A surveillance radar should have as low a frequency as the minimum object size of interest will allow. At 200 MHz the wavelength is 1.5 meters, so at this frequency, the radar would have good sensitivity to 3/4 of a meter. The world's first artificial satellite, Sputnik, which the Soviet Union launched in 1957, was 58 cm in diameter combined with four long appendages (2.4 and 2.9 meters long), which is consistent with the frequency choice for the AN/FPS-133.

A GEO- or MEO-focused SDA radar can dwell on an object for a long time to increase SNR. The lack of rapid motion makes long dwells easier. There are no GEO-focused SDA radars at this point, but the NASA Deep Space Network (DSN) has similar design constraints and serves as an interplanetary radar. The DSN's primary use is to communicate with and track interplanetary satellite probes. It also creates ISAR images of asteroids (Figure 5-22 [NASA, 2015]). The DSN comprises large steerable dish antennas, as shown in Figure 5-23 [NASA, 2013].

DSN's very long range of operation (interplanetary) requires these large steerable dish antennas with high power transmitters. A similar architecture would be reasonable for dedicated radar designed for GEO SDA operations. Building a large reflector should be less expensive than a large phased array, and in the case of both GEO and deep space operations, the beam agility that a phased array provides is not necessary.

5.2.3 RF/comm

The work of Claude Shannon in the foundational paper "A Mathematical Theory of Communication" [Shannon, 1948] established the theoretical basis for communications and communications theory. Figure 5-24 shows the general form of a communications link. It starts with information (a message) to be transmitted. This is sent to the transmitter which modulates the transmitted signal to encode the message. The signal then passes through the

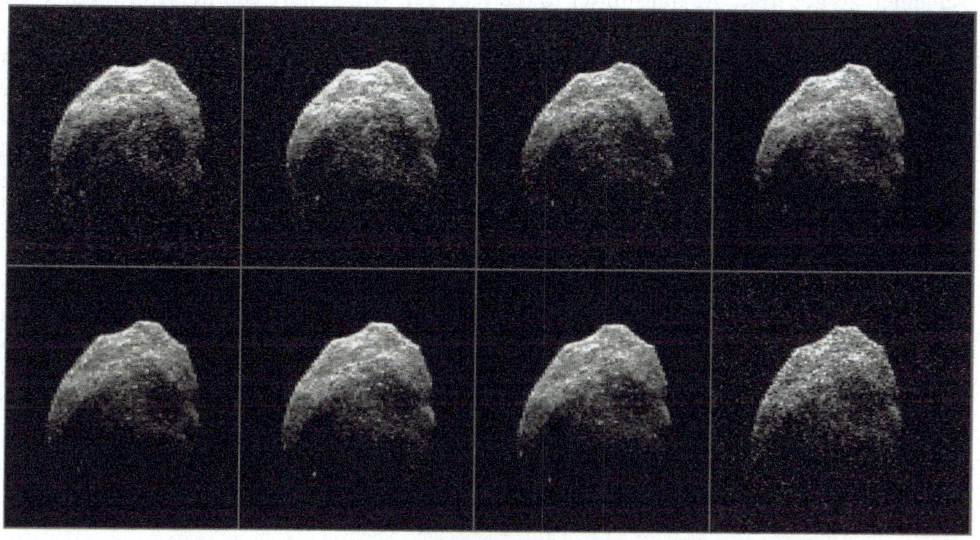

2015 TB145

FIG. 5-22. NASA Deep Space Network Images of an Asteroid. NASA and astronomers use ISAR to image asteroids that pass close to the Earth. While very far away, they are also large, and the combination of the NASA Goldstone transmitter and the Greenbank Radio Telescope creates an exceptional bistatic radar.

FIG. 5-23. NASA Deep Space Network Antennas. Dawn in the Apollo Valley. Example of a few very large antennas capable of high transmit power. These are primarily designed for communications to deep space interplanetary probes, but can also be effective radar transmitters. They are expensive, and can only point in one direction at a time, and cannot be moved very fast due to their large size.

"channel" to get from the transmitter to the receiver, which may add noise or modify the signal. This signal will then be demodulated into an estimate of the original message (with the addition of the noise). The key point of Shannon's work was determining the maximum theoretical information content that could be transmitted by such a link (how close to the original message the estimated message could be). Shannon quantified information in terms of bits. RF communication is the use of RF signals to encode the message and transmit it from the transmitter to the receiver. Other forms of communications signals are also common (e.g., acoustic/voice, optical/laser).

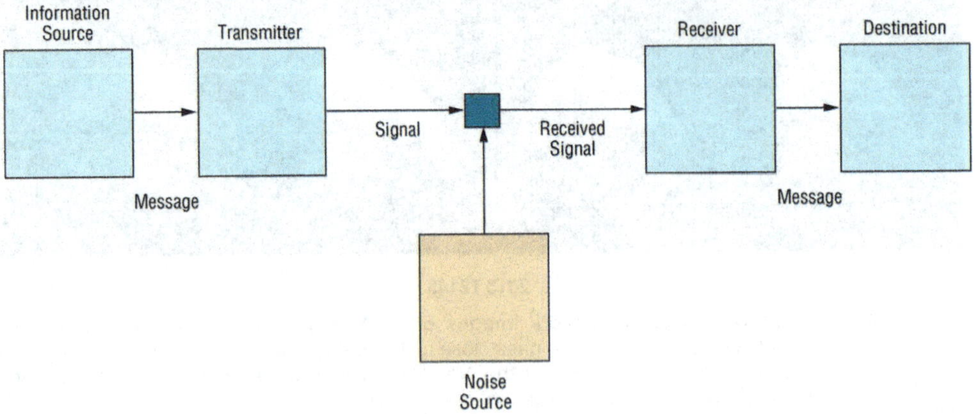

FIG. 5-24. General Form of a Communications Link. The communications path from the information source to the destination. The transfer is corrupted by noise (e.g., from the receiver, and possibly other interfering sources in the transmission path between transmitter and receiver.)

Space communications are usually via RF signals. The general types of information transmitted are: satellite commands uplinked from the ground controllers, satellite health and status information downlinked to the satellite controllers, and payload data downlinked to the satellite ground station. Sometimes the satellite is part of the message "channel" when a ground-based user uses it to relay a message to a different ground-based user. Geosynchronous communications satellites link users in this way across continents.

Communication System Performance

Not surprisingly, the key metric for a communications link is the *bit error ratio* (BER). This quantifies how often a bit of information will likely have an error, or the probability that any single bit will be estimated incorrectly. American National Standard (T1.523-2001) [ATIS Telecom Glossary] defines BER as:

> *The number of erroneous bits divided by the total number of bits transmitted, received, or processed over some stipulated period. Note 1: Examples of bit error ratio are (a) transmission BER, i.e., the number of erroneous bits received divided*

by the total number of bits transmitted; and (b) information BER, i.e., the number of erroneous decoded (corrected) bits divided by the total number of decoded (corrected) bits. Note 2: The BER is usually expressed as a coefficient and a power of 10; for example, 2.5 erroneous bits out of 100,000 bits transmitted would be 2.5 out of 10^5 or 2.5×10^{-5}.

The major system parameter that affects a communications link BER is the system signal-to-noise ratio (SNR).

The relationship between SNR and BER depends on the choice of signal modulation. Figure 5-25 depicts this relationship. It shows three different phase modulations (binary shift key (BPSK), quad shift key (QPSK) or sometimes (4-PSK), and 8-PSK). The BPSK modulation encodes a single bit (0,1) as two different phases, nominally 0 and π phase (blue highlights). QPSK adds a second bit (0,1,2,3) encoded at four phases separated by $\pi/2$ radians (brown). 8-PSK encodes 3 bits (dashed lined circles). We see that as the number of bits encoded increases, the spacing between the encoded values in the complex plane becomes less. The receiver additive noise results in each value having a spread in the complex plane. Because noise is Gaussian in real and imaginary components, the noise-induced spread is a Gaussian blur centered on the nominal noise-free location. As the noise power increases, the spread of the Gaussian blur increases. At a high enough noise level, significant overlap occurs, resulting in waveform decoding errors. As the signal power increases, the separation of the encoded values from the origin increases, which allows for higher noise power before achieving the same overlap. The 8-PSK encoding is less noise-tolerant than the BPSK coding. The equation for BPSK encoding is a function of the average SNR of the link (average receive power/average noise power):

$$BER = \left(\frac{1}{2}\right) erfc\left(\sqrt{\frac{S}{N}}\right), \qquad (5\text{-}37)$$

where *erfc* is the complementary error function. The equation also applies to the BER for QPSK. However, with the number of bits doubled, the SNR per bit is half of a BPSK waveform of the same total power. Hence to have the same BER, a QPSK signal needs twice the transmit power.

$$BER = \left(\frac{1}{2}\right) erfc\left(\sqrt{\frac{S}{2N}}\right) \qquad (5\text{-}38)$$

The BER for 8-PSK does not have a simple equation like BPSK and QPSK. The interested reader should consult communications texts that describe modulations and BER relationships. Sometimes textbooks and research papers use the term E_b/N_0 (E_b is energy per bit, N_0 is noise spectral density—noise energy per Hertz). For thermal noise, $N_0 = kT$, or Boltzmann's constant times temperature in Kelvin. The conversion from E_b/N_0 to SNR is:

$$SNR = \frac{R_b * E_b}{B * N_0} \qquad (5\text{-}39)$$

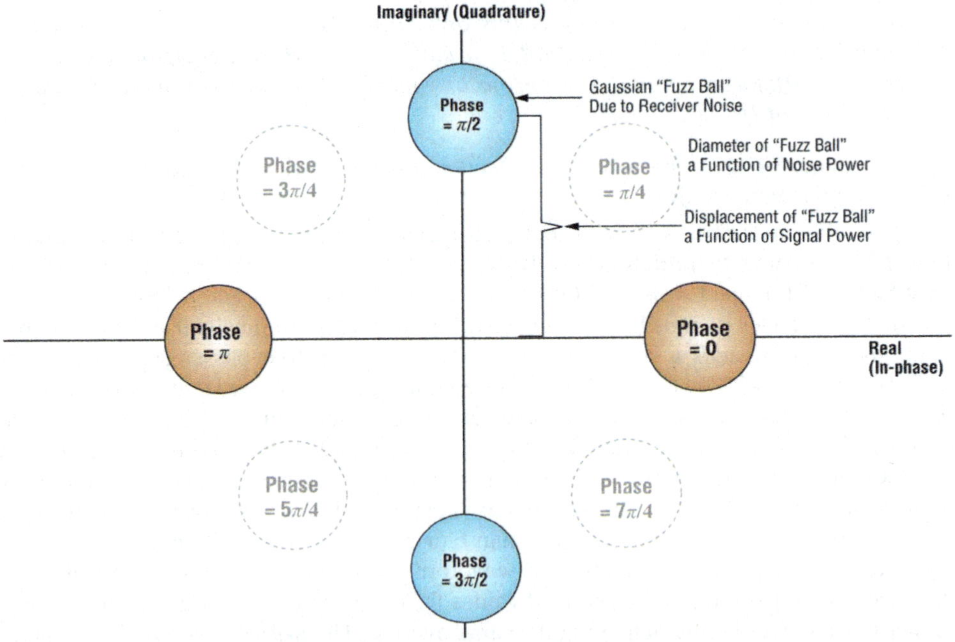

FIG. 5-25. Multiple Phase Shift Key (MPSK) Signals in the Complex Plane. Communications information uses symbols encoded in amplitude and phase, which can also be transformed to real and imaginary numbers. The symbol values can be plotted on the complex plane. Noise results in circular "fuzz balls" surrounding the nominal symbol values. The higher the noise, the larger the noise circle around the symbol values. At high enough noise the probability of a noisy symbol value being "closer" to a different symbol becomes too high, resulting in an excessive bit error rate (BER)

where R_b is the communications bit-rate, and B is the bandwidth of the communications signal. The signal power clearly must rise if we hold the Bandwidth constant and increase the bit rate. Changing from a BPSK modulation to a QPSK modulation would result in 2x the bit rate for the same bandwidth as an example. If the number of bits doubles, and the energy per bit is the same, then 2x the transmit power is needed. (we recall that noise power = kTB).

An old-fashioned modulation (on-off key—OOK) used in Morse code is seeing renewed interest in laser communication. On-off pulsing a laser does not require optically coherent modulation and demodulation, which is much simpler to implement. The BER of OOK is:

$$BER = \left(\frac{1}{2}\right) erfc\left(\frac{1}{2\sqrt{2}}\sqrt{SNR}\right) \tag{5-40}$$

A more general form of OOK is pulse amplitude modulation (M-PAM) which can encode multiple bits as a function of the amplitude transmitted, and is an amplitude version of MPSK. The BER for this as a function of the number of possible amplitudes transmitted (M) is:

$$BER = \left(\frac{1}{2}\right) erfc\left(\frac{\sqrt{SNR \log_2 M}}{2\sqrt{2}(M-1)}\right) \tag{5-41}$$

For M = 2 the BER is the same as for OOK—as it should be. A different version of multi-bit amplitude encoding becoming very popular in laser communications is pulse position modulation (L-PPM) where L is the number of time positions used to encode different values. The BER for this version is:

$$BER = \left(\frac{1}{2}\right) erfc\left(\frac{1}{2\sqrt{2}}\sqrt{SNR\left(\frac{L}{2}\right)\log_2 L}\right) \tag{5-42}$$

In L-PPM, a shorter, higher energy pulse is transmitted in a specific time slot to encode the desired value. The BER drops as L increases, due to the instantaneous power increase of the very short duration communication symbol encoding.

One of the important takeaways is that the system power density on the ground is a function of the system transmit data rate for workable communications systems. The form of the communication modulation affects what a reasonable system SNR needs to be for the transmitted data rate. If we have a ground-based receiver listening to a communications downlink, we should have sufficient SNR for simple signal detection from a small receiver (i.e., we are not attempting to demodulate the signal, but only to detect the signal's existence and possibly determine its location). More sophisticated communication systems spread the energy of a single bit in time over a larger system bandwidth, thus greatly decreasing the signal energy at any given frequency. These are *spread-spectrum* systems or *low probability of intercept* (LPI) systems. This type of modulation can have very low SNR if the spreading code is not known. Spread spectrum systems are beyond the scope of this book, but the interested reader will find many books on the subject.

In the case of a laser downlink, the transmit beam is normally very small. The commonly used optical wavelength is 1550 nm (short wave infrared). Even a small 1-cm aperture will have a very tight beam compared to a typical RF system. We calculate the 3-dB beamwidth of a 1-cm aperture at 1550 nm using Equation (5-35): ($\lambda/D = 1.55*10^{-4}$ rad = 0.0089 degrees).

General Types of Comm Systems

There are three main types of space communications links: space-to-ground, ground-to-space, and space-to-space. The ground station of a space-to-ground receive link and the ground-to-space transmit link will have different design requirements for a LEO versus a GEO satellite. For LEO satellites, the ground station must be able to track the satellite as it transits overhead. The duration that a LEO satellite is in view of a ground station is measured in minutes, often less than 10 minutes. A ground station for a GEO communications satellite can be stationary once the satellite is in its assigned "slot." A GEO satellite has a slow drift of a few (tens of) degrees per day as it drifts to its assigned location, therefore requiring significantly less pointing agility and slew rate.

Space–to–space cross-links in general require agile steerable antennas. An example is a LEO satellite that communicates with a GEO relay satellite. A typical LEO orbital period is 100 minutes; therefore in 50 minutes (half its orbit), it will need to slew the antenna pointing to the GEO relay satellite by approximately 180 degrees. If a LEO satellite has a crosslink to another satellite in the same orbit (common for a constellation like Iridium), then the crosslink pointing remains stable, except for small orbital perturbations of each satellite due to Earth's spherically non-symmetric gravity field.

From an SDA perspective, the space-to-ground link is most interesting, because it is easier to put a receiver on the ground that can detect a communications link directed at the ground.

Detecting a ground-to-space link is best done either in the air or in space. An aircraft needs to be flying quite close to the uplink to detect the signal. The uplinking ground station needs to be visible from the aircraft. A satellite-based receiver would be even more complicated. Due to orbital dynamics and orbital velocities, the space-based receiver needs to remain close to the uplink target, implying an orbit very close to the receiving satellite. If we want to detect signals from a ground station to many different satellites, we need either a satellite with tremendous fuel to rapidly move between orbits, or many individual satellites. Neither of these options is reasonable, primarily for cost reasons. Detecting space-to-space crosslinks is even more complicated if the transmitting satellite uses a directional beam, which compounds the problem that both the transmitting and receiving satellites are in motion. Some examples of space-to-space crosslinks are Iridium, with crosslinks between satellites in the constellation, and the French SILEX system. The latter is an example of a GEO-based laser crosslink relay between a ground station and other satellites in lower orbits that are in view of the GEO relay satellite, but not necessarily in view of the ground station [Gagnon et al., 2012].

Satellite Telemetry and Commanding RF Links

One of the major uses of communications links between satellites and ground stations is to support satellite commanding and receipt of satellite telemetry. These are *telemetry tracking and commanding* (TT&C) links. The telemetry downlinks have two broad purposes: to report on health and status of the spacecraft, and to downlink the acquired data or information of the payload – the whole reason for the satellite being in orbit.

Space Mission Support

Telemetry that downlinks payload data is *space mission support*. The payload performs the mission for which the satellite was created, launched, and operated. The downlink of the payload data is therefore a major part of supporting the mission of the payload, and hence the spacecraft. In many cases payload data volume is large. One example is imagery satellites in low Earth orbit (LEO). They take many images and store them in onboard memory; then when they are in view of a satellite ground station, they downlink all the stored image data. Because the visibility time of a LEO satellite to a ground station is short (on the order of 10 minutes), the images must be downlinked rapidly. A representative example is the following. We assume that a satellite has 100 images to downlink in 10 minutes, so it needs to downlink 10 images per minute. Modern commercial imaging satellites can resolve down to 30 cm in color. We further assume each image pixel has 16 bits each of red, green, and blue, for 48 bits per pixel. If the image size is 3 km square, then that image contains 9,000 pixels on a side, hence $9{,}000^2$ pixels; for the total number of bits:

$$\text{Image bit total} = 48 * 9{,}000^2 = 3.9 * 10^9, \tag{5-43}$$

which is 3.9 Gb per image. To downlink 10 images a minute, each image must be downlinked in 6 seconds, requiring a data rate of 648 Mbps. For reasonable BER, the link must have a sufficient SNR. If we assume we need a BER of 10^{-6}, then we can determine the needed SNR for a QPSK coded waveform using the following equation:

$$BER = \left(\frac{1}{2}\right) erfc\left(\sqrt{\frac{S}{2N}}\right) \tag{5-44}$$

The required SNR is 16, or 12 dB. Using the equation relating E_b/N_0 to SNR, and knowing that the bandwidth (BW) for QPSK is 1/2 the data rate we can calculate E_b:

$$E_b = SNR\left(\frac{N_0}{2}\right) = 16\left(\frac{kT}{2}\right) \tag{5-45}$$

If we assume the receiver noise temperature is 500 K, we then determine what the received energy per bit must be:

$$E_b = 16\left(\frac{6.9*10^{-21}}{2}\right) = 5.52*10^{-20} J \tag{5-46}$$

At a data rate of 65 Mbps, that is a received power of 3.58×10^{-12} W (–114 dBW). This is a very small amount of received power; however, to determine the transmit power, we need to include both the size of the receive antenna and the power loss due to range. First we calculate the required signal power density at the receive antenna. If we assume a ground station antenna radius of 5 m, the total antenna area is 78.5 m^2. Therefore, the required power density (P_d) at the receive antenna is 4.56×10^{-14} W/m^2. The reasonable maximum range for a LEO satellite is 3500 km. From the radar section we calculate the power density at the antenna using the following equation:

$$P_d = PG/4\pi r^2,$$

where P is the transmit power and G is the Tx antenna gain.

With this we can calculate PG:

$$PG = P_d 4\pi r^2 = (4.56 \times 10^{-14}) 4\pi (3.5*10^6)^2 = 7.0 \tag{5-47}$$

This implies that a fairly low-power transmitter (~1 W) with a small gain antenna (7 linear = 9 dB) will work with this downlink and meet reasonable performance requirements. If the ground receive antenna that this was designed to connect to was smaller (1 m diameter), then the antenna area for this smaller antenna is 100 times smaller, requiring a power density 100 times larger, which would require that a *PG* of 700. This downlink would require a combination of a reasonably high gain antenna and increased Tx power.

Similarly, if the downlink data rate is 10 times larger, the payload downlink requires a *PG* 10 times as large. We see that by knowing or estimating a few things about the payload, we can determine the rough transmit power and antenna gain necessary for the system. This information helps determine what kind of performance an SDA-focused system would need to detect the signal.

In the case of a system with a low gain Tx antenna, the beamwidth would be large and the SDA receiver location could be over a large area. In the case of a high gain antenna, the beamwidth would be very narrow, limiting the location of the SSA antenna to the main beam. It is possible to have the SSA antenna not in the main beam and hence receive the Tx antenna's sidelobes, with a much smaller power density. This would require a larger SSA antenna in order to have the same sensitivity as one in the main beam. Sidelobes are commonly 15-30 dB down from the main lobe.

If we are merely attempting to detect a signal, then long integration time is possible. We calculate the detection performance of an SDA antenna as a function of size and signal dwell time. We again assume a Rx noise temperature of 500 K. For the signal we just looked at, the BW will be 65/2 MHz or 32.5 MHz. The total noise power is k*T*B = $2.24*10^{-13}$ W. The total signal power is simply the antenna area times the power density, so the SNR is the signal power/noise power:

$$SNR = \frac{P_d(\pi(D/2))^2}{kTB} = \frac{4.56*10^{14}*\pi(D/2)^2 W}{2.24*10^{-13} W} \qquad (5\text{-}48)$$

This SNR equation assumes unit dwell time (1 s). If we wish to dwell longer to acquire more signal, the SNR will improve linearly with time. This is because the signal power is coherent and will increase as the square of time, while the noise power is incoherent and increases linearly with time.

$$SNR = \frac{tP_d(\pi(D/2))^2}{kTB} = \frac{t4.56*10^{14}*\pi(D/2)^2 W}{2.24*10^{-13} W} \qquad (5\text{-}49)$$

As Figure 5-26 shows, we can trade antenna size for dwell time to detect the signal. A moderately small antenna might detect the signal if we are willing to dwell for three seconds. If we want a very rapid detection in 1/4 second, we would need a 20-m antenna.

One of the other major functions of a communication link is satellite tracking. If the antenna is pointed accurately at the spacecraft, the signal power is maximum. As the pointing error increases (the line from the ground receiver to the satellite increases in angle and experiences antenna gain reduction), the signal power drops (Figure 5-27). The fact that the signal power varies depending on pointing accuracy lets us determine the time-varying angle to the spacecraft during a contact with the ground station. Referring back to Figures 5-14, 5-15, and 5-16, we can use monopulse to determine the angle error from the commanded pointing angle for the satellite.

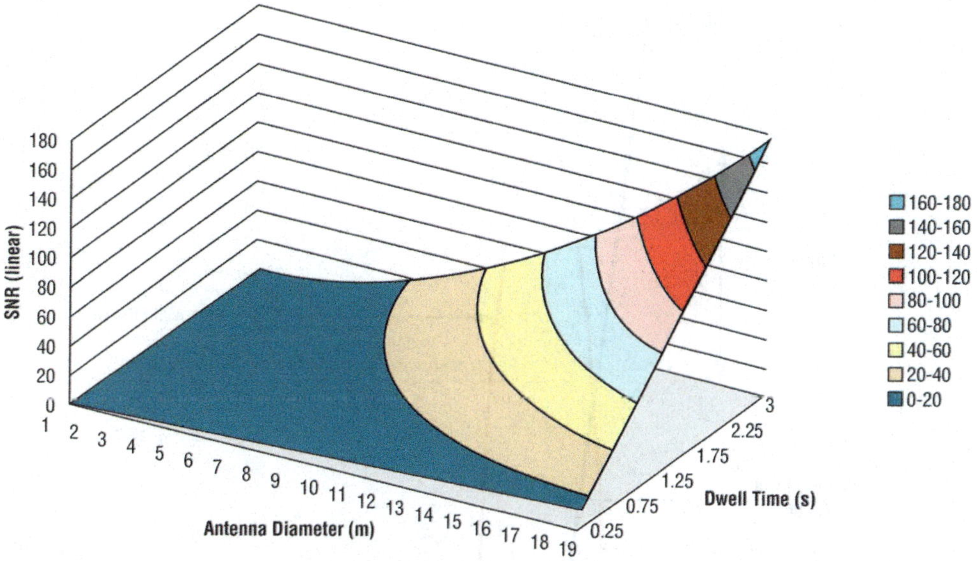

FIG. 5-26. SNR For an Example Link as a Function of Antenna Diameter and Dwell Time. The relationship between antenna diameter and dwell time and their effect on SNR. The SNR increases linearly with dwell time and with antenna area.

Telemetry and Communications (T&C)

In addition to payload data downlink, satellites also use a combination of an uplink to the spacecraft for commanding and downlink for spacecraft health and status telemetry. T&C links have comparatively low data rates, and nominally have both transmit and receive beams that cover a full 4π steradians (spherical) around the spacecraft. Thus, these links effectively have zero antenna gain. This pattern is typically achieved by two separate antennas on opposite sides of the spacecraft, each transmitting 2π steradians, with transmit power split evenly between them. This spherical coverage is to support attitude anomaly recovery, as well as orbit initialization operations that may have the spacecraft tumbling in an unknown and possibly uncontrolled manner. Without full spherical coverage, recovery from an anomalous situation that causes the satellite to lose attitude control is nearly impossible.

The lack of antenna gain on transmit results in a lower power density on the ground, based on the previous discussion of the PG relationship to the power density, with $G = 0$ db (or linear 1). Fortunately, telemetry uplinks and downlinks are commonly low data rate (10 kbps or less), and low bandwidth signals. Most T&C links are in S-band Space Ground Link System (SGLS) for DoD [Fitzgerald and Witherspoon, 1968], and Unified S-band (USB) for NASA and most commercial satellites worldwide. The SGLS downlink has three carriers, with carrier 1 having three sub-carriers for different communications types, with different modulation types. The telemetry downlink is sub-carrier 1 (1.024 MHz offset from carrier 1) using PCM modulation. From the referenced DTIC report (page 3-3):

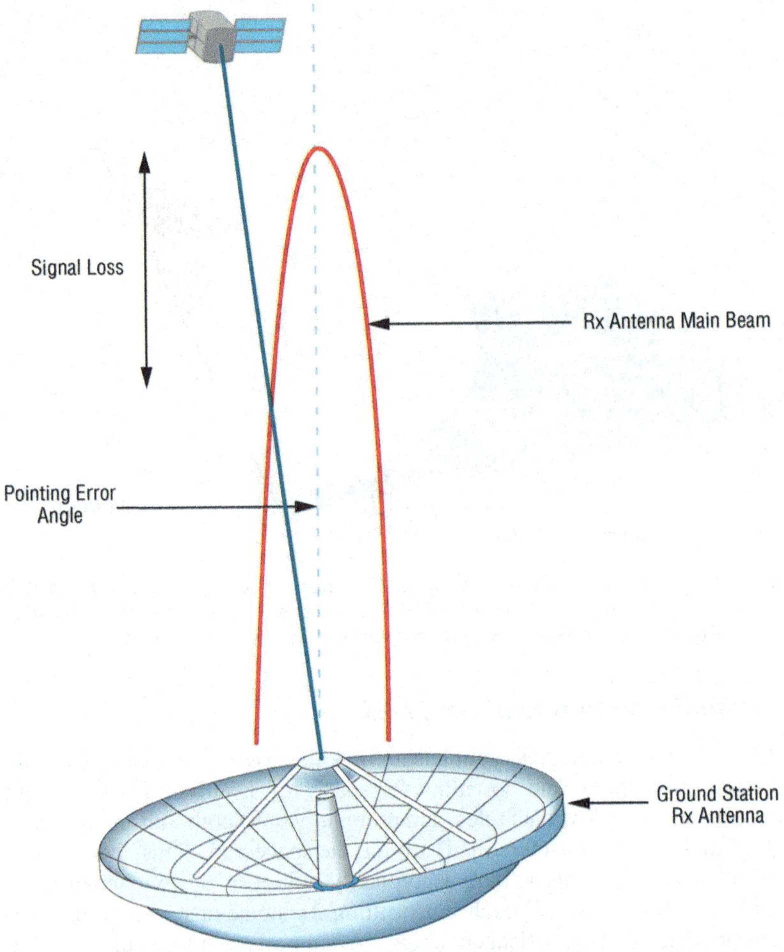

FIG. 5-27. Signal Loss Due to Antenna Pointing Error. The gain of the Rx antenna decreases as the difference between the antenna pointing direction and transmitting signal source increases.

> *"The 1.024-MHz subcarrier transmits low-rate PCM with bit rates from 7.8 bps through 128 kbps...These PCM data phase modulate the subcarrier +90 or −90 degrees in each bit interval."*

Therefore, the specific phase-coded modulation for typical T&C telemetry downlink is BPSK. We can use the previous analysis for BPSK modulation to determine reasonable transmit powers to achieve useable bit error rates on the downlink. This helps determine the necessary receiver capability to detect the telemetry (TLM) downlink.

Tracking

In addition to telemetry and commanding, we often add tracking. The combination of all three functions on a single up/downlink is *telemetry tracking and commanding* (TT&C). The SGLS system explicitly includes tracking as a function. The SGLS link adds a ranging pseudo random noise (PRN) signal to the uplink, which the SGLS receiver on the spacecraft then transmits on the downlink. The delay between the uplink and the downlink is then used to calculate the range to the spacecraft. This range data, combined with the angle information from the receive antenna monopulse angle tracking system, provides 3D information (range, azimuth, elevation) of the satellite position over time. From the referenced DTIC report (page 3-3):

> *"The PRN ranging signal is either a 1-Mbps PRN code or a 500 kHz clock signal that is extracted from the uplink carrier by the spacecraft receiver and inserted in the Carrier 1 downlink baseband. This signal…is required for range determination by the ground station."*

5.2.4 Space-based User Communications

Space-based communications most commonly serve to connect multiple terrestrial users. A space link provides the easiest way to link people across the globe, with minimal requirements on long-distance communications infrastructure (e.g., microwave relays, fiber-optic communications links, etc.). Long-distance mobile communications are one example, where due to the mobile user, land-lines are not possible (imagine a "cell-phone" with a long wire, plugged into your house), and RF communications would require potentially many distributed RF relays to cover the area where the mobile user operates. The cell phone industry has made a huge investment in very large-scale RF relays to provide mobile communications across large areas. For parts of the world that are sparsely populated, this is not a viable option, nor is it reasonable for poor countries. Additionally, for military operations, even if the region has a commercial cell phone system, this would not be a secure or reliable way to communicate. Use of geosynchronous or world-wide LEO communications constellations (e.g., Iridium) to relay communications from mobile users may be the only viable method of long-range communications.

There has also been great recent commercial interest in LEO-based high bandwidth communication to support digital data communications (including voice). Some of these recent commercial systems are: OneWeb, Telesat LEO, Space-X Starlight, and LEOsat. It is doubtful that all of these will become operational, but a few almost certainly will. They will have significantly higher transmit power to support higher data rates. Many of them also operate at high frequencies (e.g., K-band) in order to have sufficient bandwidth to support the higher data rates.

Overview of Possible RF Signals—What Could We Possibly See?

Significant SDA-relevant information is obtainable through RF links broadcast by satellites. We have two broad classes of RF signals transmitted by satellites: radar and communications. Position, navigation, and timing (PNT) signals like GPS are a specific type of communication system, broadcasting PNT relevant messages.

Communications signals are by far the most common signals broadcast from spacecraft. We previously described how to detect satellite communications signals, and monopulse makes fine angular tracking possible. This provides detection and tracking information, which is the first-order SSA information necessary to build and update an orbital catalog. Additionally, a spacecraft with an active RF downlink must be powered and is probably operational. If the downlink has a high data rate, then it likely has a directional antenna, and therefore must have a functional attitude control system to enable proper downlink pointing.

A ground-based receiver that can detect and angularly track satellites overhead must be able to scan the beam either by physically moving the antenna, or electronically steering the beam without physical motion. The antenna size combined with the receiver noise temp must be sufficient for detection. Also, the higher the SNR, the finer the monopulse angular accuracy that the system can achieve. The antenna and receiver must be sensitive to the satellite's downlink frequency. Having the receiver sensitive to S-band provides sensitivity to a very large number of satellites that use S-band for telemetry downlinks. First we need sufficient SNR to detect the signal. We assume that we have two cases: the received signal contains a satellite downlink plus noise, or the received signal contains only noise. This is called the binary hypothesis test. The standard detection approach uses Neyman-Pearson (NP) criteria to decide between the two hypotheses [Van Trees, 2001a]. NP enables calculation of a signal threshold where if the signal power is above the threshold, we declare signal detection, and below that threshold, we declare only noise. The lower the threshold, the higher the probability of detection (Pd); however this also results in a higher likelihood of a false detection, or false alarm (Pfa). The typical way to set a threshold is to determine the maximum allowable Pfa, and then the Pd can be calculated. We can plot the Pd against the Pfa for various receiver SNR values, yielding a *receiver operating curve* (ROC).

A reasonable SNR threshold for a tolerable Pfa is 12 dB. The details of the derivation of this threshold are beyond the scope of this chapter. We assume that the TLM downlink is 10 kbps, and that we require a BER of 10^{-6}. If the TLM downlink is BPSK modulation, we can use the BER equation (5-39) to calculate the required signal SNR, which comes out to 3.4 (5.3 dB). If we assume a receiver noise temperature of 500 K, we can calculate the required signal power necessary to achieve the required BER. The signal bandwidth required for 10 kbps using BPSK is 10 kHz. Therefore, the noise power (kTB) is 6.9×10^{-17} W. The required signal power is 2.37×10^{-16} W. If we assume that the satellite transmit power is one watt, the effective power density from a GEO satellite with a fully hemispherical antenna pattern at ~40,000 km range is 4.97×10^{-17} W/m^2. This requires a communications receive antenna of ~5 m^2. We now use this to calculate the required antenna size to achieve the desired detection probability SNR (12dB = 16). The needed antenna area is 22 m^2, or 5.3 m diameter. The angular resolution of this antenna at S-band (~2 GHz) where λ is 15 cm, is λ/D, or 0.028 radians. If we are using a monopulse system, the monopulse angular accuracy from a 12 dB SNR signal is 0.00707 radians (0.4 degrees). The system can achieve this angular accuracy measurement every second.

A longer signal dwell time results in a higher SNR and more accurate angular accuracy—assuming the satellite's motion is slow enough not to "blur," implying that the motion in the time measurement period is less than the angular measurement accuracy. In our example, the angular rate would need to be less than 0.4 degrees/s. If we are looking at a LEO satellite with a velocity of 7 km/s and an altitude of 1000 km, then the predicted angular rate at zenith is .007 radians/s (0.4 degrees). So our notional system will experience a small amount of blurring. To avoid the blurring we could decrease our signal integration time to 1/2 second, which would decrease our SNR to 8. That would decrease the angular resolution via monopulse to 0.57 degrees while the satellite would only traverse 0.2 degrees. The angular position measurements calculated every 1/2 second could then be fused in an orbital estimation algorithm (e.g., Kalman Filter) to determine the satellite's orbital parameters. Orbit determination is beyond the scope of this chapter.

We see that a moderate size tracking antenna with a monopulse feed is an effective SDA sensor to detect and track satellites via their telemetry downlink. The presence of a downlink provides information that the satellite is not "dead" but operational. This information is of significant importance to the overall SDA mission.

5.3 Human-based (Soft) Inputs

Of importance to space domain awareness, which is really "decision-making knowledge" for the domain, is context. Context cannot result solely from an interpretation of physics. Although most of the tracked resident space objects (RSOs) are defunct, there is a subset of the population that is actively human-controlled. Moreover, humans have valuable information to provide a system that is attempting to quantify, assess, and predict the behavior of objects in space.

We can learn much about an upcoming launch from human-based information, such as semantic corpora like "tweets" and other online media outlets. Detecting and tracking newly launched RSOs, for example, has proven to be challenging even for the most expert space surveyors. This would be easier if we could fuse or couple physics-based inputs (i.e., measurements from physical sensors such as radars and telescopes) with human-based inputs (e.g., an opinion or human-made observation). One such example is the use of so-called *two line elements* or TLEs. These are provided by USSTRATCOM but have no measure of precision or uncertainty associated with them. Are TLEs useful? They can be, as long as we treat them as a human-based input and not as a physical sensor measurement. Prior work by Delande et al. [2018] demonstrates how to fuse these two sources to achieve improved insight into RSO trajectories. It is beyond the scope of this section, but we refer the reader to Chapter 7 for more details. In essence, it is possible to model the uncertainty of an RSO as driven by systematic effects or ignorance (epistemic) rather than assuming that all uncertainty is driven by randomness (aleatory).

One of the salient challenges in exploiting human-based information is in properly curating and normalizing it. We see this in the following example:

> We assume an unmanned aerial vehicle (UAV) in an enclosed room with a ranging sensor in one corner, but this ranging sensor is unable to detect objects in the whole room. Moreover, the room contains obstacles such as furniture. We also assume that there are three windows allowing three humans to see into the whole room. Each human has a microphone to provide input into the UAV guidance,

navigation, and control, along with the physics-based (range) sensor. At a given moment, the UAV is on an almost assured path to colliding with a chair in the room. The range sensor can weakly observe the UAV, the first human says, "Watch out," the second says, "Ummmmm, that's going to hurt," and the third says, "You have an obstacle three meters ahead of you." The range sensor's input is straightforward to implement, but what about the three humans? Not only do they have information that could be useful to the UAV, but they say three very different things, which all attempt to convey the same predicted event, a collision. So, these human data need to have an associated measure of uncertainty, and we need to map the semantic inputs to a *lingua franca* or "normalize" them to minimize redundancy and confusion.

While radar and telescopes are examples of physics-based sensors, natural language processing (NLP) is an example of a human-based sensor, as its objects are semantic. Finding relevant information content in a physics-based sensor can be "trying to find a needle in a haystack" but finding relevant information content in human-based inputs can be "trying to find the needle in the needlestack."

5.4 Structured and Unstructured Information

For the purposes of this textbook, structured data are those in specific formats, often repeatable and adhering to standards (e.g., ISO). Their formatting makes them ideal because we can develop software and algorithms to consistently and appropriately interpret and exploit such data. As for unstructured data, we can think of these as serendipitous, not necessarily repeatable, or not even having any meta-data or ancillary information to provide meaningful context. For example, a query "scraping" the internet for semantic data regarding launches will result in some tweets, some online media articles, etc., and these do not have a common format or structure to them, making them more difficult to interpret. Moreover, as in the example of the UAV, many different things can allude to a common event or process, and this must also be inferred.

5.5 Standards, Calibration, and Meta Data

True and meaningful SDA cannot occur in the absence of properly engineered, modeled, and curated data. Something that makes this more straightforward is the use of and adherence to standards. For space we have quite a few, such as ISO, CCSDS, RINEX, EOSSA, and others. An interpreter can be developed for each, so that when we obtain data in any of them, it can be readily "ingested" and mapped or transformed to a common and appropriately labeled framework (e.g., a graph database). This can in turn be available for query by a variety of users. If users have to interpret each type of data to achieve their knowledge needs for space domain decision-making, this will not be timely, actionable or consistent. We need to make this interaction with "raw" information sources transparent to the user and allow the user to query some form of database or knowledge graph. This readily provides the information required to perform analytics or other tasks. Standards make this process simpler.

Information content only exists in the presence of differences or rates of change. How does one know whether or not a given information source is "credible"? Biases are commonplace among information sources, some of which are also noisy and corrupted by systematic or random errors. If all we have is a single clock we may not actually know the local time. However, if we have a time reference for comparison, we can quantify how accurate and precise our clock is. This underscores the importance of calibration within SDA. We can think of information as evidence, and we need to assess each piece of evidence for its value, relevance, and accuracy or precision. We must know how to interpret and "weigh" each source. This is the goal of calibration and is a critical element of meaningful data curation.

When interpreting information from a given source, it's not enough simply to have measurements absent context. For example, several thousand observations from an unknown telescope or radar are almost impossible to exploit. For the data to be actionable, we require details regarding where they were collected, the precision of the sensor, and perhaps other information such as how they were collected, assumptions made, etc. This context or ancillary information is *metadata*. The word "meta" means "beyond." Hence, metadata would be "beyond the information" but really is context or information regarding the information. Metadata are critical in giving the SDA practitioner guidance on how to use the data.

5.6 The Importance of "Independent Observations" and Big Data

How do we know if we have the most accurate clock in the world? We must have access to as many clocks as possible, of all types, to weigh all the independent evidence and find its "barycenter." We can think of this term as the "center of mass" of multi-source information or evidence, when all are appropriately weighed and compared. Only then can we know because we can then compare our clock to this information barycenter. How precise is our clock? Well, along with determining the information barycenter, we can also determine how scattered or spread apart these clocks are from each other, and this "variance" can provide insight into our clock's precision. Let's explore why independent observations are critical for SDA. We assume that we have a satellite near the GEO region become concerned if any other RSO comes within one kilometer of it. What if someone predicts that another RSO will be within 100 meters of our satellite this time tomorrow? Would we maneuver the satellite to avoid the risk? Suppose the person had arrived at that conclusion based on a single sensor. Would our decision change? We might want an independent source of information to either confirm or refute that hypothesis.

For evidence to be actionable, it must be corroborated, and this can only happen if there is at least one additional and independent source of information observing the same event or process. It's of little use to the global community if India makes one of their satellites explode on orbit, stating that the resultant debris will all re-enter Earth's atmosphere in just over one month and the United States says that they are tracking debris in higher orbits that will take decades to re-enter. We have two entities with opposing hypotheses and no ability to corroborate either of them publicly. This is a global problem: it is only a matter of time before two entities have a dispute to resolve. What body of evidence can satisfactorily resolve the dispute?

Moreover, having lots of telescope data to observe the debris would be welcome. Yet it is much more useful to have a lot of disparate data, for example, having several telescopes, radars, observers from the ground, observers from space, etc., all adding to the pool of evidence. A lot of data is just that, a lot of data. However, a lot of disparate data is *big data*. Why is big data important? Because it provides additional perspectives and "observability" into a common event or process via mutual information that a single type of data cannot provide due to limitations in the sensor itself. No single type of sensor can observe in all wavelengths and frequencies and resolutions. Is it more powerful to see a hazard or does hearing it and perhaps even smelling it also help make a decision? Disparate sensing allows for more quickly abducting the underlying system or behavior from the data. It helps eliminate wrong answers more quickly. We should always evaluate evidence not only for its ability to support a hypothesis but perhaps even more so for its ability to discard a hypothesis.

References

20th Space Control Squadron. 2009. USAF Fact Sheet, http://www.peterson.af.mil/About/Fact-Sheets/Display/Article/1060309/20th-space-control-squadron, accessed 1 Mar 2017.

Adamy, D. L. 2001. *EW 101: A First Course in Electronic Warfare*. London, UK: Artech House.

Arecibo Observatory. Available: http://www.naic.edu/ao/landing

ATIS Telecom Glossary. Available: http://www.atis.org/glossary/definition.aspx?id=5818

Chandrasekhar, S. 1946. On the radiative equilibrium of a stellar atmosphere. XI, Astrophysical Journal, 104, 110-32.

Cherniakov, M. 2008. *Bistatic Radar: Emerging Technology*. West Sussex, UK: John Wiley & Sons Ltd.

Cigal, J.C. 2002. "A Novel Spectroscopic Ellipsometer in the Infrared." PhD dissertation, Technical University of Eindhoven.

Collett, Edward. 1993. *Polarized Light, Fundamentals and Applications*. New York, NY: Edward, Marcel Dekker Press, p. 34.

Czerwimski, M. G. et al. 2014. "Development of the Haystack Ultrawideband Satellite Imaging Radar." Lincoln Laboratory Journal, vol. 21, no. 1. https://www.ll.mit.edu/publications/journal/pdf/vol21_no1/21_1_3_Czerwinski.pdf

Delande, E. et al. 2018. "Physics and Human-Based Information Fusion for Improved Resident Space Object Tracking." *Elsevier Advances in Space Research*, vol. 62, Issue 7, pp. 1800–1812. https://doi.org/10.1016/j.asr.2018.06.033. Last accessed 28 July 2017.

DoD Instruction 3100.11, October 24, 2016. "Management of Laser Illumination of Objects in Space." http://www.esd.whs.mil/DD/

Ewell, G. W. Defense Technical Information Center. 1972. "Investigation of Target Tracking Errors in Monopulse Tracking Radars." Defense Technical Information Center, AD748428, http://www.dtic.mil/dtic/tr/fulltext/u2/748428.pdf

Falchi, Fabio et al. June 2016. "The new world atlas of artificial night sky brightness." *Science Advances*, Vol 2, No 6. http://advances.sciencemag.org/content/2/6/e1600377.full

Fitzgerald, P. and J. Witherspoon. 15 November 1968. Space Ground Link Subsystem Ground Station System Analysis Summary Report." Defense Technical Information Center (DTIC), report number AD853122, http://www.dtic.mil/dtic/tr/fulltext/u2/853122.pdf

Freniere, Edward R. et al. 1999. "Polarization models for Monte Carlo ray tracing." Optical Design and Analysts Software, Proceedings of SPIE, Vol. 3780.

Gagnon, S. et al. Oct 2012. "Recent developments in satellite laser communications: Canadian context". Proc. *International Conference on Space Optical Systems and Applications (ICSOS)*. 1-2, Ajaccio, Corsica, France.

"Geosynchronous Space Situational Awareness Program (GSSAP) Fact Sheet." 15 April 2015. US Air Force Space Command. Accessed September 2016. http://www.afspc.af.mil/About-Us/Fact-Sheets/Display/Article/730802/geosynchronous-space-situational-awareness-program-gssap

Hayes, Eugene Nelson. 1963. "The Smithsonian's Satellite-tracking program: Its History and Organization, Part II," *Smithsonian Institution Annual Report for 1963*, pp. 331–357. http://siarchives.si.edu/collections/siris_sic_3188

http://astria.tacc.utexas.edu/AstriaGraph

http://astria.tacc.utexas.edu/compliance

Huygens, Christiann. 1690. *Traite de la Lumiere*. Leiden: Pieter van der Aa.

International Laser Ranging Service (ILRS). September 2016. "Satellite and Lunar Laser Ranging Contributions to Earth and Lunar Science." http://ilrs.gsfc.nasa.gov/science/scienceContributions/

Jim, Kevin T. C. et al. 11–13 September 2021. "Daytime GEO Satellite Characterization with the HANDS-IONS Camera." AIAA SPACE 2012 Conference & Exposition, Pasadena, California. AIAA 2012-5101. http://dx.doi.org/10.2514/6.2012-5101

Jones, R. A New Calculus for the Treatment of Optical Systems I and II. 1941. Opt. Soc. Am. 31,488 and 500.

Jones, R. A New Calculus for the Treatment of Optical Systes IV. 1942. J. Opt. Soc. Am. 32, 486.

Lacomme, P. et al. 2001. *Air and Spaceborne Radar Systems: An Introduction*. New York, NY: William Andrew Publishing.

Lederer, S. M. September 2016. "NASA's Orbital Debris Optical and IR Ground-based Observing Program: Utilizing the MCAT, UKIRT, and Magellan Telescopes." AMOS Technologies Conference, 20–23 September 2016. http://www.amostech.com/TechnicalPapers/2016/Orbital-Debris/Lederer.pdf

Lee, Chris H. et al. September 2016. "Infrared Photometry of GEO Spacecraft with WISE." AMOS Technologies Conference, 20–23 September 2016. http://www.amostech.com/TechnicalPapers/2016/Poster/Lee.pdf

Malus, E.L. 1809. Nouveau Bull. d. Sci., par la Soc. Philomatique, volume 1. Paris.

McCue, Gary A. et al. August 1971. "Optical characteristics of artificial satellites." *Planetary and Space Science*, Vol 19, Issue 8, pp 851–868. http://dx.doi.org/10.1016/0032-0633(71)90137-1

McGinn, J. W. Sept. 1966. "Thermal Noise in Amplitude Comparison Monopulse Systems." *IEEE Transactions on Aerospace and Electronic Systems*, vol. AES-2, no. 5, pp. 550–556.
doi: 10.1109/TAES.1966.4501931,
URL: http://ieeexplore.ieee.org/stamp/stamp.jsp?tp=&arnumber=4501931&isnumber=4501922

Molotov, I. et al. 2008. "International Scientific Optical Network for Space Debris Research." *Advances in Space Research* 41, pp 1022–1028. http://dx.doi.org/10.1016/j.asr.2007.04.048

Mosca, E. March 1969. "Angle Estimation in Amplitude Comparison Monopulse Systems." *IEEE Transactions on Aerospace and Electronic Systems*, vol. AES-5, no. 2, pp. 205–212. doi: 10.1109/TAES.1969.309906, URL: http://ieeexplore.ieee.org/stamp/stamp.jsp?tp=&arnumber=4103299&isnumber=4103287

Mueller, H. J. 1947. Opt. Soc. Am. 37 110.

Murray Krezan, J. et al. 2015. "GEO Collisional Risk Assessment Based on Analysis of NASA-WISE Data and Modeling." AMOS Technologies Conference. http://www.amostech.com/TechnicalPapers/2015/Orbital_Debris/Howard.pdf

NASA. 18 December 2013. https://www.jpl.nasa.gov/spaceimages/?search=DSN&category=#submit.

NASA. 3 November 2015. https://www.jpl.nasa.gov/spaceimages/?search=2015&category=Asteroids+and+Comets#submit.

"'Neighborhood Watch' supports Navy operations." 17 August 2016. 50th Space Wing Public Affairs. http://www.schriever.af.mil/News/Article-Display/Article/917972/neighborhood-watch-supports-navy-operations. Accessed February 25, 2017.

Otten, David D. et al. 1980. "GEODSS: Heavenly Chronicler." *Quest, New Technology at TRW Defense and Space Systems Group.*

Pellerin, C. Sept 2016. "DoD's Technological Superiority Depends on Out-Innovating Adversaries." https://www.defense.gov/News/Article/Article/947041/dods-technological-superiority-depends-on-out-innovating-adversaries

Phipps, Claude R. et al. 1 May 2012. "Removing orbital debris with lasers." Advances in Space Research, Vol 49, Issue 9, pp 1283–1300. https://doi.org/10.1016/j.asr.2012.02.003

Poincarè, Henri. 1892. *Theorie Mathematique de la Lumiere*. Gauthiers-Villars, Paris, Vol. 2, Chap 12.

Rhodes, D.R. 1980. *Introduction to Monopulse*. London, England: Artech House.

Rork, W. W. et al. 25 May 1982. "Ground-Based Electro-Optical Detection of Artificial Satellites in Daylight from Reflected Sunlight." MIT Lincoln Laboratory Project Report ETS-62. http://www.dtic.mil/dtic/tr/fulltext/u2/a117413.pdf

Scott, Robert (Lauchie) and Alex Ellery. July-August 2015. "An approach to ground based space surveillance of geostationary on-orbit servicing operations." *Acta Astronautica*, Vol 112, pp 56–68. http://dx.doi.org/10.1016/j.actaastro.2015.03.010

Shannon, C. E. July 1948. "A Mathematical Theory of Communication." *The Bell System Technical Journal*, vol. 27, no. 3, pp. 379–423. doi: 10.1002/j.1538-7305.1948.tb01338.x URL: http://ieeexplore.ieee.org/stamp/stamp.jsp?tp=&arnumber=6773024&isnumber=6773023

Sibert, D. et al. September 2016. "Collaborative Commercial Space Situational Awareness with ESpOC-Empowered Telescopes." *2013 Advanced Maui Optical and Space Surveillance Technologies Conference*, 20–23 September 2016, Maui, Hawaii. http://www.amostech.com/TechnicalPapers/2013/POSTER/KELSO.pdf, Accessed September 2016.

Skinner, Mark A. et al. 2015. "Comparison of Thermal IR and Visible Signatures of Graveyard Orbit Objects." 66th International Astronautical Congress, Jerusalem, Israel. IAC-15-A6.1.4.

Solar Spectrum. 2000. "2000 ASTM Standard Extraterrestrial Spectrum Reference E-490-00." http://rredc.nrel.gov/solar/spectra/am0/

Speicher, Andy. 2015. "Identification of Geostationary Satellites Using Polarization Data from Unresolved Images." PhD dissertation, The University of Denver.

Speicher, Andy et al. 17-21 August 2014. "Calibration of a system to collect visible-light polarization data for classification of geosynchronous satellites." *SPIE Optics + Photonics Symposium Proceedings*, San Diego, CA.

Stead, R.P. 1967. "An Investigation of Polarization Phenomena Produced by Space Objects." M.S. Thesis, GSP/PH/67-7, Air Force Institute of Technology, Wright-Patterson Air Force Base, Ohio.

Stokes, George Gabriel. 1852. "On the Composition and Resolution of Streams of Polarized Light from different Sources." Trans. Camb. Phil. Soc. 9, part III, pp 399–416. Reprinted in Mathematical and Physical Papers 3, 233-58 Cambridge University Press, 1901.

Stokes, Grant H. et al. 1998. "The Space-Based Visible Program." *Lincoln Laboratory Journal*, Vol 11, No 2, pp 205–238.

Van Trees, H. L. 2001a. *Detection, Estimation, and Modulation Theory Part I*, 19-46. New Jersey: John Wiley & Sons, Inc.

Van Trees, H. L. 2001b. *Optimum Array Processing: Part IV of Detection, Estimation, and Modulation Theory*. New Jersey: John Wiley & Sons, Inc.

Very Large Array. Available: https://public.nrao.edu/visit/very-large-array/

Chapter 6

Information Organization, Storage, and Management

Ramona L. Walls and Peter Jansen,
University of Arizona
Moriba Jah, *University of Texas at Austin*
Maria Esteva, Weijia Xu, and Amit Gupta, *University of Texas at Austin's Advanced Computing Center*
Nevan Simone, Shiva Iyer, *Oden Institute for Computational Engineering and Sciences at University of Texas at Austin*

Data form the basis of scientific knowledge and discovery, but data cannot reach their full potential unless they are well organized, understandable, and discoverable for future use, following good data management practices. Data science provides many tools for making the most of data, including semantics and natural language processing, as described in more detail in this chapter. Together, data management and data science generate and organize space domain awareness data and support information fusion for knowledge discovery.

6.1 Data Science and Data Management

Data and **information** are the raw material for scientific and applied research and form the basis of knowledge [Rowley, 2007]. As the speed and volume at which data and information are produced increase, new fields of research for dealing with **big data** and generating information from them have evolved in concert. This chapter highlights methods from data and information science that are useful for SDA, with a focus on how to apply them to the use cases of this book and the Space Domain Information Fusion (SDIF) paradigm. It includes information on how to integrate data science and data management techniques to support the SDIF model of SDA (Figure 6-1). As a paradigm for fusing information, SDIF depends on sophisticated data management techniques to support data generation, integration, queries, and decision-making. SDA practitioners draw on both well established and cutting-edge methods from data, information, and library sciences to organize, store, and manage their data and information.

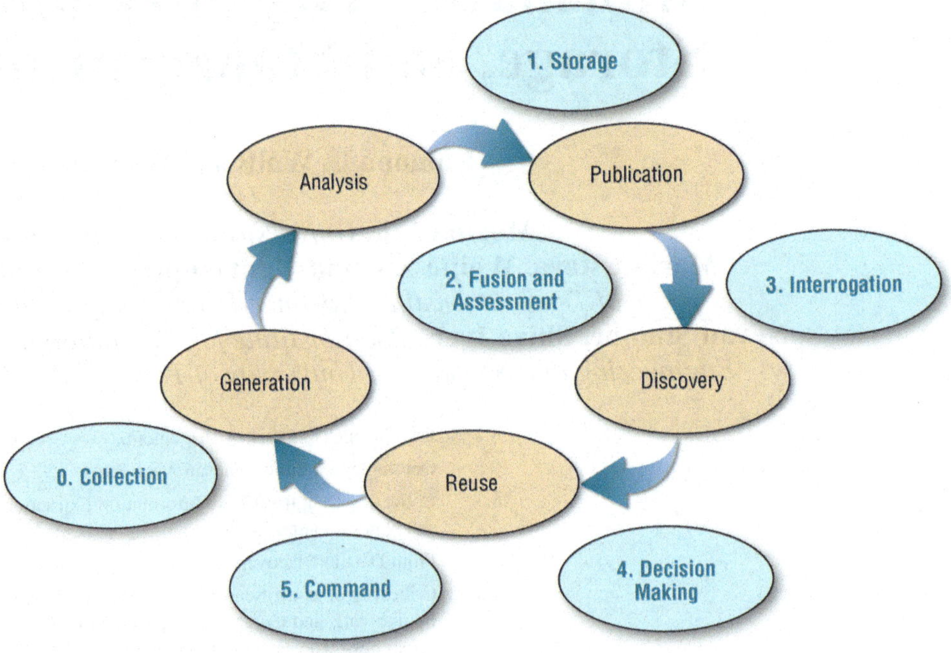

FIG. 6-1. The Different Levels of the SDIF Paradigm (Blue Outer Ovals, 0-5) and How They Map to the Data Life Cycle (Tan Ovals). Mapping is not one to one, because the SDIF is about more than data.

Data science uses various techniques to extract knowledge or insight from data, usually big data [Carmichael and Marron, 2018]. **Data management** studies how best to collect, store, preserve, and use data—how to get the most value from them [Briney, 2015]. This chapter is primarily about data management, but includes an introduction to some data science methods relevant to SDA, such as ontologies, natural language processing, and data mining. Data management and data science work together synergistically, and both are needed to meet the challenges of generating and fusing information for SDA.

One of the fundamental principles of data management is that data are not static, but rather follow a life cycle (tan ovals, Figure 6-1). The data life cycle represents a series of processes (e.g., analysis, discovery) for working with data, that, as the name suggests, are cyclical rather than linear. Individual researchers and practitioners rarely work with the same dataset throughout the whole cycle, and we have no guarantee that any individual dataset will follow the whole cycle. However, most researchers encounter each part of the data life cycle as part of their work. The key take-home of the data life cycle is to make researchers aware that their data may participate in processes outside their sphere of influence. Therefore, they must manage their data with consideration for how others may need to interact with them at some future time. Although some disciplines lack a culture of data publication (sometimes intentionally, for security reasons), the general trend in science is for funders to require that data be published and shared. The data life cycle reiterates the importance of managing the data generation, analysis, and publication processes to facilitate future data discovery, reuse, and generation.

The SDIF paradigm is also cyclical, with decisions and command driving future data collection and interrogation. However, the subtle distinction that SDIF deals with *information* rather than *data* means that the SDIF paradigm focuses on a different set of processes (blue ovals, Figure 6-1). Because SDA scientists and decision makers require data-based information, a strong correspondence exists between the data life cycle and the SDIF information cycle. "Collection" in SDIF corresponds closely to data generation, which also involves the reuse of existing data from external sources. Information storage happens at every point in the data life cycle.

In SDIF, "storage" deals specifically with how information is stored so it can support such other levels as interrogation or decision-making. Therefore, it corresponds most closely to data publication, as the two processes share similar goals of preserving data and making data accessible to others. The "fusion and assessment" level in SDIF cuts across many aspects of the data life cycle, including analysis, publication, and interrogation, so it appears inside the cycle in Figure 6-1. "Decision making" depends heavily on data discovery and is a type of data reuse: it uses data to generate information and knowledge, which guide decisions. Likewise, knowledge and decisions help guide commands, and so "command" is also a type of data reuse.

For scientists and those making evidence-based decisions, managing information requires the same tasks and resources as managing data. Table 6-1 describes some of the resources available for those tasks.

6.1.1 Basic Vocabulary for Data Fusion

Data fusion is an interdisciplinary process that combines techniques for managing big data with methods from data science for extracting knowledge from data, with **semantics** that define the meaning of data and metadata.

Metadata, or data about data, is essential for effective data management. It takes many forms, but is often specified as combinations of data properties (aka attributes) and values, with an optional unit field for a given attribute. A familiar example is the metadata attribute "author" associated to an article, with the author's name as a value. There is no strict distinction between metadata and data. As the saying goes, "One person's metadata is

TABLE 6-1. Resources for Data Management for SDA. These resources also apply to information management.

		Useful for				
Resource	**Main features or benefits**	**Generation**	**Analysis**	**Publication**	**Discovery**	**Reuse**
Metadata	"Data about data," supports reuse and discovery			x	x	x
Data Standards	Promote exchange, interoperability	x	x	x	x	x
Resource Description Framework (RDF)	World Wide Web Consortium (W3C) compliant, consists of "triples," flexible, many supporting technologies and tools			x	x	x
Controlled Vocabulary	Promotes consistent use of terminology			x	x	x
Thesaurus	Specified as simple knowledge organization system (SKOS), W3C compliant, hierarchical			x	x	x
Ontology	Specified as web ontology language (OWL), W3C compliant, hierarchical, built-in semantics			x	x	x

TABLE 6-1. Resources for Data Management for SDA. (Cont.) These resources also apply to information management.

Resource	Main features or benefits	Useful for: Generation	Analysis	Publication	Discovery	Reuse
Data Mining	Extracts previously unknown patterns and information from big data	x	x			x
Machine Learning	Teaches computers to learn to recognize and classify	x	x			x
Natural Language Processing (NLP)	Extracts information from large corpora of natural language text	x	x			x
Workflow Managers	Support automatic collection of provenance and other metadata	x	x	x	x	x
Databases	Aggregate and add value to data			x	x	x
Repositories	Long term preservation of data			x	x	x

another's data." To facilitate sharing of data across platforms and applications, metadata should be specified in machine readable languages such as **JSON** (Java Script Object Notation) or **JSON-LD** (JSON linked data), **RDF** (Resource Description Framework), or **XML** (eXtensible Markup Language), although XML is an older format. In RDF and similar formats, attributes are properties (relations) that link the data being described to values. Metadata, whether serialized (i.e., written) as RDF or not, can be made more widely understandable through the use of controlled vocabularies and standards.

Controlled vocabularies (CVs), thesauri, and ontologies are all collections of terms used in data management to improve data harmonization and discovery, but they vary greatly in their structure and level of complexity. CVs restrict the language used to describe things. Often a CV takes the form of an enumerated list specifying the allowable values for a particular metadata attribute. Because CVs are simple lists of terms, without identifiers and generally without definitions, their utility is restricted to the applications or communities that generated them. Thesauri and ontologies, in contrast, are developed specifically for use with the Internet, and therefore are better suited for data fusion.

A **thesaurus**, in the context of information management, is a vocabulary specified as a simple knowledge organization system (SKOS). Like a traditional thesaurus, a **SKOS** thesaurus contains sets of related terms, but SKOS is structured according to a standardized format. A SKOS vocabulary consists of a set of "concepts" with unique identifiers (see e.g., http://www.w3.org/2004/02/skos/core#Concept) related by properties such as "has exact match," "has broader match," or "has narrower match." SKOS vocabularies add very basic semantics or meaning. An **ontology** includes more complex semantics. An ontology (Box 4.1), in data and information science, is a set of terms—and relations among those terms—specified in a formal language, usually **OWL** (Web Ontology Language). Both SKOS and OWL ontologies feature extensively in **semantic web** applications and are standardized by the World Wide Web Consortium or W3C. Both can be serialized in RDF or related formats for use on the World Wide Web or in other data exchange applications. More information on using ontologies for SDA is in Section 6.2.2.

The primary source of data for SDA is observations of objects (e.g., optical, radar; see Chapter 5) and modeling outputs (see Chapter 7). However, to fully assess risks and threats in SDA, additional information is needed on causes and motivations. This information often comes from "soft" data sources like news articles, blog posts, and Twitter. **Natural language processing** provides methods for automated extraction of information from text. The ultimate goal of natural language processing is to be able to automatically read and perform human-level tasks with natural language, and the field has made tremendous progress in many tasks that support **machine reading**. For example, **Information Extraction (IE)**, allows researchers to mine large **corpora** (collections of documents) for a set of domain-specific patterns using a variety of statistical, machine learning, and hand-made rule-driven techniques.

Workflow managers or **workflow management systems** provide a way to set up, perform, and monitor a sequence of tasks. In data science, workflow managers generally manage complex analysis pipelines. But they also support data management by automating metadata generation and collection and tracking **provenance**, which is essential for reproducibility and estimations of trust and reliability. Workflows are represented graphically in the form of a **directed acyclic graph** or DAG. For big data analysis, workflow managers are integrated into **high performance computing** (HPC) systems used for compute tasks that require large amounts of memory or highly parallel processing (performing the same task many times on different processors). There are also HPC systems optimized to store and access large datasets.

By structuring related data based on a **knowledge model, databases** (DBs) provide single access points to related data generated by multiple sources. A DB's structure reflects an interpretation of the underlying relationships among different pieces of data, in order to increase the efficiency of data storage, query, and retrieval. The most common types of DBs are relational DBs and graph DBs. Relational DBs organize data using tables, records (rows in a table), and columns. Each row in a table has a unique key specified in a designated column, allowing that row to link to rows in other tables. A graph DB (Figure 6-2), in contrast, comprises one or more graphs consisting of a set of nodes (pieces of data) and edges (directional relationships among the data). Both types have different strengths and weaknesses, making them suitable for different applications. Graph DBs are well suited for use with ontologies, which are also a type of graph.

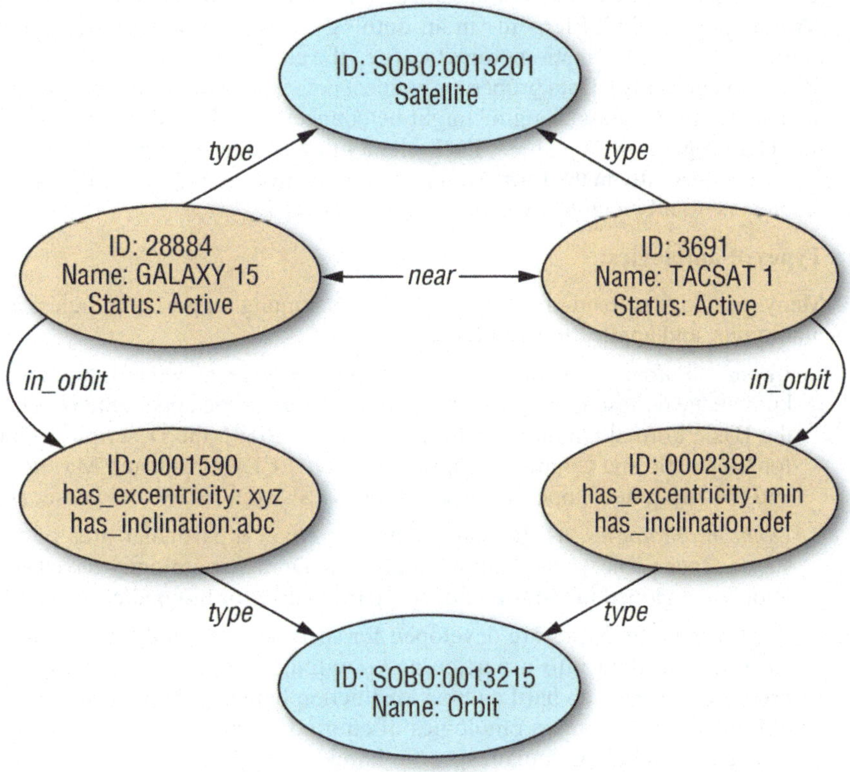

FIG. 6-2. **How Data is Organized in a Graph Database.** Each node (oval) has a unique identifier and can have various data properties (e.g., Name). Nodes may be unique individuals or types (also known as classes) and are related to one another by edges, also known as properties (arrows).

We can combine tools such as metadata, ontologies, and workflows into an integrative data management method, as described more in Section 6.4. The following are basic ontology terms:

- **Elements of an ontology:**

  The key elements of an ontology are classes, individuals, and properties (i.e., relations). Classes (also known as universals), are collections of all the individuals (also known as instances) that fit the definition of the class (see e.g., http://ontogenesis.knowledgeblog.org/1418, OWL spec). For example, Abraham Lincoln is an instance of the class "human being." Classes and individuals are related to one another by object properties that form the logical structure of the ontology. For example, an ontology might specify that every instance of the class "country" "has_citizen" some "human being," which at the instance level could be satisfied by asserting the "United States of America" (an instance of "country") "has_citizen" "Abraham Lincoln." Elements in an ontology may also be linked to additional information using annotation properties (i.e., the class label is generally stored as an annotation property). Data properties connect data values to instances. For example, an instance of the class "length" might be connected to the value "20 cm" through the data property "has_value." Each element of an ontology must have a unique identifier, generally in the form of an international resource identifier (IRI), which in practice is usually a uniform resource locator (URL).

- **Types of ontologies:**

  Many ontologists recognize the distinctions among upper ontologies, domain ontologies, and application ontologies.

  - *Upper ontologies* are domain neutral and provide a general framework for knowledge representation. Commonly used open source upper ontologies include the Basic Formal Ontology (BFO) [Arp et al., 2015], the Descriptive Ontology for Linguistic and Cognitive Engineering (DOLCE) [Borgo and Masolo, 2010], and the Suggested Upper Merged Ontology (SUMO) [Niles and Pease, 2001].
  - *Domain ontologies*, as the name suggests, cover a particular area of research such as space objects, astronomy, or physics. Domain ontologies have been most widely developed in the life sciences, particularly for biomedical research.
  - *Application ontologies* are developed for specific data handling purposes, such as ingesting data into a particular knowledge base or supporting an NLP program. There is no hard and fast distinction between domain and application ontologies, with domain ontologies often used for applications and application ontologies including domain specific knowledge. However, a key difference is that domain ontologies are generally reused widely, and thus are often developed as open, community efforts.

- **Reasoning and data:**

  A major benefit of ontologies is their ability to infer additional knowledge beyond what is explicitly recorded in a dataset. They do so by using a reasoner. A reasoner uses a given logical profile (e.g., DL in the Hermit reasoner [Glimm et al., 2010] or EL in the ELK reasoner [Kazakov et al., 2014]) to create new statements about classes or individuals. For a simple example, if an ontology asserts that all "presidents" are "human," and includes the assertion that "Abraham Lincoln" is an instance of the class "president," then a reasoner will infer that "Abraham Lincoln" is also a member of the class "human." In order to reason over data, it must first be converted to a form that the reasoner understands. One can "triplify" data by converting spreadsheets or other formats into RDF triples that can then be ingested into a graph DB, along with the ontology, to be reasoned over together.

6.2 Ontologies for Space Domain Awareness

Data for SDA come from disparate sources and in a variety of formats. Since it is often not possible to make a decision based on data from a single source, information fusion is an integral part of SDA and is only efficient if good data management practices are followed throughout the data lifecycle. No easy shortcut exists to the hard work of gathering, cleaning, and integrating large sets of data, but the data science and management practices mentioned above make the process traceable, repeatable, and (assuming limited resources) feasible.

6.2.1 Roles of Ontologies in Information Fusion

Ontologies and their accompanying semantics fulfill several roles in data fusion. At their most basic, term labels from ontologies, just as SKOS terminologies, serve as a controlled vocabulary, ensuring that researchers use the same words to describe the same entities. In this role, ontologies supply the values for metadata attributes. For example, if a metadata standard requires researchers to supply an "orbit name," values from an SDA ontology should serve as a pick list of names of different types of orbits. Ontologies can also handle synonymy. While it may be easy for a human to know that "LEO" and "low Earth orbit" refer to the same thing, a computer must be told this. If researchers are allowed to use free text strings in metadata values, ample evidence indicates that tens or even hundreds of different strings will end up referring to the same entity.

Using a controlled vocabulary solves an important hurdle in information fusion, but it is not without cost in terms of researcher effort (e.g., having to use a specialized tool to ensure the terminology matches and having to check definitions to make sure the chosen term is appropriate for the data). If the vocabulary comes from an ontology, the reward for the effort of using a controlled vocabulary increases greatly, because the meaning of metadata and data can be understood and reasoned over by a computer. Semantics allow a computer—depending on the complexity of the ontology—to infer facts that were not directly recorded in the data. For example, if a dataset contains parameters for an instance of a low Earth orbit, a reasoner and the appropriate ontology will be able to infer that the dataset also contains parameters about an Earth orbit and a planetary orbit. This inference is possible because the ontology specifies that low Earth orbit is a subclass of Earth orbit, which is a subclass of planetary orbit. While this inference is simple for a human, the advantage of having it coded in an ontology is that anyone who searches for data on planetary orbit parameters will be able to find this dataset, without any additional effort on the part of the data generator or curator.

Ontologies are vital for information fusion not only as CVs, however, but for providing the logical relations among entities that allow new knowledge to be generated. Returning to the LEO example, suppose that LEO was defined in the ontology based on the values of precise parameters (e.g., distance from earth < 2000 km). In that case, a reasoner would be able to infer that a dataset was about a LEO without a curator specifying an ontology annotation, simply based on the values in the dataset. Automatic classification (using data to specify the type of an instance) is a key use of ontologies in information fusion. Because SDA covers multiple domains, effective integration of SDA data will require ontologies for multiple domains (Figure 6-3), which could be independent, interoperable ontologies or modules in a single SDA ontology.

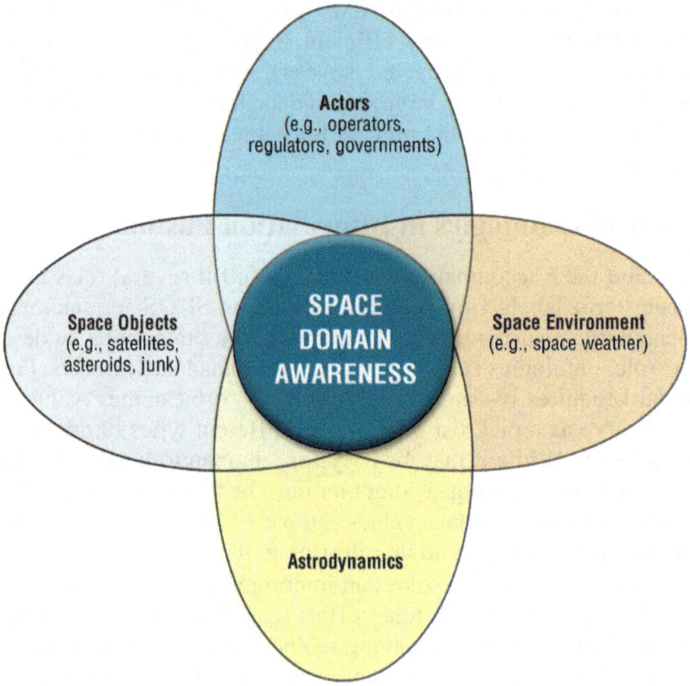

FIG. 6-3. SDA's Multiple Overlapping Domains Drive Ontology Requirements.

While domain ontologies provide the semantics needed to describe the space domain, they are not sufficient for handling data and tracking provenance. This task often goes to application ontologies custom built for specific data ingestion pipelines, but several general purpose data integration ontologies exist in other fields, and the semantics of those ontologies could be reused in SDA. At the highest level, the Provenance Ontology (PROV-O, https://www.w3.org/TR/prov-o/) is a W3C standard that expresses the PROV Data Model (https://www.w3.org/TR/2013/REC-prov-dm-20130430/) in OWL2. It provides a set of classes, properties, and restrictions that represent and interchange provenance information and can also be specialized to model provenance information for different applications and domains. The Semantic Sensor Network Ontology (SSN) (https://www.w3.org/TR/vocab-ssn/) was developed by the Earth science community [Cox, 2017]. It describes sensors and their observations, the procedures involved, the studied features of interest (i.e., the things being sensed), the samples used to do so, and the observed properties of the features of interest, as well as actuators. SSN supports a wide range of applications and use cases, including satellite imagery, large-scale scientific monitoring, industrial and household infrastructures, social sensing, citizen science, and observation-driven ontology engineering. In life sciences, the BioCollections Ontology [Walls et al., 2018] and Extensible Observation Ontology [Madin et al., 2007] support data integration for biodiversity and ecology, respectively. In additional to observation-based ontologies, open source ontologies exist for general domains like time or units (https://www.w3.org/TR/owl-time/, https://www.w3.org/2005/Incubator/ssn/ssnx/qu/qurec20.html), people and their relations to one another (http://www.foaf-project.org/), and the recently released Common Core Ontologies (https://github.com/CommonCoreOntology/CommonCoreOntologies).

6.2.2 Limitations of Ontologies

Ontologies written in OWL and the reasoners used with them are derived from the description logic (DL) family of knowledge representation languages [Baader et al., 2003; Hinzler et al., 2009]. While DL languages have advantages in decidability and efficiency, like all knowledge representation languages, they have limited expressivity. For efficiency, large ontologies often rely on the ELK reasoner (https://github.com/liveontologies/elk-reasoner/wiki, [Kazakov et al., 2014]). ELK uses the EL profile of DL, which greatly increases reasoning speed, and thus scalability, but it even further restricts the types of statements that can be made in an ontology (e.g., it allows only existential, not universal restrictions). Many knowledge representation limits can be circumvented with ontology engineering methods (e.g., reification overcomes the inability to express ternary relationships), but others are best dealt with by combining ontologies with other methods. For example, in OWL, it is only possible to say that an individual is or is not an instance of a class. To express uncertainty, error, or probability, it is better to use methods such as Bayesian or frequentist statistics combined with ontologies as part of an information fusion system (Chapter 7).

A major limitation currently encountered in ontology-based data systems is the lack of suitable domain ontologies or incompatibility among ontologies used to describe different domains. Although automatic ontology generation is an active area of research, is still far from ready for practical applications. If ontologies do not exist for a needed domain, the only current alternative is to develop them. Reuse of existing ontologies is generally a better practice than creating a new one, so minor incompatibilities should be overcome with mapping axioms whenever possible. In the life sciences community, the Open Biological and Biomedical Foundry has developed a set of principles that allow ontologies developed following those principles to easily interoperate with one another ('The OBO Foundry,' 2017). The SWEET family of ontologies, developed by the Earth science community, was intended to provide a suite of ontologies covering all domains relevant to Earth science. During a period of inactivity, the SWEET ontologies fell out of use, but they were recently adopted by the Earth Science Information Partners (ESIP), http://www.esipfed.org/ and may be further developed (https://github.com/ESIPFed/sweet).

6.2.3 Tools for Building and Using Ontologies

A number of tools exist to make it easier to build and use ontologies for information fusion, from graphical ontology editors to reasoners to graph database systems (Table 6-2). Because new tools and technologies for working with ontologies are constantly being developed, this table focuses on key resources that are relatively stable. Nonetheless, there are likely to be new tools and changes even before this book is published.

TABLE 6-2. Tools for Building and Using Ontologies, Organized by Category. These tools come from a wide range of disciplines.

Name	Description	URL/Reference
Ontology editors		
		See also https://www.w3.org/wiki/Ontology_editors
OWL API	A web service for loading, modifying, saving, and checking ontologies	https://owlcs.github.io/owlapi/
Protégé	Open source desktop and web tools for creating and editing ontologies	https://protege.stanford.edu/
TopBraid	Commercial tool for creating ontologies and linked data applications	https://www.topquadrant.com/tools
RDF		
SPARQL	Query language	https://www.w3.org/TR/rdf-sparql-query/
Index/search		
ElasticSearch	Indexing and search tools	https://www.elastic.co/
Lucene/SOLR	Open source search platform	http://lucene.apache.org/solr/
Graph database and RDF tools		
Apache Jena	Open source Java framework for building semantic web and linked data applications	https://jena.apache.org/
BlazeGraph	Open source graph database software	https://www.blazegraph.com/
Fuseki	A SPARQL server providing REST-style SPARQL over HTTP	https://jena.apache.org/documentation/fuseki2/

TABLE 6-2. Tools for Building and Using Ontologies, Organized by Category. (Cont.) These tools come from a wide range of disciplines.

Name	Description	URL/Reference
GraphDB	Graph database software with free and enterprise versions	http://graphdb.ontotext.com/
Neo4J	Suite of tools for building and editing graph databases. Uses its own language.	https://neo4j.com/

6.2.4 Gaps and Needs

The biggest gap inhibiting the use of ontologies for SDIF is the lack of ontologies and community-based standards in the SDA domain. Several mature vocabularies exist for astronomy (IVAO, http://www.ivoa.net/rdf/; UAT, http://astrothesaurus.org/), but they do not sufficiently cover space objects in the way needed to reason over SDA data. According to Cox et al. [2016], a set of interoperable ontologies for SDA was developed, but is not publicly available for use beyond the research group that developed them. The open source movement that has gained support in other scientific domains is less firmly established in SDA, perhaps due to security concerns and the large role played by private industry. A move toward open data and shared standards and ontologies, including infrastructure to support sharing and publishing relevant data, will go a long way toward facilitating SDA information fusion.

A shortage of machine readable "soft" data in SDA (i.e., data not directly derived from physical observations), also inhibits the advancement of SDA. The next section describes how NLP can help to overcome this limitation.

6.3 Collecting Data from Text: Information Extraction and Natural Language Processing

Although the primary data collection methodology of SDA is physical observations of spacecraft, the field is beginning to explore how the vast amount of textual information available on the web can augment physical observations, including identifying motivations, events of interest, or behavioral explanations for physical observations. For example, a news article describing how a spacecraft has suffered a loss of communication provides predictive information for how it might behave in the future if the loss of contact is sustained, or explanatory information if that spacecraft is already behaving erratically. Some of the information contained in text may already be available to physical observations—such as a news brief (or social media post) describing a maneuver to change the orbit of a given spacecraft. But the text may provide additional information including the behavioral motivation for that maneuver, from routine operation to moving the spacecraft into a graveyard orbit in preparation for decommissioning.

The ultimate goal of natural language processing is to be able to automatically read and perform human-level tasks with natural language. Today, "machine reading", the idea of having a computer read, understand, and act on arbitrary text, is still well beyond the capability of modern computational and cognitive science. However, the field has made tremendous progress in many supporting tasks, such as information extraction (IE), which allows us to mine large corpora for a set of domain-specific patterns (such as a loss of communication in satellites), using a variety of statistical, machine learning, and hand-made rule-driven techniques. In general, the field as a whole performs well when very large amounts of textual data are available for the domain in question and when the data contain a large amount of redundancy—that is, the same information is mentioned in different ways in the text. Cases that tend to perform poorly are tasks that face few mentions of relevant information in a large database of text, tasks that require deep or human-level understanding or inference capabilities to integrate or explain information, and situations that require world knowledge (the kind of knowledge a person usually acquires before the age of five) to solve (Figure 6-4).

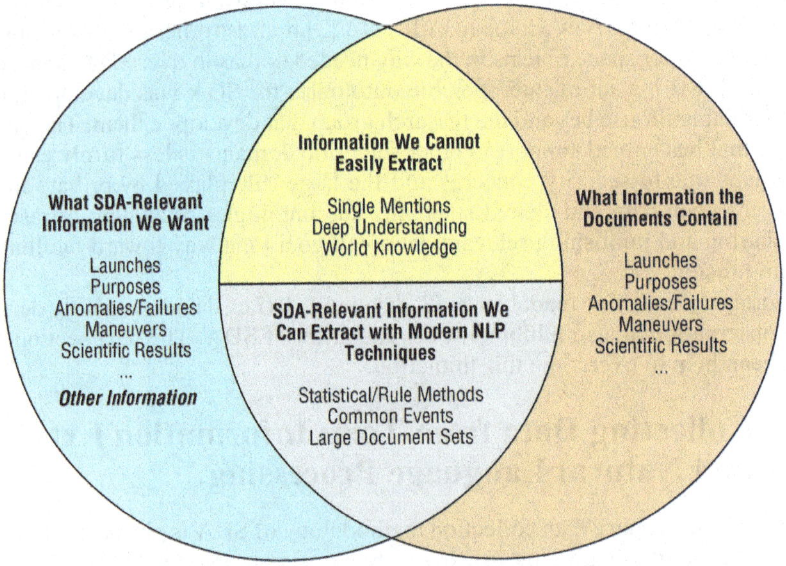

FIG. 6-4. **A Venn Diagram Illustrating the Overlap Between the Information Required for a Given SDA Task, and the Information Available in a Given Corpus.** The intersection represents task-specific information, and can be further subdivided into information that can and can not be easily automatically extracted using NLP techniques (SOBS).

6.3.1 Natural Language Processing Tasks and Pipelines

Natural language processing tasks typically exist on a continuum from high-level tasks that produce output directly useful to a human-level task (like answering a question), to low-level linguistic tasks—like part-of-speech tagging or syntactic parsing—that are critical for enabling higher-level tasks. The latter typically depend on processing from lower-level modules, and these modules (like syntactic parsing) may further depend on the output from even lower-level modules (such as tokenization). Because of this, natural language processing toolkits that integrate many different low-level processors are usually packaged together, reducing the burden on the developer from assembling each of these fundamental tools independently when they would like to build a system to perform high-level tasks, such as information extraction. In research and production settings, the Stanford CoreNLP toolkit (https://stanfordnlp.github.io/CoreNLP/) typically includes state-of-the-art components written in Java, while both spaCy (https://spacy.io) and the Natural Language Processing Toolkit (NLTK, https://www.nltk.org/) are common for Python developers. Table 6-3 includes examples of common natural language processing tasks, sorted from low-level to high-level, with examples of each from the space domain.

TABLE 6-3. Example Natural Language Processing Tasks, from Low-level to High-level. This table includes examples relevant to SDA.

Name	Description
Tokenization	Splitting a sentence string into distinct words and punctuation. e.g. SpaceX's Dragon spacecraft lifted off today
Lemmatization	Normalizing the specific instances of words into their lemmas, or the concepts that they represent. 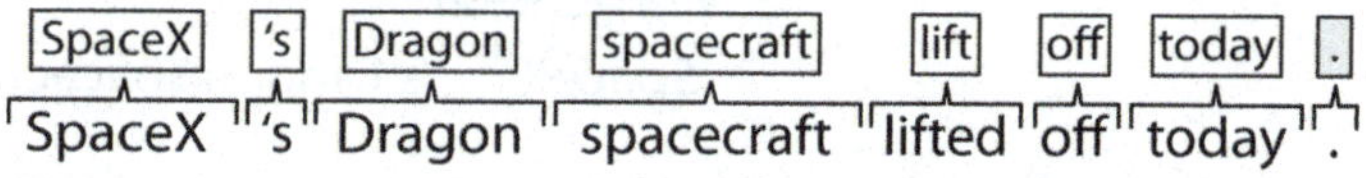
Part-of-speech Tagging	Identifying the part-of-speech category that each word in a sentence belongs to (e.g., noun, verb, adjective, determiner, possessive,...). 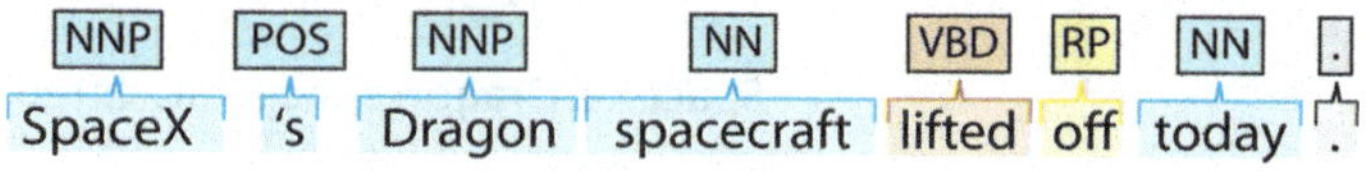
Syntactic Parsing	Parsing a sentence into a syntax tree, which consists of hierarchically organized high-level phrases (e.g., noun phrase, verb phrase) and lower-level terminal nodes. 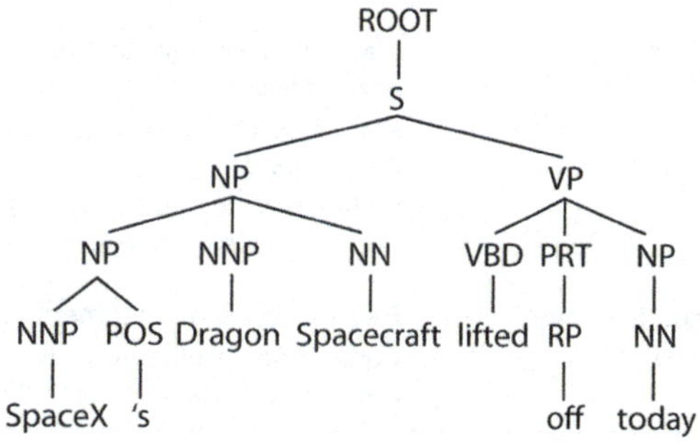

TABLE 6-3. Example Natural Language Processing Tasks, from Low-level to High-level. (Cont.) This table includes examples relevant to SDA.

<table>
<tr><th>Name</th><th>Description</th></tr>
<tr><td>Semantic Parsing</td><td>Identifying common semantic roles in a sentence.

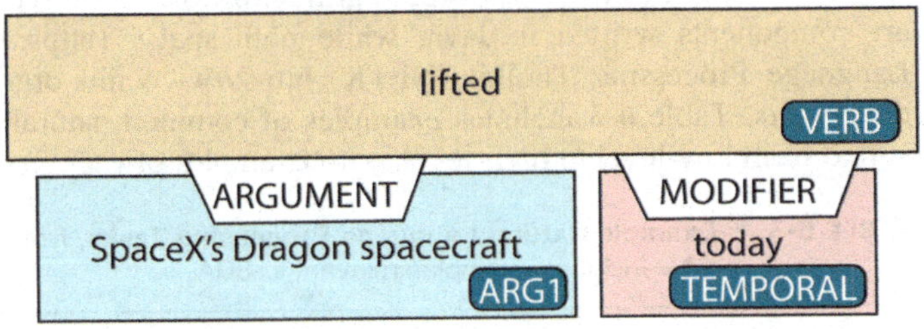
</td></tr>
<tr><td>Named Entity Recognition</td><td>Identifying common entities mentioned in text, such as people, organizations, locations, countries, or dates. These categories can be extended to identify domain-specific information, such as names of spacecraft, or names of launch vehicles.

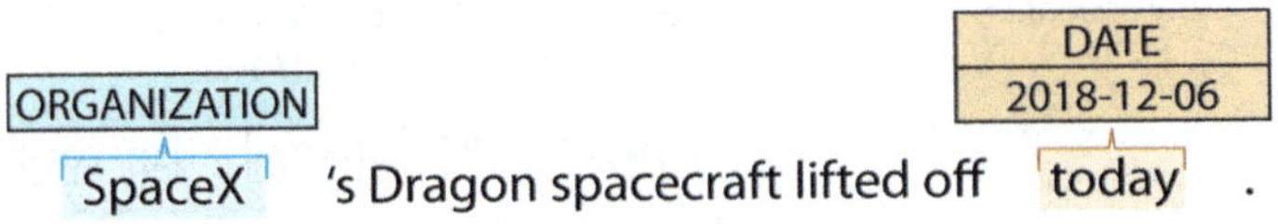
</td></tr>
<tr><td>Co-reference Resolution</td><td>Identifying which mentions refer to the same entity, especially in cases of ambiguity such as pronouns (e.g., "it").

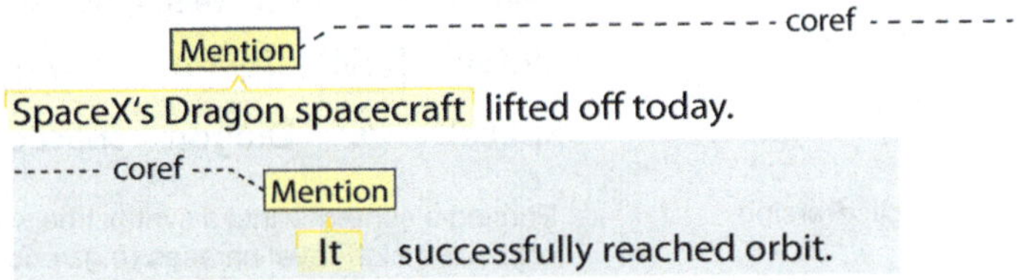

"It" -> refers to SpaceX's Dragon spacecraft]</td></tr>
<tr><td>Sentiment Analysis</td><td>Identifying whether text appears to be conveying something positive or negative.
e.g., "The launch vehicle failed to achieve orbit, resulting in the loss of the payload"
Failed and loss both have negative sentiment. This text likely describes a negative event.</td></tr>
<tr><td>Information Retrieval</td><td>Retrieve a list of documents that are relevant to a given topic.
For example, retrieving a list of documents that mention the launch of a specific spacecraft, sorted by relevance.</td></tr>
</table>

TABLE 6-3. Example Natural Language Processing Tasks, from Low-level to High-level. (Cont.) This table includes examples relevant to SDA.

Name	Description
Relation extraction	Relations (from open information extraction): SpaceX -> 's -> Dragon spacecraft Identifying relationships that exist between two entities in text. 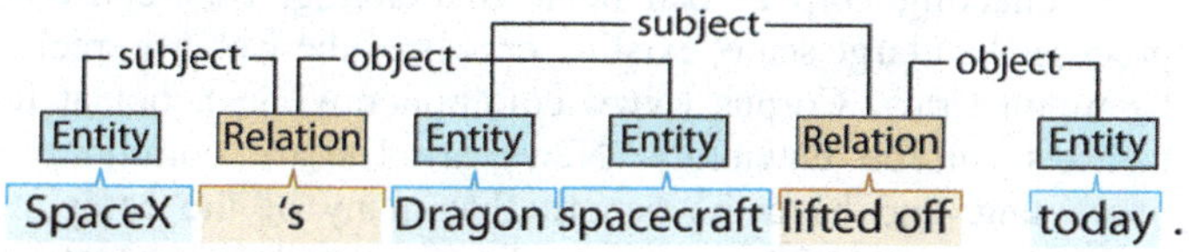 SpaceX's Dragon Spacecraft -> lifted off -> today
Information extraction	Identifying specific events in text. For example, extracting instances in a set of documents of spacecraft lifting off. Event: Spacecraft Liftoff Organization: SpaceX Spacecraft: Dragon Destination: orbit time: today
Question Answering (look-up)	Asking a natural language question that can be answered by looking up facts. For example, "What spacecraft have been launched in the last week?"
Question Answering (complex inference)	Asking a natural language question that requires combining facts through complex inference to arrive at an answer. For example, "What are the possible causes of the anomaly on Spacecraft X?"

6.3.2 Corpora and Collecting Raw Documents

Pragmatically, natural language processing operates on collections of documents, or corpora. Corpora are important both for the development process and for deployment. End-users typically want to build an information extraction system that can (for example) extract events of interest from a large, unseen, and continually expanding set of documents, such as news articles, web pages, or documents from some other source. Similarly, the development of such an extraction system normally requires having a large collection of documents to serve as training data to build and evaluate the performance of either a machine learning system, or a system of rules built by a human designer.

Different types of documents have different availabilities, depending both on the properties of the documents and the domains of interest. For example, tens of thousands of news articles are written in English and published to the web each week, making it possible to collect corpora of millions of news articles. Restricting the domain to only English articles that mention active satellites in orbit reduces this considerably, to approximately 50,000 articles available at the time of writing. Restricting this further to space-domain articles written in other languages can either increase or substantially decrease this number, causing some domains to be labeled **low-resource domains**. If a topic of interest is either rarely written about, or information on it is difficult to collect, it may be a low-resource domain. This is important because many modern machine learning techniques depend on having a large amount of training data available to achieve good performance, meaning that a system designer is likely to choose different methodologies depending on whether their domain has tens, thousands, or millions of relevant documents available.

Collecting corpora can be a challenging, time-consuming, and sometimes costly process. At a large scale, existing crawls of the web are freely available in the form of the Common Crawl Corpus (www.commoncrawl.org), but at hundreds of terabytes, these datasets can be unwieldly. News articles are commonly used in natural language processing, and have the benefit that many of the component systems (like syntactic parsers) are trained on corpora of news articles, suggesting the performance of these systems may be more accurate when using within-genre text. Collecting news articles is currently possible using the Microsoft Bing News Search API (https://azure.microsoft.com/en-us/services/cognitive-services/bing-news-search-api/). For the analysis below, a corpus of 61,409 articles was collected by independently searching for each of the approximately 1,700 satellite names listed in the Union of Concerned Scientists (UCS) Satellite Database, and retrieving (at most) the top 100 documents for each query.

6.3.3 Preliminary Content Analysis

The success of a given task depends not only on the number of documents available, but also on how often the information we want to use is mentioned in that text. Certain genres of text tend to describe some types of events while rarely describing others, independent of how often those events happen in the world. For example, spacecraft launches occur with a given frequency, and also tend to be reported by news articles, suggesting that the news genre of text may be a suitable source of documents to extract information about launch events.

Similarly, detailed information about a spacecraft's instrumentation specifications and capabilities are unlikely to be reported in news articles, suggesting another source of text may be more appropriate for this information, though such a source may be challenging to find.

One means of getting a sense of whether a corpus is likely to contain content of interest is to perform a *frequency analysis*, counting the number of documents in the corpus that contain a certain linguistic pattern related to the events that one would like to extract. A common surface pattern that minimizes the linguistic processing involved is to look for the frequency of specific **n-grams**. An n-gram is the name given to a series of *n* words (or lemmas) in a definite order, where a *unigram* contains a single word, *bigrams* contain two words, *trigrams* contain three words, and so forth. Such a frequency analysis for a number of generic SDA unigrams (e.g., launch, failure, maneuver, anomaly) and bigrams (launch failure, power failure, engine failure,...) in the satellite news corpus is in Table 6-4.

Examining the unigram frequencies, what we observe is that nearly 52% of the news articles in the satellite news corpus mention the word "launch" at some point in the article. This suggests that a large number of the news articles may either be about spacecraft launches, or include historical information about a launch when discussing a given spacecraft.

TABLE 6-4. Frequency of Example Unigram and Bigram Patterns Found in the News Corpus. Bigrams are significantly less likely to occur than unigrams.

N-gram	Frequency (Number of Documents)	Frequency (Proportion of Documents)
Total documents (news articles) collected		61,409
launch	31,736	52%
launch failure	640	1%
failure	4,784	8%
power failure	194	0.3%
engine failure	132	0.2%
system failure	125	0.2%
partial failure	99	0.2%
maneuver	1411	2%
anomaly	1321	2%
lost contact	220	0.4%
lost control	208	0.3%

We also observe that almost 8% of the articles mention the word "failure," suggesting that nearly one tenth of news articles about satellites discuss failures of some kind.

N-grams suffer from *sparsity*, or the notion that as we increase the length of the string we're matching, the chance of observing that string generally decreases dramatically. As the size of the n-grams in the frequency analysis for Table 6-4 increases from unigrams to bigrams, the number of observations also decreases. Whereas "failure" was observed in 8% of articles, the bigram "power failure" was observed in only 0.3% of articles, i.e., wherever the word "failure" was observed, in only 4% of cases did the word "power" precede it. Similarly, "engine failure", "system failure", and "partial failure" were each observed in about 0.2% of the news articles in this corpus, demonstrating that moving from a coarse-grained analysis of considering all mentions of *failure* to more precise kinds of failure (like *power failures*, or *engine failures*) greatly reduces the number of times these specific cases will be observed in a corpus. Because the number of training examples available to a machine learning model is often a concern, moving from a unigram (e.g., "failure") to a bigram (e.g., "power failure") may not be sufficient to acquire a hypothetical pattern to extract these kinds of events, depending on the specific information extraction task to be learned. In general, except for corpora of extreme size, n-grams longer than bigrams often suffer from sparsity unless they capture common domain-specific noun phrases or other high-frequency patterns involving function words (e.g., "on the <word>").

6.3.4 Extracting Events of Interest

The above content analysis suggests that news articles frequently discuss at least a number of SDA events. Consider the case of being interested in launch events. A next likely step would be the desire to automatically monitor a news feed for articles that mention launches, and to extract information from these articles, such as *what satellite was launched*, and *who launched the satellite*, to serve as subjects and objects in the triples of an ontology-based knowledge base—e.g.,*<organization>* launched *<spacecraft>*. Intuitively, searching for **specific patterns** of words or surface patterns in the text seems like an attractive alternative. Consider the case of building a **named entity (NE)** recognizer able to identify space-domain *organizations* and *spacecraft*. The pattern:

<NE:organization> launched <NE:spacecraft> (1)

may be sufficient to extract a large number of these events, such as extracting the tuple *<United States> launched <Explorer 1>* from the sentence "The United States launched the Explorer 1 satellite on January 31st, 1958". Unfortunately, surface form extraction patterns tend to be brittle to both minor and major variations in the presentation of the same knowledge. Consider the following sentences, which all express the same knowledge, but with different surface forms:

"The United States launched the Explorer 1 satellite on January 31st, 1958." (2)

"The United States, through a program called Project Orbiter, launched the Explorer 1 satellite on January 31st, 1958." (3)

"Explorer 1 was a satellite launched by the United States in 1958." (4)

"1958 marked the beginning of the US satellite program with the successful launch of the Explorer 1 satellite." (5)

The surface extraction pattern <NE:organization> launched <NE:spacecraft> would fail to identify this relation in all but the first example, because additional content words occur between the arguments of the extraction pattern, or the arguments are reversed, as in (4). This means that surface patterns generally suffer from low **recall**, or an inability to extract specific knowledge from a corpus, even when that knowledge is frequently expressed, because of variations in the surface form of the expression of the same ideas. Many techniques and systems have been developed to make extraction systems more robust to surface form variations, to increase recall. Often recall is improved by performing deeper linguistic processing on the input sentences, by subjecting them to deeper syntactic or semantic processing, then extracting patterns using these more complex syntactic or semantic representations. The two methods examined below make use of traversing a graph of relations to capture long-distance relationships between words in text. They are thus more robust than searching for adjacent words.

6.3.5 Extracting Relations Using Open Information Extraction (OpenIE)

Open information extraction (OpenIE) represents a family of information extraction algorithms for extracting *<entity, relation, entity>* triples without having to rely on first constructing domain-specific tools like named entity recognizers. OpenIE has the benefit of being able to more robustly extract relation triples than surface forms, allowing triples to be extracted from even very large corpora. The triples extracted from the above examples include:

<The United States, *launched*, the Explorer 1 Satellite> (2)
<a program, called, Project Orbiter> (3A)
<The United States, *launched*, the Explorer 1 Satellite> (3B)
<Explorer 1, was, a satellite launched by the United states in 1958> (4A)
<a satellite, *launched*, by the United States> (4B)
<1958, marked, the beginning of the US satellite program with the successful launch of the Explorer 1 satellite> (5)

In this case, patterns from two of the four variations of our sentences were correctly extracted supplying the knowledge in (2) and (3B), increasing recall over the surface pattern. In the final two cases, the extracted triple was not the triple of interest for our task of identifying launches, with the relation *was* in (4A), and *marked* in (5). These mismatches highlight a potential limitation of this approach, but also a possible benefit, if the additional knowledge is useful. When large amounts of redundant data are available, such as multiple news articles mentioning the same event using different linguistic patterns, this may help improve task performance by allowing the event to be successfully extracted from at least one document.

6.3.6 Extracting Known Events Using Rule-Based Extraction

While using tools intended for **open-domain extraction** can reduce development time, these methods may be suboptimal both in terms of precision and recall. When a specific set of relations or events of interest is known ahead of time, it may be possible to observe some subset of these event mentions in a training corpus, and hand-craft rules to extract these events. Typically these rules couple manually-identified trigger phrases that tend to signify certain events, with syntactic extraction patterns that search for arguments to the trigger phrase using domain-specific named-entity recognizers. The development cost of a human authoring rules for a given domain is higher than fully-automatic methods, but with the benefit that these methods tend to achieve higher precision, even in scenarios with little data available.

Consider a hypothetical user interested in automatically monitoring for events where organizations lose contact with spacecraft, such as in the following example from a news article in spaceflightinsider.com:

"The Israeli company Spacecom Communications Ltd. announced on Saturday, Nov. 21, 2015, that it had lost contact with its Amos-5 satellite."

News Article: spaceflightinsider.com, November 23rd, 2015

Processing this text through an NLP pipeline with named entity recognition, co-reference resolution, and dependency parsing, might yield the results shown in Figure 6-5.

Rule generation usually begins by identifying a set of *trigger phrases* that signify a given event may be present in text. For a loss-of-contact event, these trigger phrases might include "lost contact" (example in Figure 6-5), "lost communications," "loss of contact," and other related forms. By specifying the relation in terms of dependency relations, the

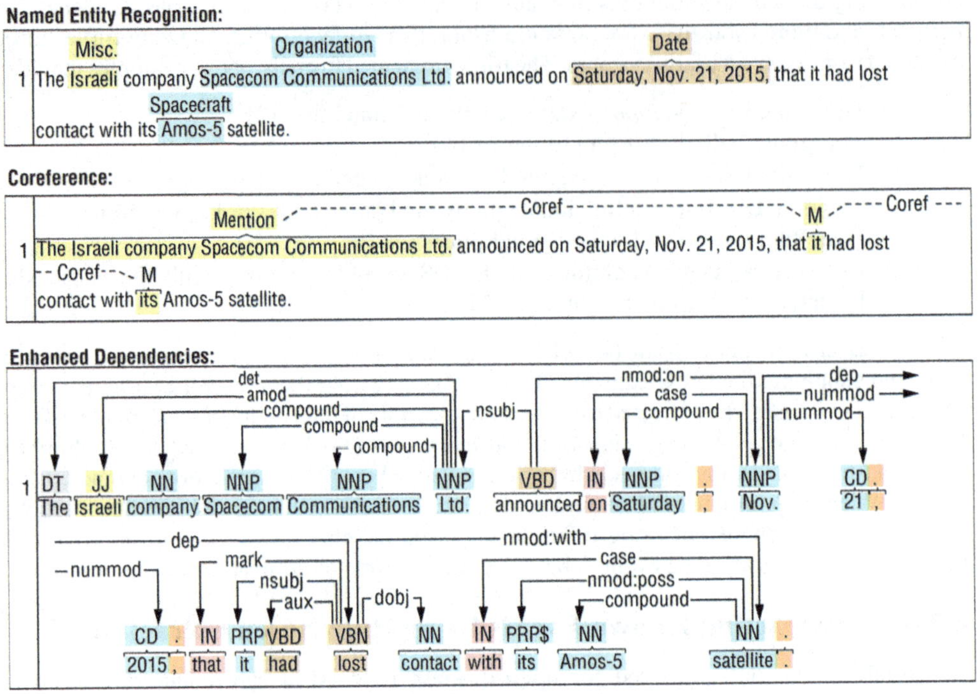

Fig. 6-5. **Partial Output From an NLP Pipeline that Serves as Input for Information Extraction.** This pipeline example uses the Stanford CoreNLP (https://stanfordnlp.github.io/CoreNLP/), with hypothetical named entity recognition augmented for the SDA domain to identify spacecraft.

rule can capture long-distance arguments for a given event while being somewhat robust to surface-form changes. In this example, the *nominal subject (nsubj)* dependency of the "lost contact" event is the entity who lost contact—in this case, "it," which refers to the organization "Spacecom Communications Ltd." Similarly, the *nominal modifier (nmod:with)* dependency points to the entity that contact was lost with—here, the spacecraft "Amos-5." This event, implemented in the Open Domain INformer (ODIN) rule-based extraction language, might take the form:

Extracted Event:

Event: Loss of Contact
Organization: "Spacecom Communications Ltd."
Spacecraft: "Amos-5"

Rule:

- name: loss_of_contact
label: LossContact
pattern: |

trigger = [lemma=/loss contact/]
company: Organization = nsubj
ship: Spacecraft = nmod: with}

6.3.7 Learning Extraction Patterns Automatically With Machine Learning

While it is possible to use open-domain tools to extract relations/triples, or build high-precision hand-crafted rules to extract particular event patterns, it is also possible to use machine learning techniques to automatically learn extraction patterns that tend to express given events. Using a set of pre-extracted training examples, such as *<Spacecom Communications Ltd, lost contact, Amos-5>*, and *<NASA, lost contact, IMAGE>* these methods can search a training corpus for all the instances where, e.g., *Spacecom, lost contact*, and *Amos-5* are mentioned in the same sentence. They then observe the dependency or other patterns between these named entities, and learn to apply them to extract other unseen instances of the loss-of-contact event, such as that <ISRO, lost contact, GSAT-6A>. When large amounts of training data are available, these methods can achieve high recall, but may *overfit* to the training examples, causing erroneous relations to be extracted, reducing precision.

6.4 From Data to Discovery

Data for SDA come from sensor networks, astrodynamic observations and models, and text sources with natural language processing. To "fuse" all this data, generate knowledge, and support decision-making, the resources described above and in other chapters of this book must be combined into a coherent information system. Such a system must include tools processing raw data, analyzing data of all sorts, ingesting data into a knowledge base, querying the knowledge base, and making decisions based on query results. Walls et al. [2016] describe one such system. A high-level workflow for applying the SDIF paradigm to SDA is shown in Figure 6-6, encompassing four basic steps: data collection, data analysis, data integration, and knowledge generation. These are typical steps in workflows using integrated data, and we must bear in mind that each step may be performed by separate and multiple actors.

Integrating good data practices early—as soon as data are generated or collected—ensures that data will be reusable and discoverable. This can be done by building metadata collection into the data collection process and using standard, non-proprietary data formats. Clean-up and pre-processing steps often happen concurrently with data collection, and these activities must be well documented, including: noting if outliers were removed and the removal method, disturbances that occurred during the data collection process and the times of the disturbances, and parameters used in pre-processing steps. Scientists are of course used to collecting such information and even publishing it in methods sections, but it is necessary to maintain that information as metadata associated to the original and processed data. Such metadata are crucial for tracing data **provenance**.

Analysis is the most familiar data-handling step for many scientists, but as the size and complexity of data analyses become larger, data management becomes an ever more important

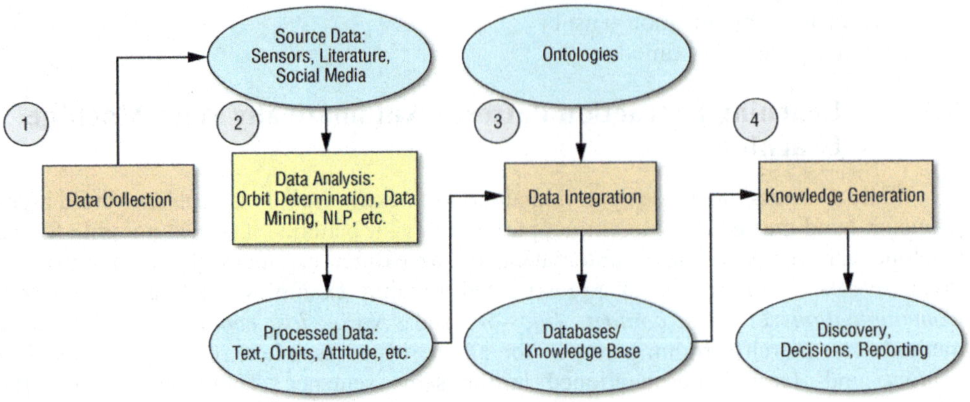

FIG. 6-6. **A High-level Workflow for SDA Information Fusion.** Source data are collected from multiple hard (physics-based) and soft (human-based) sources (1) and serve as input to analyses (2). The use of automated workflows in step 2 assures that the processed data are correctly annotated with metadata keyed to ontologies, facilitating data fusion, or ingestions into a graph DB or knowledge base (3). Information is extracted from the knowledge base and used in knowledge generation processes such as modeling or Bayesian Network analysis (4), to produce useful information that guides decision making and command and control.

part of analysis. Planning for data reuse (even when you are the only re-user) means planning for metadata and provenance capture as part of analysis. Ontologies, thesauri, and controlled vocabularies are all valuable for data integration, but annotating data with standard terminology post-facto is time-consuming and error prone. It is better to include standardized metadata during data collection and analysis to facilitate later data integration. Open source console or notebook-based tools such as R-studio (https://www.rstudio.com/) and Jupyter (http://jupyter.org/) provide a way for scientists to record their methods as part of analysis, as do commercially available laboratory information systems (LIMS). Cloud analysis platforms such as the CyVerse Discovery Environment (https://www.cyverse.org/) are useful to scientists, in part because they automatically record provenance metadata—inputs, outputs, and parameters—for all analyses. Workflow managers [Talia, 203] such as Pegasus (https://pegasus.isi.edu/), JMS (http://dx.doi.org/10.1371/journal.pone.0134273), Taverna (www.taverna.org.uk/), or Kepler (https://kepler-project.org/) can also help capture important provenance metadata. If automated workflows are set up to use ontology-based metadata, then data integration is further facilitated.

Data integration is the process of combining data from multiple sources to provide a unified view of the data. Likewise, information fusion combines multiple types of information with the goal of discovering new knowledge, but no standard procedures for data integration or information fusion yet exist. There are two main challenges: discovering appropriate data or information to integrate (which includes determining if discovered data are fit for the purpose) and getting all of the data into compatible formats. Ontologies, CVs, and metadata—if they were included during data collection and analysis—help overcome these challenges by providing semantic context, allowing the data's utility to be evaluated automatically by computers. In the absence of metadata, techniques such as data mining, machine learning, and NLP contribute to data integration, but these methods rely on large datasets and often require manual annotation of training datasets. Integrated data are stored in one or more databases, where they are available for knowledge generation. Such a semantically encoded database is known as a **knowledge base**.

The goal of an SDIF workflow is to generate knowledge about the space domain. Although this task has traditionally been a human activity, scientists increasingly seek to automate knowledge generation through processes such as machine learning and artificial intelligence. Human knowledge may be based on familiarity and understanding, which arise from facts, experience, or any number of factors. Scientific knowledge is a specialized type of human knowledge based on data, but is also influenced by familiarity (e.g., trust in the researcher who generated the data) and experience (e.g., how often similar data have been recorded). Therefore, the creation of scientific knowledge requires decisions about the reliability of a data source or repeatability of an experiment. Tracking provenance is key to weighing different lines of evidence for decision-making because in science something is "known" only after multiple lines of evidence have supported the same conclusion, and there is always the possibility that future evidence will change what is known. Here the importance of looking at data as part of a life cycle comes into play because knowledge can feed back into the cycle to generate new data, leading to revised knowledge.

6.5 Criticality of Multi-Source Information Modeling, Engineering, and Curation for Space Domain Awareness: The Case for ASTRIA

To make the best decisions as possible regarding the space domain, we must establish a robust common space domain awareness operating picture that fully leverages the information (opinions and beliefs) provided to it from a variety of sources, hard (physics based) and soft (human based). ASTRIAGraph was initially funded by the U.S. Federal Aviation Administration (FAA) to develop a framework that could serve as a demonstration of a common operating picture (COP). Creating a COP is challenging because of conflicting objectives and constraints. This is to be expected. The definition of common operational pictures varies but we use the following Wikipedia definition:

> A common operational picture (COP) is a single identical display of relevant (operational) information (e.g., position of own troops and enemy troops, position and status of important infrastructure such as bridges, roads) shared by more than one Command.

A COP offers a standard view, thereby providing information that enables decision-makers to make effective, consistent, and timely decisions. Compiling data from multiple sources and disseminating consistent and harmonized information ensures that all users have the same understanding and awareness of the complete situation and circumstances when conducting space operations.

One could also consider a user-defined operational picture (UDOP), in which the user selects what information to include or exclude. As such, a given user would only be visualizing information for specific needs, as opposed to the COP being suitable for situational awareness of the big picture and making decisions from that information.

What follows are general concepts taken from the Big Data Science and Analytics community that are recommended for use as a foundation for space domain awareness and creating a common/user defined operating picture. We also leverage the material presented earlier in the chapter as a foundation and show how these principles can be successfully implemented with real data and relevant user-defined scenarios and inquiries.

In general, one is interested in information acquisition, organization/management, analysis/exploitation, and decision-making. Figure 6-7 shows these same processes in an Observe-Orient-Decide-Act (OODA) model. Much of the space domain awareness community believes that the key to knowledge is mostly just in gathering observations. However, this is naive. Observations by themselves are not useful unless they are appropriately "oriented." By this we mean that the observations or measurements or data should be transformed into a *lingua franca* that allows for curation and common understanding/interpretation and exploitation of such data. The most critical aspect of space domain awareness is to have all the space domain beliefs and opinions mapped into this harmonized set of information. Earlier in the chapter, we presented the use of ontologies (or schema) as a mechanism to store and curate these data.

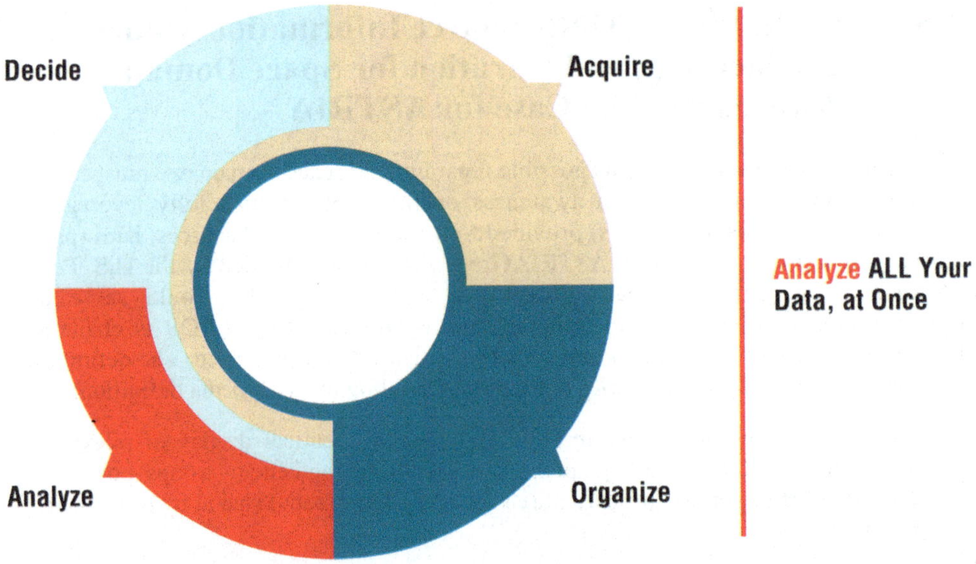

FIG. 6-7. **From Information Collection to Decision-Making.** A Recursive Process where all available data must be analyzed concurrently to maximize mutual information. (Image from Oracle Online Presentation.)

An analogous model was proposed by John Boyd, called an OODA (Observe-Orient-Decide-Act) process illustrated in Figure 6-8.

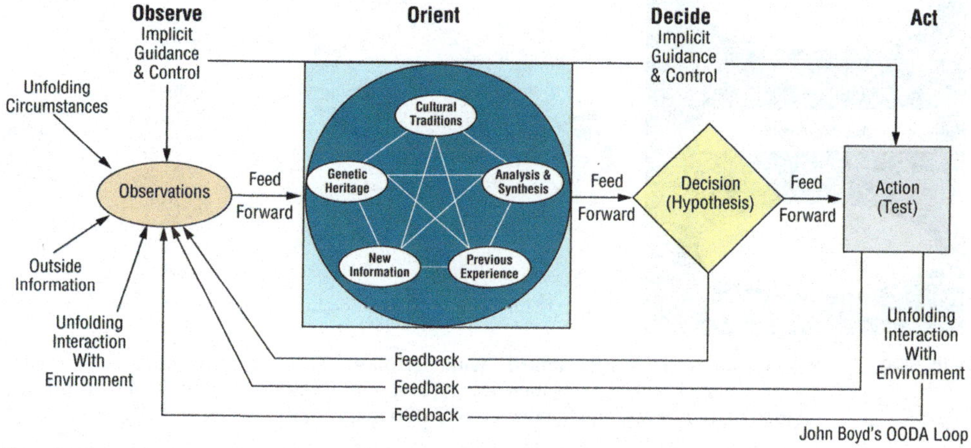

FIG. 6-8. **Observe Orient Decide Act Loop.** The OODS Loop describes the generalized process of going from observations to decisions. (2880px-OODA.Boyd.svg.png)

A key tenet of Big Data is to analyze all sources of available information simultaneously so as to get the maximum mutual information on desired space domain awareness criteria and thus enable going from data to discovery. A "best practice" in organizing and curating information is to do so in an ontology or a data model which can be implemented through the Resource Description Framework (RDF) standard and within in a triple store. In AstriaGraph we use a semantically consistent data model which is implemented in a graph database that does not use a triple store. The AstriaGraph data model is designed based on research inquiries posted by researchers. The model grows in a modular way, as researchers come up with new questions to solve using data and as more data sources are ingested. The data model can be exchanged as an RDF document and turned into an ontology or a SKOS.

The power in organizing information in a semantically connected fashion is that it enables the linking of disparate sources of information, creating vast trees of descriptions of many different elements and encouraging "discovery" from this data linking. As an example, in Figure 6-9, more semantic links and descriptions could be defined for "Andrew" and "Susan", such as their children any, etc. Although their relationship might be unknown, by seeing correlations between them, it could facilitate "discovering" that they are indeed related to each other (i.e., discovery of a causal relationship).

Equivalently, in the ASTRIAGraph data model (Figure 6-10), we could "discover" a consistent behavior in which a set of space actors that launch their anthropogenic space objects into orbit do not comply with debris mitigation guidelines, and are significantly latent in implementing the treaty for registering their objects with the United Nations.

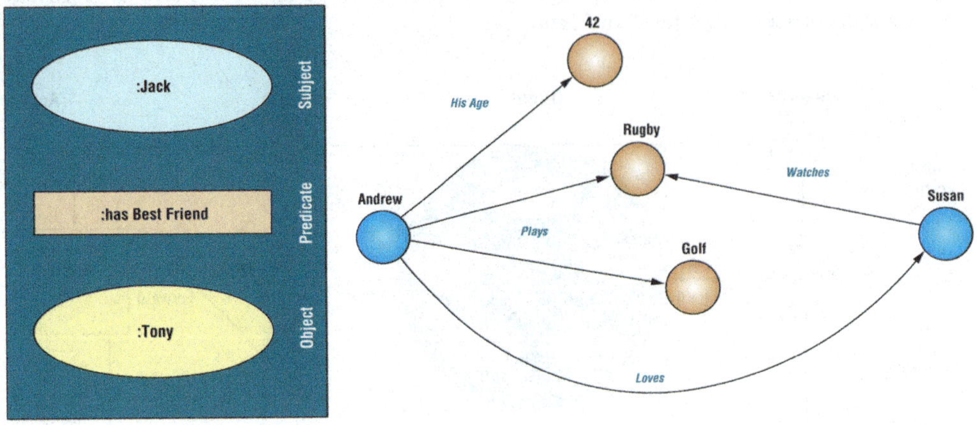

FIG. 6-9. **Knowledge Graph Structure with Triples.** This figure illustrates semantic descriptions of entities and their relationships in a given domain: Subject-Predicate-Object. (http://www.thefactmachine.com/rdf-basics-post/).

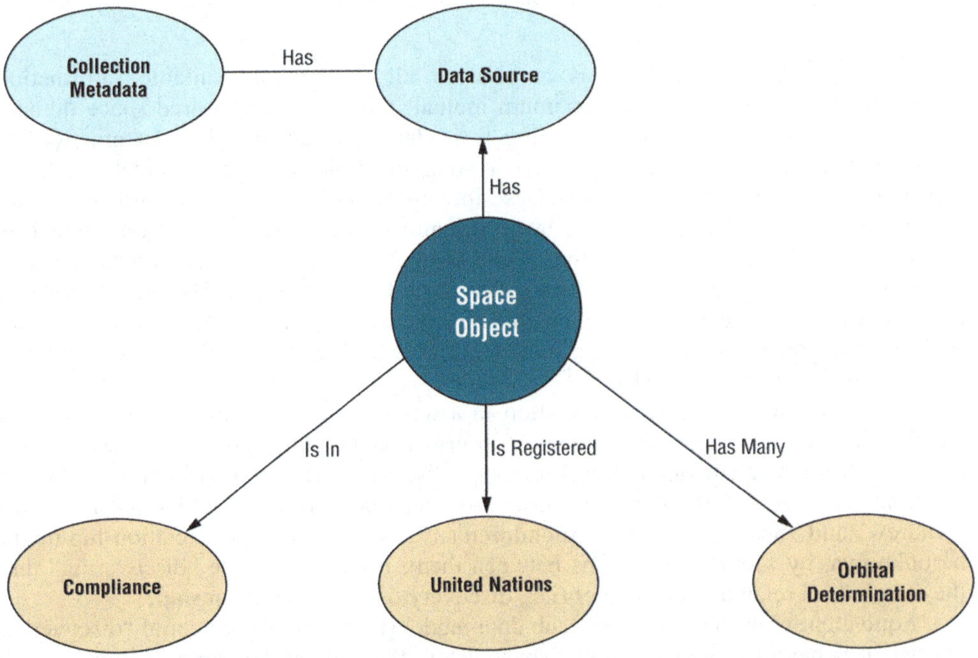

FIG. 6-10. **ASTRIAGraph Data Model Schematic.** This shows some relevant classes and their semantic relations. This is an alternative to a Triple Store but is an equivalent semantic model by which to describe entities in a given domain and their mutual relationships.

The power in organizing information in a semantically connected fashion is that it enables the linking of disparate sources of information, creating vast trees of descriptions of many different elements and encouraging "discovery" from this data linking. As an example, in the figure above, more semantic links and descriptions could be defined for "Andrew" and "Susan" such as their children if they had any, etc. and although their relationship might be unknown, by seeing correlations between them, it could facilitate "discovering" that they are indeed related to each other for instance (i.e., discovery of a causal relationship). Equivalently, in the ASTRIAGraph data model, we could "discover" a consistent behavior in which a set of space actors that launch their anthropogenic space objects into orbit do not comply with debris mitigation guidelines, and are significantly latent in implementing the treaty for registering their objects with the United Nations. See Figure 6-10.

It should be straightforward to envision creating relations for elements within the space domain, starting with known anthropogenic space objects, their characteristics (physical, functional, kinematic, etc.), whom they belong to, who built them, etc. and by linking disparate sources of information in this knowledge graph structure, being able to discover otherwise unknown causal relationships between space actors, the objects and their environments, etc. We imagine having a user bring an application that queries the knowledge graph using natural language searching for any ASO behavior that is correlated with any measured space environment event. For instance, perhaps there are several hundred RSOs in LEO that have a change in their nominal states roughly 14 hours after a solar flare of a specific magnitude. One would then want to know which objects and the common elements amongst them. Perhaps they have all been built with a specific type of Kapton. One might then attempt to find out what physical processes would cause this effect upon that type of Kapton. This is but one example of a significant discovery that could be made simply by organizing, managing, and exposing information in a smarter way. Previously in this chapter, this was explained via curating information that could be exploited by a *reasoner*.

The attributes expressed as nodes and relationships as edges are used in a graph database and are a middle layer in a holistic SDA framework. The information to be used gets acquired by a variety of methods and sources into so-called "instances" as seen below. These can be structured (data from sensors) or unstructured (data reported by humans online, via tweets, etc.). The desire is to have automated processes that are always acquiring information and storing them as instances. The information would have a so-called "landing zone" with metadata generated to provide a timestamp on this process. The information is stored as is and absolutely no changes or alterations are made to any instance data. The data are in their "raw" or acquired form.

There must be a separate process that takes the information from the instance data and maps these into the knowledge graph according to an agreed upon standard and semantically consistent vocabulary (for ASTRIAGraph, these are schema). The knowledge graph should also have a dictionary or thesaurus that can be readily accessed by those who interact with the knowledge graph. Given the international nature of the desired C/UD-OP, cultural competency is desirable in defining the knowledge graph vocabulary where the user is not asked to or forced to know this vocabulary. Instead, the knowledge graph developers have established which country- and sector-specific terms map to which knowledge graph entities, somewhat like a machine-enabled translation service.

The knowledge graph is where the current knowledge or beliefs and opinions of the space domain reside, with descriptions concerning all of these elements of information. The way in which the information within the knowledge graph can get updated or reconciled given evidence is via the top layer of applications that run or perform queries on the knowledge graph. One can envision one application being Space Object Orbit Determination and Prediction, for instance. Another would validate the knowledge graph data for semantic, physical, and statistical consistency. Related content and relations can be discovered by navigating connected entities and reasoning across them.

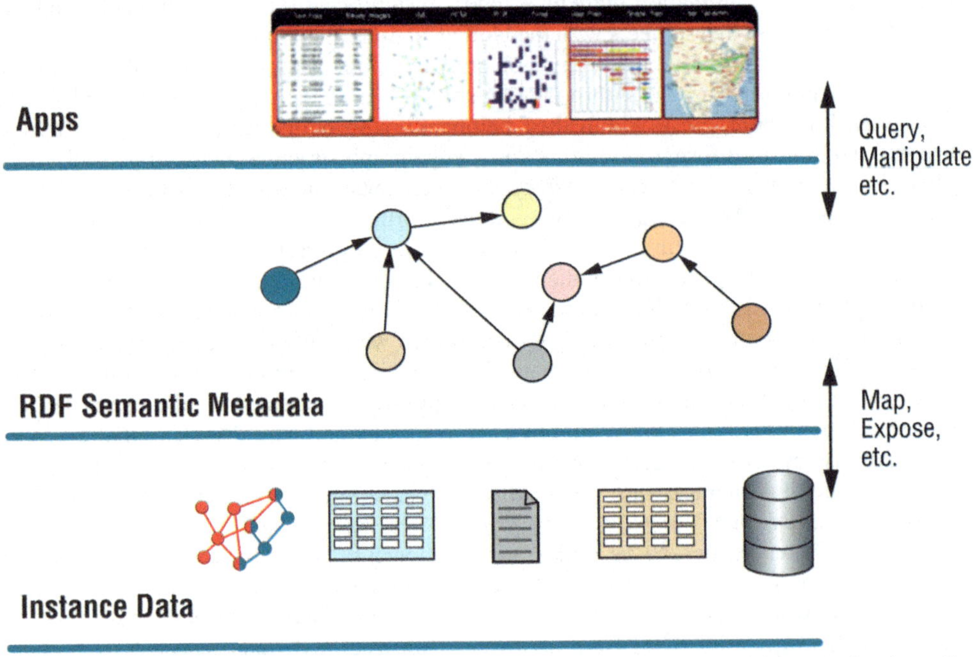

FIG. 6-11. **Graph-based Big Data Framework.** This is a three tiered framework where the "ground floor" is where information comes from the external world in whatever format, a middle (mezzanine) layer comprising the actual knowledge graph database that models and curates these data, a top layer (penthouse) for applications that can query the graph database directly. Applications can trust to query data that is curated instead of unknown raw information.

Figures 6-11 and 6-12 provide examples of the various functions that come together to go from data to discovery with this big data framework. The "customers" can be exchanged for "space objects." The demographics for a space object can be defined as its characteristics (physical, operational, functional, kinematic, ownership, etc.).

This knowledge graph technology is still in its infancy regarding space domain awareness, and it remains to be seen if it can be made relevant to a Space C/UD-OP. Some salient research issues involve how to make the knowledge graph able to handle dynamic

Problem Classification	Sample Problem
Anomaly Detection 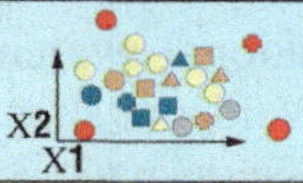	Given demographic data about a set of customers, identify customer purchasing behavior that is significantly different from the norm
Association Rules 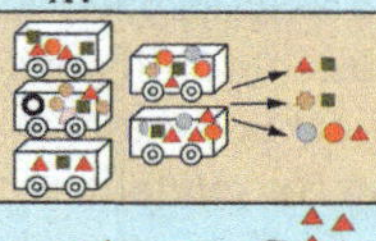	Find the items that tend to be purchased together and specify their relationship – market basket analysis
Clustering 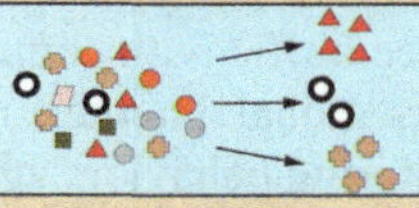	Segment demographic data into clusters and rank the probability that an individual will belong to a given cluster
Feature Extraction 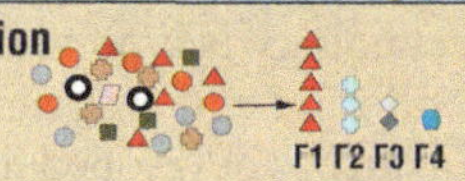	Given demographic data about a set of customers, group the attributes into general characteristics of the customers

FIG. 6-12. Examples of Big Data Analytics.

information, which in turn may be uncertain or random. Which queries are relevant? There must be some competence in mapping instance data (independent information sources) to the knowledge graph framework. A vocabulary and dictionary must be developed. There is much yet to be done to make this into a near real time, robust, and resilient architecture that could underpin a space C/UD-OP. Initial peer-reviewed research exposes the current status on these developments*.

A real world example has been developed by Prof. Moriba Jah at The University of Texas at Austin, and it is called ASTRIAGraph, which can be accessed here: http://astria.tacc.utexas.edu/AstriaGraph.

The current capability of ASTRIAGraph is as follows:

- Various sources of "instance data" are autonomously retrieved online and deposited into a "landing zone"
  - Government, industry, and open-source
- Relevant information from "instance data" are autonomously mapped into the knowledge graph in the graph database platform NEO4J based upon developed schema
  - Apps update the knowledge graph
- An app autonomously queries the knowledge graph, retrieves observational data, and processes these to yield an informed orbital estimate product, and updates the appropriate knowledge graph entities

* Esteva, M., Xu, W., Simone, N., Gupta, A., Jah, M., (2020) *"Modeling Data Curation to Scientific Inquiry: A Case for Multimodal Data Integration"* Proceedings of the 2020 IEEE Joint Conference on Digital Libraries (JCDL2020), Xi'an, China, August 1 – 5 (accepted – in Print)

- Currently all Planet Flock Cubesats are processed autonomously, daily, along with Iridium 64, and Maxar (DigitalGlobe) satellites

Interestingly, most of the attention on SSA (space situational awareness) focuses upon space hazards, namely conjunction analyses and collision avoidance support. However, this is easy to assess compared to being able to monitor and assess space actor behavior. The information and analytics requirements for predicting possible collisions is dwarfed by the requirements to develop a body of evidence of space object behaviors that can be assessed for compliance or non-compliance with space laws, rules, or guidelines.

By and large, given the significant increase in space traffic with commercial operators launching thousands of satellites for services such as global internet, many people are concerned about malicious behavior on orbit and monitoring behaviors against such norms as space sustainability guidelines. Knowledge graph technology is a powerful information construct that can significantly enable this activity: monitoring and assessing space object behaviors. As an example, The University of Texas at Austin has developed the first instance of space object behavior monitoring and assessment against soft information inputs such as the UN COPUOS long-term sustainability guidelines, specifically, GEO disposal for space debris mitigation, which is demonstrated here: http://astria.tacc.utexas.edu/compliance. What follows is a deeper description of how this has been implemented to assess a large number of important space policies, laws, and guidelines (e.g., United Nations Committee On peaceful Uses of Outer Space Long-Term Sustainability) [https://www.unoosa.org/oosa/en/ourwork/topics/long-term-sustainability-of-outer-space-activities.html].

6.5.1 ASTRIAGraph as a Knowledge System

ASTRIAGraph is the only public multi-source compendium of space domain information [ASTRIAGraph, 2020]. Using multiple data sources and analysis methods, ASTRIAGraph enables exploration of scientific problems. Information resulting from the analyses is presented through a publicly available web interface allowing users to make inferences about the location and compliance of near space objects at a given time.

Knowledge systems use data and analysis processes for representing, exploring, and understanding domain problems. As bigger and more diverse datasets are aggregated in these systems, their transparency, and consequently the results that they produce, can be seriously compromised. Not only are different datasets created, structured, and maintained differently, but across them, the same fields may be inconsistent and or discontinued. In addition, different datasets may contain duplicate and even conflicting information of diverse levels of resolution, scale, quality, and uncertainty. A fused estimate requires achieving some form of information consensus.

Currently, ASTRIAGraph ingests data from eight external sources, including public and private entities. Because information about an anthropogenic space object is reported by many sources, each source may contribute multiple records about the same object over time. Sources differ in the number of fields and in the format of the data they report. The volume of accumulated files per source also varies.

In many cases the labels of the fields are inconsistent across data sources as well. Significantly, some sources assign their own unique identifiers to anthropogenic space objects. This makes it challenging at best to distinguish when two or more sources report about a common object. In Figure 6-13, a query result in the ASTRIAGraph interface illustrates how three data sources report different information about the same object. In the example, all three records correspond to the same satellite, named *FLOCK 2k 28*. However, each source indicates a different position in Euclidean space, marked as 1, 2, and 4 in Figure 6-13. In summary, information disagreements about same anthropogenic space objects are common, and the differences range from subtle to completely contradictory. In many cases, it is not possible to identify if an object is the same over time. Hence, the differences between records from different sources need to be reconciled or assessed to determine which are more accurate. Therefore, a data curation challenge is managing multiple datasets concurrently: organizing, describing, keeping track of, and assessing relevant quality features across multiple datasets. In ASTRIAGraph we undertake these tasks within the knowledge system and consider all its functions as well as the scientific inquiries at stake through what we denominate ASTRIA Data Design.

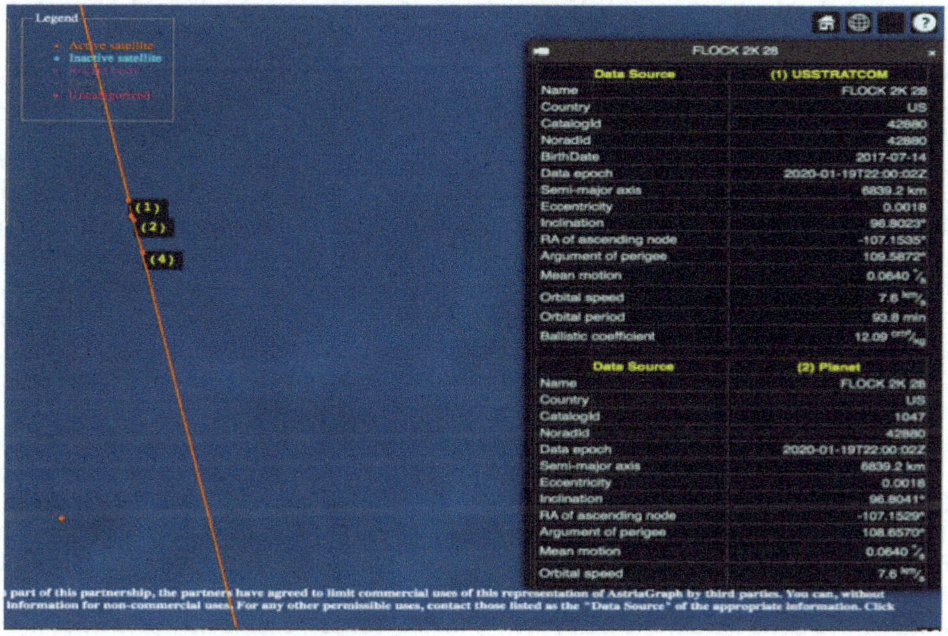

FIG. 6-13. **ASTRIAGraph Interface Snapshot.** This interface shows the results of a query in which different data sources report different fields and position values for the same anthropogenic space object (1, 2, 4).

6.5.2 ASTRIA Data Design

Our work is grounded in data modeling, an effort that is analogous to ontology (schema) development. Ontologies are hierarchical semantic representations of a knowledge system as classes, their properties, and their relations. Designing ontologies is an involved process, requiring the consensus of domain experts and tight mechanisms for approving classes and properties. We are pursuing ASTRIA Data Design in a modular and applied way, through data modeling and testing of relevant scientific use cases that build progressively towards a comprehensive SSA knowledge system and eventually an ontology. Because of this methodology, we can evolve our vocabulary and the relationships between classes dynamically and test their application promptly. Scientific inquiries are implemented within the ASTRIA Data Design.

Data Modeling

Data modeling entails describing entities that are important to a system and how they relate to each other [West, 2011]. The resultant model conforms to a semantic structure in which self-contained classes are related to each other to form a natural context representation [Perttunen, 2009]. In our project, data modeling is a transdisciplinary task with participation from aerospace domain scientists, an information scientist/curator and a computer scientist. It begins with the domain scientists proposing goals for the knowledge system and defining the SSA research problems that they want to answer through it. It is also an ongoing process, as research questions evolve and new ones emerge. Throughout the discussions, the information scientist identifies main entities within the research propositions as classes representing physical objects, research problems, data sources, collections, and curation activities. The scientists then determine the properties that characterize the classes and map them to corresponding fields within the datasets that are available to the knowledge system. The connections between classes are drawn in relation to how data will be used to solve the research propositions. The computer scientist oversees how the relationships are established for efficiency and scalability of the graph database.

Figure 6-13 shows the main classes in ASTRIAGraph representing the near-Earth anthropogenic space objects and space traffic management domain knowledge. The Space Object class is the central focus of the knowledge system, and it is connected to the rest of the classes. Through its properties, it describes the characteristics of anthropogenic space objects. The Sensor class represents the physical instruments from which the observational data derives, and is measured data. The Object Resolution, Orbital Determination, and Compliance classes use data to resolve scientific problems through different algorithms. The rest of the classes (Provenance, Collection Metadata)- are used to manage the data from multiple sources efficiently and transparently by linking. As new scientific inquiries emerge (e.g., http://astria.tacc.utexas.edu/ingestion_queries_demo), the team continues working to expand the knowledge system with new classes, adding new properties in the existing classes, and establishing new relationships among classes.

We maintain and share a diagram of the complete data model online, and we use webProtege [Protégé, 2020] to maintain classes, properties, and their relations. To name classes and properties we use terms from the aerospace engineering domain, using authoritative vocabularies when possible [Unified Astronomy Thesaurus, 2020]. For interoperability purposes we use standard metadata terms to label properties in the Provenance and the Collection Metadata classes [DCMI: DCMI Metadata Terms, 2020]. Managing multimodal collections within knowledge systems such as ASTRIAGraph presents complex data management and curation challenges that we detail in the next sections.

Data Ingestion and Management

We developed scripts to download and to parse data files from multiple and disparate data sources, and to import the fields and values of interest into the Neo4j graph database. Data in Neo4j is managed based on the ASTRIA Data Design [Webber, 2012]. The interconnected nature of ASTRIA Data Design makes the graph database a good choice for its implementation. Nodes are created with the names of the classes, and each class is given properties that correspond to fields in the data. Nodes are connected to other nodes with edges, the latter reflecting the relationships designed in the model. Iterating through the data model design by adding classes and properties, or changing their relationships, is very easy to do in Neo4j. If a class is updated with a new property in our model, changes are simply made in the import scripts.

Previously, we mentioned the discrepancies between the names of similar fields across data sources. Field label reconciliation is also done via the scripts that perform data ingest. For example, if data source A has a field called "ECC" and source B has one called "Eccentricity," the related import scripts are written to put those values into a property called "Ecc" in the Orbital Determination class.

Data Curation: Collection Metadata and Provenance

Data curation is supported through the Collection Metadata and the Data Source classes and throughout the research project's lifecycle [Pouchard, 2015]. We gather descriptive and administrative metadata for each data collection that is ingested and aggregated into ASTRIAGraph. This is performed both manually for static information, and automatically from the data files as their contents (appropriate fields) are ingested. Different from institutional repositories that focus on curation of individual collections, we use the metadata to curate multiple collections. Descriptive metadata standards used in digital repositories include elements to record information about the data sources known as provenance. In our data model we separate provenance from other descriptive metadata as two individual but related classes. This increases efficiency dealing with collections' updates and growth, decreases data storage, and facilitates tracking the provenance of individual fields within the data collections. The Data Source class contains properties to record the name and contact information of the source as well as licensing agreement. The Collection Metadata class combines static and dynamic information about the files contributed by a given source. Static information includes data formats, collection ID, and number and type of fields. Dynamic properties are processes that monitor aggregation and the inclusion of time stamps required to track changes reported in the data files over time (See Table 6-1). Fields in the Collection Metadata class correspond to elements in the extended Dublin Core standard, including format, size, frequency, method of accrual, and date of collection.

The queries in Table 6-5 reflect how the data model enables relating field values to their provenance information as well as to collection's metadata, providing transparency and authenticity to the results. Designed collaboratively between domain and information scientists, the queries' structure reflects the relevance of provenance information both to data curation and to research inquiries. The query format conveys our research inquiries through a common syntax and style. The FROM clauses indicate the classes from which these queries pull information. The WHERE clauses further refine the searches according to specified conditions, and the SELECT clauses choose and organize the returned fields. For these queries, the number of returned rows is dependent on how many sources report the requested inquiry. This process demonstrates how data is integrated as queries are resolved in the knowledge system.

TABLE 6-5. Example Data Curation Questions and Their Corresponding Queries, Results, and Performances.

Question	Query	Results	Time (sec)
What data sources provided range/range rate measurements on a given day?	**FROM** data sources, measurements **SELECT** data source **WHERE** {measurement date = day}	1	0.1
What space objects were updated by a given source on a given day?	**FROM** data sources, space objects **SELECT** space object **WHERE** {space object date = day, data source = source}	235	0.6
How many records of space objects have been accumulated from a given source?	**FROM** data sources, space objects **SELECT** COUNT space object **WHERE** {data source = source}	267	15
Do multiple sources report the same value for a field, such as orbital eccentricity for the same object?	**FROM** data sources, space objects, orbits **SELECT** source, orbit eccentricity **WHERE** {space object identifier = id}	8072	50

To illustrate the performance of our design, in Table 6-5 we include the time spent and the size of the return result for each query execution. The first query only involves a small number of nodes and returned results in less than a fraction of second. The rest of the queries involving the Space Object class took longer to retrieve. The last query also involving the Orbital Parameters class took 50 seconds to query over 15 million nodes.

Synchronic Multimodal Data Assessment

When integrating multiple multimodal datasets, a challenge is maintaining a clear record of which sources report which fields. This is required for a Common Operating Picture (COP). To demonstrate our solution, we consider the eight external data sources, which altogether have 38 different fields. Based on provenance information and on individual collection's metadata such as number and type of fields, we generated an Information Record Matrix (IRM) that shows which data sources report which fields. Figure 6-14 below shows the IRM including visual cues to understand which sources report the same or different fields.

	2	0	1	4	9	3	7	8
P1	✓	✓	✓	✓	✓	✓	✓	✗
P2	✓	✓	✓	✓	✓	✓	✓	✗
P3	✓	✓	✓	✓	✓	✓	✓	✗
P4	✓	✓	✓	✓	✓	✗	✓	✗
P5	✓	✓	✓	✓	✓	✓	✗	✗
P6	✓	✓	✓	✓	✓	✓	✗	✗
P7	✓	✓	✓	✓	✓	✓	✗	✗
P8	✓	✓	✓	✓	✓	✓	✗	✗
P9	✓	✓	✓	✓	✓	✓	✗	✗
P10	✓	✓	✓	✓	✓	✓	✗	✗
P11	✓	✓	✓	✓	✓	✓	✗	✗
P12	✓	✓	✓	✓	✓	✓	✗	✗
P13	✓	✓	✓	✓	✓	✓	✗	✗
P14	✓	✓	✓	✓	✓	✓	✗	✗
P15	✓	✓	✓	✓	✓	✓	✗	✗
P16	✓	✓	✓	✓	✓	✓	✗	✗
P17	✓	✓	✓	✓	✓	✓	✗	✗
P18	✓	✓	✓	✓	✓	✓	✗	✗
P19	✓	✓	✓	✗	✓	✗	✓	✗
P20	✗	✓	✗	✓	✗	✗	✓	✗
P21	✗	✓	✓	✗	✗	✗	✗	✗
P22	✓	✗	✗	✗	✗	✗	✗	✓
P23	✓	✗	✗	✗	✗	✗	✗	✓
P24	✓	✗	✗	✗	✗	✗	✗	✓
P25	✓	✗	✗	✗	✗	✗	✗	✗
P26	✓	✗	✗	✗	✗	✗	✗	✗
P27	✗	✗	✗	✗	✗	✓	✗	✗
P28	✗	✗	✗	✗	✗	✗	✓	✗
P29	✗	✗	✗	✗	✗	✗	✓	✗
P30	✗	✗	✗	✗	✗	✗	✓	✗
P31	✗	✗	✗	✗	✗	✗	✓	✗
P32	✗	✗	✗	✗	✗	✗	✓	✗
P33	✗	✗	✗	✗	✗	✗	✓	✗
P34	✗	✗	✗	✗	✗	✗	✓	✗
P35	✗	✗	✗	✗	✗	✗	✓	✗
P36	✗	✗	✗	✗	✗	✗	✓	✗
P37	✗	✗	✗	✗	✗	✗	✓	✗
P38	✗	✗	✗	✗	✗	✗	✓	✗

FIG. 6-14. Information Record Matrix (IRM). The data fields contained in this IRM were contributed by eight data sources.

Each column in the IRM corresponds to an external data source identified with an ID. The column is ordered so that the source that provides more fields is shown to the left. For example, data source 2 provides the most fields with 24, and data source 8 only provides 3. Each row corresponds to a field. The rows are ordered so that fields common to more sources are shown at the top. Rows are color coded according to the classes in the data model. Brown indicates information for a space object, blue represents measurements, and yellow represents information related to orbital determination.

The IRM allows curators to understand the scope and content of the information provided by multiple data sources, reveals information coverage (number and data fields reported by each source) and the overlap between them. Since each collection has different native information organization, format, and structure, without the matrix providing a uniform overview of all of them, it would be very difficult to understand differences and similarities between them.

Using the data from the IRM we created a visualization to understand the similarity between data sources. The information from each source is mapped to a binary vector of 38 dimensions, each of which corresponds to a data field. We computed Euclidean distances among the binary vectors and mapped them into a two-dimensional coordinate system using the multidimensional scaling (MDS) method [44-46]. Figure 6-15 shows the results. Each circle represents a data source labeled with its ID. The size of the circle corresponds to the number of fields reported by the data source. For example, the smaller circle represents source 8, which only provides three fields. Proximity between the circles indicates their similarity. In the visualization we can see that data sources 0, 1, and 9 are the most similar to each other. In contrast, data sources 7 and 8 are quite different from the rest, which indicates that they provide complementary information.

Understanding the coverage, the differences, and the agreements of multiple collections informs both data curation and research decisions. Knowing about the duplication and the disagreement between fields across different collections helps researchers to assess uncertainty. For example, if multiple data sources report the same information about a space object, there is less uncertainty about that information. Instead, conflicting information reported by different sources suggests more uncertainty. Both assessments can help users make better research judgments. The results of the matrix assessment can be stored as part of the data model in the form of edges (relations) between nodes corresponding to each collection.

Implementing Compliance Inquiry in AstriaGraph

This case illustrates how our data model design helps in conducting research with data transparency. Built as a class in our data model the Space Object Compliance inquiry informs researchers whether satellites, a type of anthropogenic space object (ASO), are compliant with international guidelines. Proprietors of satellites in geosynchronous orbit (GEO) are highly encouraged to follow United Nations guidelines regarding disposal of these ASOs at their end-of-life. These were initially put in place in 1997 by the Inter-Agency Space Debris Coordination Committee (IADC) to further the prevention of orbital debris around Earth [IADC]. Compliance of a GEO satellite is represented as an algorithm,

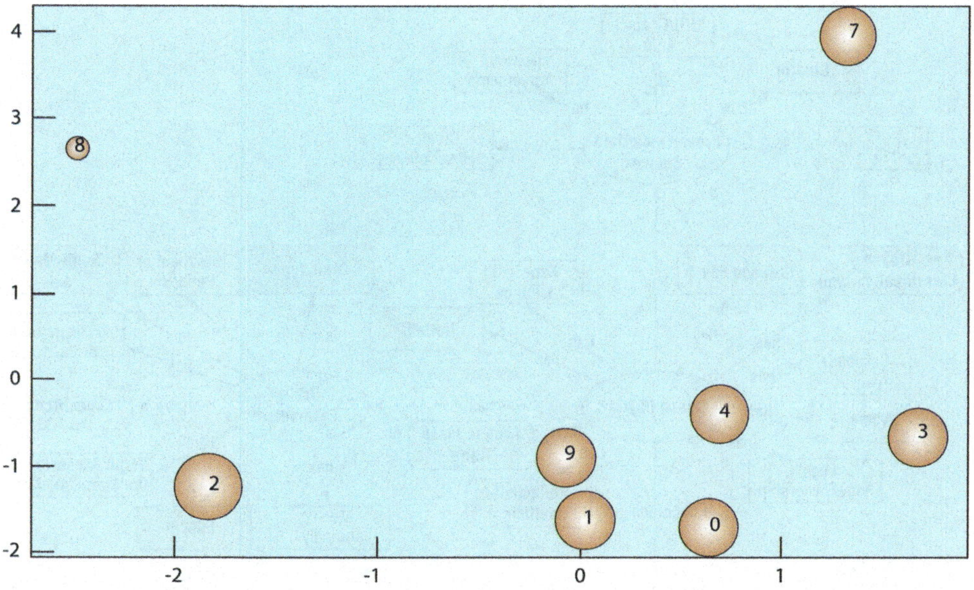

FIG. 6-15. **Information Record Matrix (IRM).** The data fields contained in this IRM were contributed by eight data sources.

which primarily relies on fields such as the altitude of the satellite at perigee, in addition to other orbital parameters and space object features like area-to-mass ratio.

The amount of space traffic is growing at an alarming rate. Experts the world over agree that the dominant problem is that approximately half of space operators are not complying with agreed upon space debris mitigation guidelines. One cannot enforce compliance if one is unaware of who is or is not compliant. Furthermore, one cannot know that which is not measured. Therefore, a relevant and needed capability we chose to quantify and assess as a use-case that could be scaled to other space traffic and awareness challenges was GEO disposal compliance.

In ASTRIAGraph we implemented the compliance algorithm in a Python script to compute and report results considering relevant classes and fields. Figure 6-16 shows the classes and the fields (in tan) required by the algorithm. Examples of queries are in Table 6-7. We note that the second query, designed by the domain experts, specifically requires information about the data sources. Currently, if two different sources report on the same object, we calculate compliance twice, one per data source report about the ASO. This is a case in which the information coverage and similarity assessments (Section 5.2), can help researchers evaluate the uncertainty of the results.

System's Architecture and Scalability

The scale of data that needs to be managed is significant. Table 6-7 shows a summary of classes and relations mapped to the proper nodes and edges in the database backend.

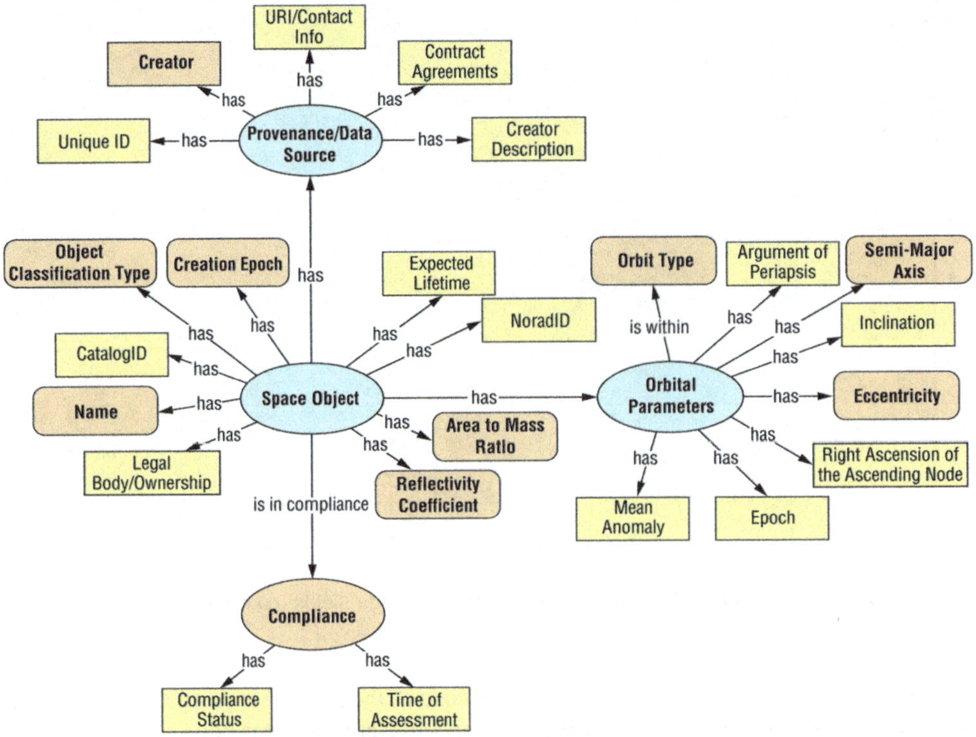

FIG. 6-16. Classes and Properties (tan) Involved in the Compliance Inquiry. This is an example of how semantically curated information can be brought to bear to both support models and scientific/policy inquiries. Analytical applications can rapidly find patterns of correlation and make these available to interested users. This example is focused on quantifying and assessing space actor compliance with anthropogenic space object end-of-life disposal guidelines.

Currently we have 8 external and 3 internal data sources. The database includes over 7 million nodes for space objects and orbital parameters. Altogether, there are about 15 million nodes and about 30 million edges in the current graph database. Performance issues are carefully considered in the data model design. For accountability and long-term preservation, all downloaded files from each source are aggregated and stored in a geographically replicated storage system at TACC [TACC Corral Storage].

The ASTRIAGraph website and its associated web services reside on a Virtual Machine (VM) hosted at the Texas Advanced Computing Center (TACC) internal VM Farm [Texas Advanced Computing Center]. The VM consists of 2 Intel Xeon E5-2660 (@ 2.20GHz) CPUs, 8 GB of RAM and 118GB of local hard disk space. The VM also has larger high-performance distributed storage areas available by remote mount points for storing larger datasets [TACC Corral Storage]. The Neo4j instance is hosted on a service node which provides ample storage space and processing power for the database to grow. The web

TABLE 6-6. Example Questions in the Compliance Inquiry, and Corresponding Queries, Their Results and Performances.

Question	Query	Results	Time (sec)
Which satellites have been compliant since a given year?	**FROM** space objects, compliance **SELECT** space object **WHERE** {space object compliance row = compliant, space object date >= year}	637	2
How many satellites are currently in GEO disposal compliance that are reported by a given data source?	**FROM** data sources, space objects, compliance **SELECT** space object **WHERE** {space object compliance row = true, data source = source}	637	2
Which satellites are compliant or noncompliant and owned by a given legal ownership?	**FROM** space objects, compliance **SELECT** space object **WHERE** {space object compliance row = compliant, space object country = legal body ownership}	133	1.3

TABLE 6-7. Summary of Nodes and Edges in Neo4J Instance for ASTRIAGraph.

	Description	Counts
Node	Provenance/Data Source	11
	Space Object	7140877
	Measurement	76053
	Orbital Parameters	7166860
	Compliance	211007
Edge	Provenance2SpaceObject	7169507
	has_orbit	7267279
	has_measurement	81591
	has_compliance	211007

service queries the database remotely from the VM via secure SSH connections. Queries to the database are written in Cypher, and external tools, such as the ASTRIAGraph's web interface, connect with the database through the Neo4j Python driver.

6.5.3 Conclusions

ASTRIAGraph is a real world implementation which encompasses innovative approaches to accomplish data curation, organization and data exploration. It is a first-of-kind framework that combines data curation with scientific inquiry in the same data model, and it expands curation from a single data collection to multiple ones. The result is a unified, expandable, and scalable data curation and scientific representation across multiple data sources. The data model design helps in efficiently conducting user-defined inquiries using large volumes of integrated data whose provenance is clear. When implementing this work in ASTRIAGraph we verified how relevant provenance and collections metadata are to the explainability and transparency of its research outputs. Monitoring for Anthropogenic Space Object Compliance with Orbital Debris Mitigation Guidelincs was shown as an example. In addition, the value of the transdisciplinary process stimulated the involvement of domain scientists and space policy-makers in data curation. By focusing the data model design in relevant space domain concerns and inquiries, we take a more focused and modular approach than is typical. The design grows as new questions are formulated and as new data are ingested. These data are continuously assessed and organized accordingly so that it is ready to be integrated as problems are being asked through queries. Always, provenance can be clearly identified. Without this end-to-end information framework, some inquiries would be impossible to address due to an absence of mutual-information content. SDA will benefit from being analyzed as a big data problem. To date, no one has been able to have a big data problem in SDA. This framework enables such a problem to be formulated and addressed.

Glossary

big data: A dataset characterized by having very high volume, velocity, and variety of data.

controlled vocabulary (CV): A selected list of words and phrases, which are used to tag units of information (documents or other work) so that they may be more easily retrieved by a search.

corpus (plural corpora): A collection of text documents from which information is extracted, e.g., by natural language processing.

data: Possible sources of latent knowledge that are intended to be a statement or opinion about something. Unlike inferred knowledge, data need not provide an answer to a question or resolve uncertainty.

database (DB): A structured set of data held in a computer, especially one that is accessible in various ways.

data integration: The process of combining data from multiple sources for use in a new analysis. Data may be of the same type or different types.

data management: The design and execution of plans and processes to manage data throughout their life cycle.

data science: A field of science that uses various techniques to extract knowledge or insight from data, usually big data.

directed acyclic graph (DAG): Finite directed graph (i.e., a collection of nodes or vertices and directed edges) with no directed cycles. There is no way to start at any vertex v and follow a consistently-directed sequence of edges that eventually loops back to v again.

high performance computing (HPC): Utilizing supercomputers that can perform calculations at much higher rates than general purpose or personal computers.

information: Any entity or form that provides the answer to a question of some kind or resolves uncertainty. Unlike data, information need not be intended to be true.

information extraction (IE): The process of extracting structured information from text.

information fusion: The process of inferring or discovering latent knowledge that only exists in a combined set of data.

JSON-LD: JSON for linking data (https://json-ld.org/). An extension of JSON that includes statements on "context" to provide additional mappings from JSON to an RDF model [JSON- LD, 2015].

knowledge: A familiarity, awareness, or understanding of someone or something, such as data, facts, information, description, or skills, which is acquired through experience or education.

knowledge base: A structured data storage system combining data with an inferencing system capable of generating new information or facts from the stored data.

knowledge model: Also known as knowledge representation. A machine readable representation of knowledge in a domain, such as an ontology or database model.

low-resource domains: A topic of interest that is either rarely written about, or for which information is difficult to collect. Low-resource domains may be difficult to study using natural language processing or machine learning approaches.

machine reading: The eventual goal of natural language processing—the automatic understanding of text by a computer.

metadata: Data about data. For example, if a dataset lists astrometric values, the metadata might describe the date collected and methods used to generate them.

named entity (NE): A set of predefined categories of entities that can be identified in text, such as a person, location, organization,date, or quantity.

natural language processing (NLP): A collection of methods used to process text with the goal of being able to automatically read and perform human-level tasks with natural language.

n-gram: The name given to a series of n words (or lemmas) in a specific order, where a unigram contains a single word, bigrams contain two words, trigrams contain three words, and so forth.

ontology: In computer and information science, a set of terms representing classes/concepts/categories and relations among them specified in a formal language, usually OWL, intended to represent the knowledge of one or more domains.

open-domain information extraction: The process of generating a structured, machine-readable representation of information from text (e.g. in the form of triples [trinary] or n-ary propositions).

OWL: Web Ontology Language. Standard language for ontologies used in knowledge management or web applications. [Hitzler et al., 2012]

part-of-speech tagging: The process of labeling each word in a sentence as corresponding to a particular part of speech, such as noun, verb, adjective, or adverb.

provenance: The chronology of the ownership, custody, or location of an entity. In computer and information science, provenance traces the history of a particular piece of information, such as a file or other digital object.

RDF (Resource Description Framework): A family of specifications originally designed as a metadata data model and used as a general method for conceptual description or modeling of information implemented in web resources. RDF uses a variety of syntax and serialization formats.

recall: In natural language processing, the proportion of something (for example, relations or events) a given method is able to extract from a corpus, relative to the total number of those events that exist in the corpus. Low-recall methods miss many events or relations, whereas high-recall methods are able to extract many of those present in a corpus.

semantic web: "W3C's vision of the Web of linked data. Semantic Web technologies enable people to create data stores on the Web, build vocabularies, and write rules for handling data." (https://www.w3.org/standards/semanticweb/)

semantics: The study of meaning. Traditionally used in linguistics, now part of computer science, to describe the meaning of information.

SKOS (Simple Knowledge Organization System) vocabulary: A set of related terms structured according to standardized format and related by properties such as "has exact match," "has broader match," or "has narrower match."

surface pattern: In natural language processing, a pattern expressed in terms of words or lemmas rather than deeper structure (such as syntax, semantics, or dependencies).

syntactic parsing: In natural language processing, automatically ascribing syntactic structure to a sentence.

triple: A tuple (ordered list) consisting of three elements. The basic unit of RDF is a triple consisting of subject-predicate-object.

workflow manager (workflow management system): In computer science, infrastructure for the set-up, performance, and monitoring of a defined sequence of tasks, arranged as a workflow.

XML (extensible markup language): A markup language that defines set of rules for encoding documents in a format that is both human-readable and machine-readable (https://www.w3.org/TR/xml/)

References

Arp, Robert et al. 2015. Building Ontologies with Basic Formal Ontology. MIT Press. Cambridge, Massachusetts.

Astria Data Model demo. http://astria.tacc.utexas.edu/ingestion_queries_demo.

ASTRIAGraph. http://astria.tacc.utexas.edu/AstriaGraph/. Accessed 22 Jan. 2020.

Baader, Franz et al. 2003. *The Description Logic Handbook: Theory, Implementation, Applications*. Cambridge University Press, Cambridge, UK. ISBN:0-521-78176-0

Borgo, Stefano and C. Masolo. 2010. Ontological foundations of DOLCE. In *Handbook on ontologies*. pp. 279-295. DOI:10.1007/978-90-481-8847-5_13.

Briney, Kristen. 2015. *Data Management for Researchers: Organize, maintain and share your data for research success*. Pelagic Publishing.

Carmichael, Iain and J. S. Marron. 2018. Data science vs. statistics: two cultures? *Japanese Journal of Statistics and Data Science*. 1:117-138. DOI:10.1007/s42081-018-0009-3

Cox, Alexander P. et al. 2016. The Space Object Ontology. *Proceedings of the 2016 19th International Conference on Information Fusion (FUSION), Heidelberg, Germany*. IEEE. ISBN: 978-0-9964-5274-8

Cox, Simon. 2017. Ontology for observations and sampling features, with alignments to existing models. *Semantic Web Journal* 8(3):453–470. DOI: 10.3233/SW-160214.

DCMI: DCMI Metadata Terms. https://www.dublincore.org/specifications/dublin-core/dcmi-terms/. Accessed 23 Jan. 2020.

Glimm, Birteet et al. 2010. Optimising Ontology Classification. In Proc. of the 9th Int. Semantic Web Conf. (ISWC 2010) 6496:225-240, November 7–11, Shanghai, China. P.F. Patel-Schneider et al., eds.

Hitzler, Pascal et al. 2012. OWL 2 Web Ontology Language Primer (Second Edition). Available at https://www.w3.org/TR/owl2-primer/. Accessed 10 Nov. 2018.

https://neo4j.com

https://www.unoosa.org/oosa/en/ourwork/topics/long-term-sustainability-of-outer-space-activities.html

IADC Space Debris Guidelines. https://www.unoosa.org/documents/pdf/spacelaw/sd/IADC-2002-01-IADC-Space_Debris-Guidelines-Revision1.pdf. International Journal of Digital Curation, vol. 4, no. 3, 3, Dec. 2009, pp. 93–103. www.ijdc.net, doi:10.2218/ijdc.v4i3.117.

JSON-LD JSON for Linking Data. 2015. Available at https://json-ld.org/. Accessed 10 Nov. 2018.

Kazakov, Yevgeny et al. 2014. The Incredible ELK. Journal of Automatic Reasoning 53:1-61. DOI:10.1007/s10817-013-9296-3

Madin, J. et al. 2007. An ontology for describing and synthesizing ecological observation data. *Ecological Informatics* 2: 279–296.

Niles, I. and A. Pease. 2001. Toward a Standard Upper Ontology. In Proceedings of the 2nd International Conference on Formal Ontology in Information Systems (FOIS-2001), Chris Welty and Barry Smith, eds. 2–9.

Perttunen, Mikko, Jukka Riekki, and Ora Lassila. 2009. Context representation and reasoning in pervasive computing: a review. International Journal of Multimedia and Ubiquitous Engineering 4, no. 4: 1-28.

Pouchard, Line. Revisiting the Data Lifecycle with Big Data Curation. International Journal of Digital Curation, vol. 10, no. 2, June 2015, pp. 176–92. www.ijdc.net, doi:10.2218/ijdc.v10i2.342.

Protégé. https://protege.stanford.edu/products.php#web-protege. Accessed 23 Jan. 2020.

Rowley, Jennifer. 2007. The wisdom hierarchy: representations of the DIKW hierarchy. *Journal of Information Science*. 33(2), 163–180. doi:10.1177/0165551506070706.

TACC Corral Storage. https://www.tacc.utexas.edu/systems/corral.

Talia, Domenico. 2013. Workflow Systems for Science: Concepts and Tools. *ISRN Software Engineering*, vol. 2013: 404525. DOI: 10.1155/2013/404525.

Texas Advanced Computing Center. https://www.tacc.utexas.edu/. Accessed 23 Jan. 2020. Unidimensional scaling: Theory and applications. Journal of Educational Measurement 40, no. 3 (2003): 277–280.

Unified Astronomy Thesaurus. Unified Astronomy Thesaurus, http://astrothesaurus.org/. Accessed 20 Jan. 2020.

Walls, R. L. et al. 2016. Assessing the IADC Space Debris Mitigation Guidelines: A case for ontology-based data management. *Proceedings of the 2016 Advanced Maui Space Surveillance Technologies Conference.*

Walls, R. L. et al. 2018. Integrating and Managing Biodiversity data with the Biocollections Ontology, in *Application of Semantic Technologies in Biodiversity Science*, Anne Thessen, editor. IOS Press. ISBN978-1-61499-853-2 (print) | 978-1-61499-854-9 (online).

Webber, Jim. 2012. A programmatic introduction to neo4j. In Proceedings of the 3rd annual conference on Systems, programming, and applications: software for humanity, pp. 217–218.

West, M. 2011. Developing High Quality Data Models. Burlington, USA: Morgan Kaufmann Pub.

Chapter 7

Uncertain Data Representation

Emmanuel Delande, *University of Texas at Austin*
Jérémie Houssineau, *University of Warwick*

Many actions, decisions, or analyses in space domain awareness (SDA) must deal with limited information about the system of study. It is relatively straightforward to determine whether two satellites are on a collision course, for example, given the *exact* dynamics of their respective geometry and of their kinematic state; yet, in an operational context, we must often assess collision risks without knowing *exactly* what these dynamics are. Likewise, we can't base data-driven policies on clear-cut strategies assuming perfect knowledge of the studied system but must rely instead on a partial description of the system supported by limited data.

Many algorithms involved in the SDA signal processing chain, by design, operate on a limited description of the system. They require, however, a *quantifiable* representation of the uncertainty induced by partial information. To this end a probabilistic representation remains the candidate of choice, adopted in the design of the overwhelming majority of modern estimation algorithms. Indeed, it is both flexible enough to accommodate the wide range of uncertain phenomena involved in an environment as rich and complex as SDA (e.g., orbital dynamics, sensor noise, or maneuvering actions), and structured enough to incorporate well-established inference rules suited to the study of dynamic systems (e.g., Kalman filters).

Despite its potency, the probabilistic paradigm does not offer a definitive answer, universal and without ambiguities, to the complex problem of representing uncertainty. The aim of this chapter is to highlight some shortcomings of the probabilistic representation in the context of SDA, discuss the notion of uncertainty in a broader context than the probabilistic paradigm, and present an alternative method of representation based on a generalization of probabilities.

Section 7.1 introduces the frame of work adopted in this chapter, namely, the representation of uncertain data involved in the construction, evolution, and exploitation of a catalog of space objects. Section 7.2 draws two examples of uncertain components from the frame of work, and highlights the dual nature of uncertainty in the SDA context. Adopting this duality, Section 7.3 then summarizes the collision assessment method in the conjunction plane and shows the limitations of the probabilistic framework for the representation of uncertainty. Section 7.4 introduces an alternative formulation, the estimation framework for *outer probability measures* (o.p.m.s), and presents its main features. Finally, Section 7.5 revisits the conjunction method with an o.p.m.-based, or *credibilistic*, representation of uncertainty.

7.1 Frame of Work

Probabilities are great at describing *random* phenomena, such as flipping coins or throwing dice. Yet we have an intuitive understanding that a typical system of study in the context of SDA—say, the *state* (kinematic state, mass, attitude, etc.) of a resident space object (RSO) at a given epoch—does not quite fit as the product of a random experiment. In order to illustrate and discuss the distinction between *randomness* and the more general notion of *uncertainty*, we focus throughout this chapter on the signal processing chain depicted in Figure 7-1.

Suppose that an operator's task is to maintain a catalog of RSOs, providing up-to-date information on the state of the identified objects to space analysts querying the catalog. In order to predict orbital trajectories, the operator exploits dynamical models established by the astrodynamic community, accounting for the forces influencing the object's motion (Earth's gravitational pull, solar radiation pressure, etc.) to varying degrees of fidelity. Since the *exact* behavior of orbiting objects is beyond the reach of the community, these models are *uncertain* (lower blue arrow in Figure 7-1). The information on the objects' states is corrected by observations collected from various sensor stations (e.g., telescopes or radars). The signal processing chain of a physical sensor always includes some form of process noise (e.g., distortions in the images produced by an optical sensor), so that the relationship between an object's observed state and the corresponding observation is *uncertain* (upper blue arrow in Figure 7-1). Since the catalog operator is provided with uncertain data, the states of the objects in the catalog, updated through sequences of data prediction and data correction steps, are *uncertain* as well (red arrows in Figure 7-1). Consequently, the state of an object transmitted to a space analyst querying the catalog is also *uncertain* (green arrow in Figure 7-1).

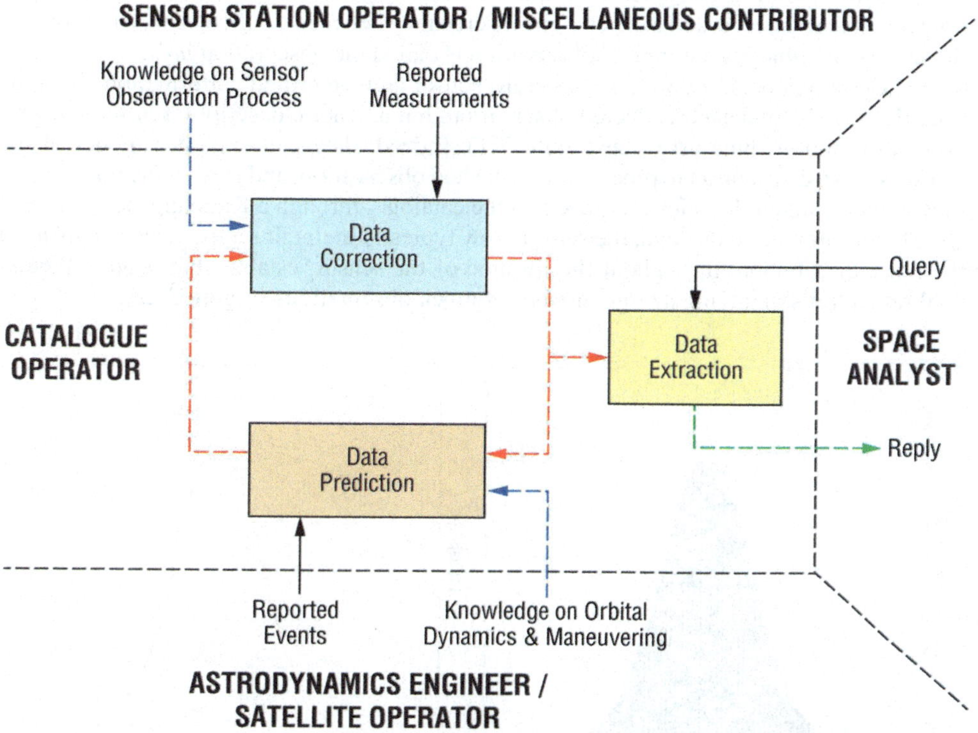

FIG. 7-1. Data Flow of an RSO Catalog. The dotted arrows represent uncertain data. The primary sources of uncertainty are the limited knowledge in the orbital dynamics and maneuvers of the RSOs, and the noise/detection capabilities of the sensors (blue lines). The modeling of the RSOs' dynamics and the sensor's observation process reflects this uncertainty, so that the information on the inferred population of RSOs is uncertain as well (red lines), and so is the data extracted from the catalog (green line).

7.2 Do We Always use Probabilities to Describe Random Systems?

7.2.1 Sensor's Observation Noise

We focus first on the integration of a new sensor to the data flow of the catalog (blue arrow in Figure 7-1). A sensor, when pointed at an RSO, provides some information on its state through a generated observation. The observations are elements of the sensor's *observation space*, some multi-dimensional space describing the physical quantities that the sensor measures—for example, a three-dimensional space describing vectors of range, azimuth, and elevation components for a radar. Assuming perfect sensing capabilities, one can map an object's state to its projection in the observation space, or theoretical observation (e.g., its true range, azimuth, and elevation coordinates). Yet measurement by any physical sensor is corrupted by imperfections in the sensing device, the storage and transmission of the recorded data, etc. The discrepancy between a collected observation and the underlying object's theoretical observation is called the observation *noise*.

The catalog operator does not expect the sensor crew to provide information about the specific noise behind each collected observation, but a generic description of the sensor's noise distribution (blue arrow in Figure 7-1). Indeed, they can exploit a model of the sensor's noise distribution to process *any* collected observation, and correct the information they maintain about the relevant objects in the catalog—through a Kalman filter [Kalman, 1960] for example. To this end, the sensor crew typically maintains a frequency analysis of the noise distribution, throughout the lifetime of the sensor, establishing it upon regular calibration trials involving a large number of mock observations (Figure 7-2a).

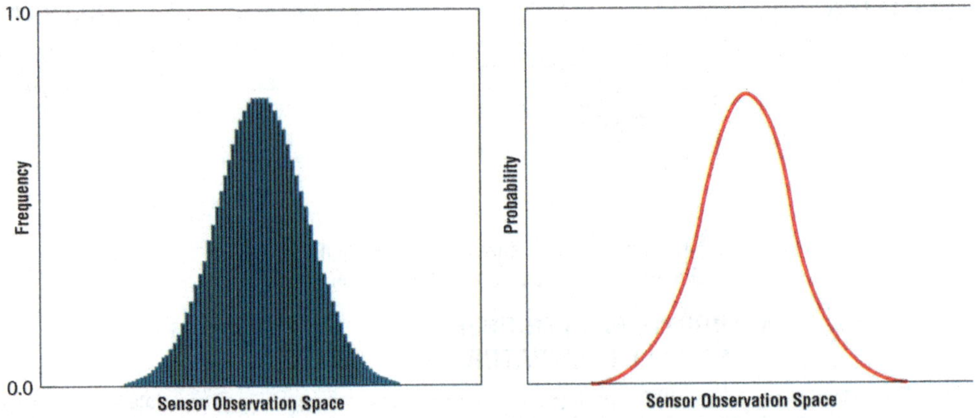

(a) Frequency pattern of the noise distribution, described by a histogram

(b) Probabilistic representation of the noise process, characterized by a probability density function (p.d.f.)

FIG. 7-2. **Probabilistic Interpretation of a Sensor's Observation Noise.** Left - frequency pattern of the noise distribution, described by a histogram. Right - probabilistic representation of the noise process, characterized by a probability density function (p.d.f.).

In this context, the noise pattern in Figure 7-2 (left) reflects some core uncertainty in the sensor's inner mechanisms, which remains beyond the reach of the sensor crew's analysis following extensive calibration trials. The interpretation of the sensor's noise generation as a *random* experiment then holds reasonably well, similar to the throwing of a die or the flipping of a coin, and the sensor crew can characterize this random process with the probability distribution reflecting the frequency pattern they have established (Figure 7-2 (right)). A probabilistic representation of the sensor noise—for example, a mean bias and a standard deviation on each component of the observation space—is an appropriate formulation for catalog operators, who needs to integrate this new sensor into their signal processing chain.

7.2.2 Object's State

We now focus on a catalog query about the state of a given RSO of interest at a given epoch, say, to assess collision risks with another spacecraft (green arrow in Figure 7-1). The purpose of space catalogs is to provide specific information on identified objects, and the expectations from the space analyst are very different from those of the catalog operator in the previous example (Section 7.2.1). Unlike a sensor's noise generation process, an RSO's dynamical process is a complex, unique, and transient experiment that no one can replicate at will to establish a frequency pattern from its recorded outcomes. Establishing frequency patterns on the *random* state of a cataloged object makes little sense to the catalog operator or to the space analyst, and the idea that a RSO's state at a given epoch is a *random* experiment characterized by such a frequency pattern appears ludicrous.

Yet the catalog operator only has limited information on the objects' states, and needs a way to quantify some form of *uncertainty* and convey it to the space analyst who produced the query. Throughout the underlying filtering algorithms (red arrows in Figure 7-1), space catalogs typically maintain a probabilistic description of the objects' states—for example, with a mean state and an associated covariance—and queries about an object's state are addressed with a probability distribution as well. That is, catalog operators and space analysts both handle p.d.f.s to describe objects' uncertain states, even though they do not lend them a random interpretation. However, since the probabilistic framework is designed to manipulate and describe *random* variables, information-based analyses set within this paradigm apply to *random* phenomena. We now illustrate, on a simple analysis of collision risks, that the lack of distinction between different sources of uncertainty may lead to confusing statements, misleading conclusions, and potentially ill-fated operational decisions.

7.3 An Example Problem: Assessing Collision Risks in the Conjunction Plane

We now describe a well-established technique to assess a collision event between two spacecrafts, the collision assessment in the conjunction plane [Alfriend et al., 1999]. In this probabilistic method, the two objects' states are described by probability distributions and a *probability of collision*, a scalar between 0 and 1 quantifying the likelihood of a collision, is estimated. The higher the probability, the more likely a collision; typically, avoidance actions will be appropriate if the probability of collision evaluated by the space analyst exceeds a given threshold.

7.3.1 Conjunction Plane and Probability of Collision

Suppose that a space analyst must assess the collision risk between two given RSOs. We assume that the situation meets the conditions for an assessment in the conjunction plane [Klinkrad et al., 2005], namely:

- The time of closest approach t_c of the two RSOs is known;

And, during a small period of time surrounding t_c:

- Both RSOs move along straight lines at constant and known velocities;

- A collision occurs if and only if the distance between the two RSOs' centers of gravity (in the position subspace) is below a known *collision radius* r_c;
- The positions of the two RSOs are described by two independent random variables, each of them characterized by a Gaussian p.d.f. with constant covariance.

Under these conditions, we can determine a collision event in the 2-dimensional conjunction plane, depicted in Figure 7-3, rather than the more complex 6-dimensional kinematic space.

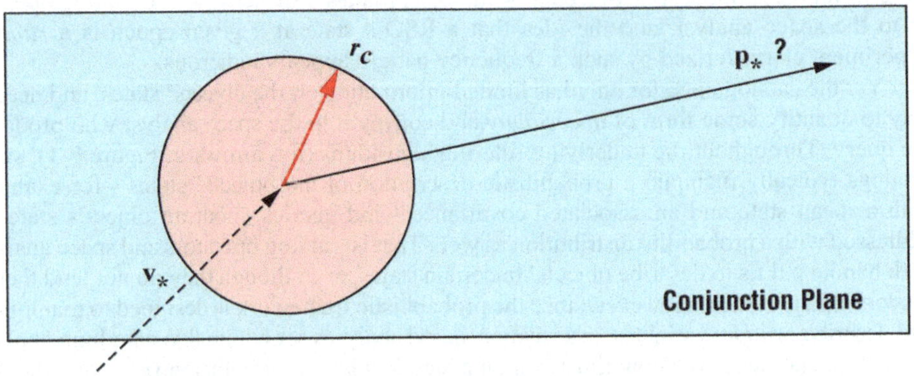

FIG. 7-3. **The Conjunction Plane, at the Time of Closest Approach t_c.** The relative velocity of the two RSOs, v_*, is known and defines the conjunction plane. The relative position of the RSOs in the conjunction plane, p_*, is unknown. If, and only if, the relative position lies within the collision region (depicted in red), a collision occurs.

The two objects come into contact and collide if their distance at closest approach is lower than the radius of collision r_c, that represents the relative geometries of the RSOs. The space analyst, then, must assess the probability of such an event. To do so, the analyst proceeds as follows. From the two six-dimensional Gaussian p.d.f.s describing the kinematic states of the objects at the time of closest approach, they produce first the three-dimensional Gaussian p.d.f. p_{p}, describing the relative position p of the two RSOs, parameterized with some mean value μ and some covariance Σ. The analyst then determines the conjunction plane and the projection of the relative position p into the conjunction plane, denoted as p_*. From the p.d.f. p_{p} they obtain the two-dimensional Gaussian p.d.f. p_{p_*}, describing the relative planar position p_*, parameterized with some mean value μ_* and some covariance Σ_*. A collision is defined by the event $\|\mathrm{p}_*\| \leq r_c$, i.e., when the drawn position in the conjunction plane is smaller than the radius of collision. This is given by the probability:

$$P_c = \int_{\|p_*\| \le r_c} N(p_*; \mu_*, \Sigma_*) dp_* \tag{7-1}$$

$$= |2\pi\Sigma_*|^{-1/2} \int_{\|p_*\| \le r_c} \exp\left(-\frac{1}{2}(p_* - \mu_*)^t \Sigma_*^{-1} (p_* - \mu_*)\right) dp_*$$

7.3.2 Risk Assessment

To ease the following discussion and without loss of generality, we suppose that the space analyst knows the state of one of the two RSOs, denoted as the *target object*. This way, the sole source of uncertainty regarding a collision event stems from the state of the other RSO, denoted as the *risk object*. We suppose that the analyst queries some information about the state of the risk object from the catalog (green arrow in Figure 7-1), and obtains a six-dimensional Gaussian p.d.f. Following the procedure described in the section above, they first produce the p.d.f. p_p of the relative position of the two objects, then produce the p.d.f. p_{p*} of the relative planar position in the conjunction plane, and finally determine the probability of collision Equation (7-1).

Figure 7-4 illustrates the probability of collision, assuming that the mean relative position μ of the RSOs is contained in the conjunction plane and has a magnitude of 10 m. The associated covariance Σ is assumed diagonal and isotropic, i.e., of the form $\Sigma = \sigma^2 I_3$, and the discussion focuses on the influence of the standard deviation σ on the estimated probability of collision.

It is formative, for the sake of this discussion, to check that the analysis above is consistent with a frequentist interpretation of uncertainty. That is, we assume first that the space analyst interprets the risk object's state as a *random* process, whose value accords with the p.d.f. queried from the catalog. If the standard deviation σ is zero (left-hand side of Figure 7-4), then the relative position takes the mean value μ with certainty. Thus, if the radius of collision is smaller than the mean distance of 10 m, the probability of collision is zero—i.e., the objects are certain to avoid each other; likewise, if the radius of collision is larger than the mean distance, the probability of collision is one—i.e., the objects are certain to collide with each other. When the standard deviation σ grows, so does the dispersion of the objects' relative position around its mean value μ.

In the case where the mean distance is larger than the radius of collision (lower plots in Figure 7-4), the probability that the drawn distance is smaller than the radius of collision increases from zero, and so does the probability of collision. Likewise, in the case where the mean distance is larger than the radius of collision (upper plots in Figure 7-4), the probability that the drawn distance is larger than the radius of collision decreases from one, and so does the probability of collision. If the standard deviation σ grows larger, the dispersion of the relative position around its mean value increases in magnitude, and so does the dispersion of the distance between the two objects.

Since the distance may be drawn with non-negligible probability from a larger range of values, the relative likelihood of drawing a value within the collision region decreases, and so does the probability of collision. At the limit case, when the standard deviation σ

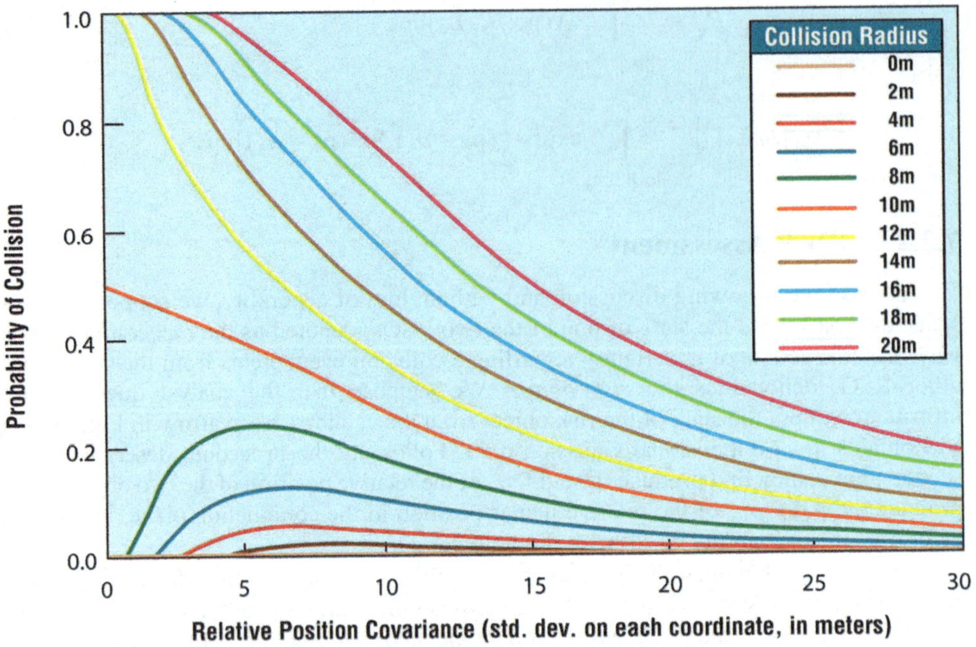

FIG. 7-4. Probability of Collision with Respect to the Standard Deviation, on Each of the Three Coordinates, of the Relative Position Between the Two RSOs [Delande et al., 2019]. The mean distance between the RSOs is fixed at 10 m. The different plots correspond to different values for the radius of collision, ranging between 0 m and 20 m.

tends to infinity, the distance can assume *any* non-negative value with equal probability: drawing a value smaller than the radius of collision becomes an event with negligible probability, and thus the probability of collision tends to zero. Broadly speaking, the more diffuse the p.d.f. transmitted by the catalog operator, the more erratic the behavior of the risk object, and the less likely a collision event becomes. This is known as the dilution effect of probability distributions [Alfano, 2003], [Alfano and Oltrogge, 2018], [Balch et al., 2018]. If the frequentist interpretation of uncertainty holds, the analysis above makes sense and is intuitively appealing. However, as discussed before in Section 7.2.2, the information content extracted from a catalog is of a different nature: in all likelihood, the p.d.f. provided by the catalog operator reflects the information gathered on the queried object so far, and is not the product of a frequency analysis.

In realistic queries, the space analyst interprets the covariance Σ as a measure of the quality of the information contained in the catalog; a low covariance, for example, suggests a high confidence that the object's state is in the vicinity of the mean value μ. In light of this alternative interpretation, the probabilistic analysis detailed above does not deliver a satisfying and safe assessment of the collision risks. Indeed, if a larger covariance Σ reflects a lower amount of information available from the catalog about the risk object, then the risks of collision should become more difficult to establish. The results in Figure 7-4, instead, lead to the counter-intuitive and hazardous conclusion that a collision becomes less likely when less is known about the risk object.

The risk assessment analysis illustrated in Figure 7-4, while sound under the premises of random uncertainty, does not bear scrutiny in a practical context and requires post-fixing by the space analysts. The dilution effect on probabilistic assessment methods is well known to the space community, and low probability collision values are not taken at face value [Balch et al., 2018]. *Unlikely* and *uncertain* collisions reflect distinct realities that may warrant different operational decisions, but a probabilistic representation does not provide such a distinction; in this context, the precise meaning of a probability of collision is a topic of discussion [Carpenter et al., 2017].

Other analyses hint at the inadequacy of p.d.f.s as sound representatives of the information we typically maintain on orbiting objects [Worthy and Holzinger, 2017]. There is merit, then, in exploring an alternative representation of the uncertain data in a space catalog.

7.4 The Estimation Framework for Outer Probability Measures

The limitations of the standard probabilistic approach to the representation of uncertainty have long been established, and alternative solutions have arisen in the last 50 years such as fuzzy logic [Zadeh, 1965], Dempster-Shafer theory [Dempster, 1967], [Shafer, 1976], possibility theory [Dubois and Prade, 1983], imprecise probabilities [Walley, 1991], fuzzy random sets [Yen, 1990], or plausibility measures [Friedman and Halpern, 2001]. They have seen wide use for applications where the information to be assimilated refers to a fixed unknown state; however, their application to the estimation of dynamical systems is less straightforward, limiting their relevance to sequential estimation problems.

The rest of this chapter introduces another solution with strong links to these alternative approaches to standard probability theory, but its formulation enables filtering equations to be derived and expressed simply for a range of Bayesian estimation problems.

7.4.1 Aleatory and Epistemic Uncertainty

We first consider a broad notion of uncertainty, encompassing the different concepts presented in Section 7.2 through the examples of the sensor's observation noise and the object's state.

We assume that any studied system possesses some inherent, irreducible core of uncertainty, representing the frequency pattern of its realizations through repeated outcomes. This random or aleatory uncertainty is characterized by a p.d.f. on the system's state space $\mathrm{X} \subseteq \mathrm{R}^n$ reflecting the frequency pattern, or aleatory profile, of the system. A deterministic system, in particular, has no random uncertainty: it takes some true state $x^* \in \mathrm{X}$ in a deterministic manner, and thus its state is characterized by the Dirac delta distribution centered on the true state x^*.

In the general case, an observer of the studied system does not have enough information to establish its frequency pattern, and thus does not *know* the p.d.f. characterizing the system's aleatory profile. In addition to the irreducible, aleatory uncertainty of the system, the observer possesses some systematic or *epistemic* uncertainty due to limited information about the rules governing the system's dynamics, limited opportunities to observe the system, or any other limiting factor. However, while naturally adapted to the representation of aleatory profiles, the probabilistic framework does not provide a straightforward tool to quantify the epistemic uncertainty of an observer.

Quantifying epistemic uncertainty is the main focus of the remainder of this chapter. It is informative to revisit the examples introduced in Section 7.2 in light of the dual representation of uncertainty into aleatory and epistemic components. The observation noise of a well-calibrated sensor, as studied by the operating crew, is a system that presents no uncertainty but a core component, established by the crew over countless calibration trials. Thus the crew has no epistemic uncertainty in the system, and their frequency analysis yields the p.d.f. characterizing the system's aleatory profile (Figure 7-2). On the other hand the information typically contained in a catalog about an object is limited, but one can expect the uncertainty on the object's state to be reducible (for the most part) through additional data. Thus the uncertainty contained in the catalog is largely epistemic in nature and, as illustrated by the risk assessment in Section 7.3, it is inadequately represented by a p.d.f. on the object's state space.

7.4.2 Quantifying the Epistemic Uncertainty

We represent an uncertain system, whose state is described by the state space X, by a uncertain variable X on X [Houssineau and Ristic, 2017]. The aleatory profile of the system is characterized by a probability distribution P_X on X, with associated density p_X. Thus the probability that the system's state lies within some region $R \subseteq \mathrm{X}$ is given by the scalar

$$P_X(R) = \int_R p_x(x)\mathrm{d}x \tag{7-2}$$

In particular, $P_X(\emptyset) = 0$ and $P_X(\mathrm{X}) = 1$, since the system must have a state in X. Probability distributions are *additive*, i.e., it holds that

$$P_X(R \cup R') = P_X(R) + P_X(R') \tag{7-3}$$

for any disjoint regions of the state space R, $R' \subseteq \mathrm{X}$, $R \cap R' = \emptyset$. In consequence, it holds that

$$P_X(R) + P_X(R^c) = 1 \tag{7-4}$$

for any region $R \subseteq \mathrm{X}$ and its complementary $R^c = \mathrm{X} \backslash R$, indicating that the system's state either lies in R or in R^c *almost surely*, i.e., with probability one.

From Probability Distributions to Outer Probability Measures

Since an observer does not have enough information to establish the system's aleatory profile, they do not *know* the probability distribution P_X or its associated p.d.f. p_X. The information they possess on the system, however, allows them to refine their knowledge on

P_X. We shall quantify the observer's uncertainty in the system with a function $\bar{P}$ on the subsets of the state space **X**, whose nature is yet to be determined.

If the observer had no evidence preventing the system from being in a given region $R \subseteq X$, then they would have no objection for the *unknown* probability $P_X(R)$ to be as high as one. A natural way to quantity their confidence of the system being in this particular region R would be to assign the weight $\bar{P}(R) = 1$. The observer may have no evidence preventing the system from being in the complementary R^c *as well*, leading them to assign the weight $\bar{P}(R^c) = 1$ too. We see that the function $\bar{P}$ cannot be a probability distribution, because the additive property Equation (7-4) precludes the assignment $\bar{P}(R) = \bar{P}(R^c) = 1$.

In order to quantity the observer's uncertainty in the system, we need to adopt a looser representation than probability distributions and replace the additive property by the less stringent *sub-additivity* property, i.e.,

$$\bar{P}(R \cup R') \le \bar{P}(R) + \bar{P}(R') \qquad (7\text{-}5)$$

for any regions of the state space $R, R' \subseteq X$, even if they are disjoint. The quantity $\bar{P}$ is, in fact, a generalization of a probability distribution and is called an *outer probability measure* (o.p.m.) [Houssineau and Bishop, 2018].

From Probability to Credibility

An o.p.m. $\bar{P}$ reflects the information possessed by a specific observer about the system, and provides some leads about the unknown, true probability distribution P_X characterizing the system. According to $\bar{P}$, the probability distribution of the system must be such that

$$1 - \bar{P}(R^c) \le P(R) \le \bar{P}(R), \text{ for any } R \subseteq X \qquad (7\text{-}6)$$

where we use the generic notation P to describe a potential candidate for the probability distribution P_X. It holds true for any o.p.m. $\bar{P}$ on X that $\bar{P}(\emptyset) = 0$ and $\bar{P}(X) = 1$ [Houssineau, 2018b], such that selecting R as X in Equation (7-6) yields

$$1 = 1 - \underbrace{\bar{P}(X^c)}_{= \bar{P}(\emptyset) = 0} \le P(X) \le \underbrace{\bar{P}(X)}_{=1} = 1 \qquad (7\text{-}7)$$

That is, the observer knows that $P(X) = 1$ and thus that the system lies *somewhere* in the state space X. If we choose $R \subseteq X$ as some specific region of interest for an analysis (e.g., a collision region), we see from Equation (7-6) that the observer cannot determine the probability $P(R)$, in the general case. However, they can provide an upper probability bound or *credibility* $\bar{P}(R)$, and a lower probability bound $1 - \bar{P}(R^c)$. Thus, according to the observer, the probability that the system is in R is no less than $1 - \bar{P}(R^c)$ and no more than $\bar{P}(R)$.

In the general case, infinitely many probability distributions P may satisfy the constraints Equation (7-6) for a given o.p.m. $\bar{P}$. Such an o.p.m. reflects a level of information insufficient to establish the aleatory profile of the system. When more information is acquired, the observer's uncertainty on the system's state reduces accordingly and the updated o.p.m. induces tighter bounds in Equation (7-6). In the limit case, the inequalities in Equation (7-6) become equalities and the corresponding o.p.m. becomes a probability distribution that characterizes—according to the observer—the system's aleatory profile.

A Special, Non-informative Case

It is instructive to build explicitly the o.p.m. $\bar{P}_u$ representing the least informative case, i.e., the absence of information regarding the system's state within X. Since the observer has no evidence against the system being in *any* non-empty region of the state space, it must hold that $\bar{P}_u(R) = 1$ for any $R \subseteq X, R \neq \emptyset$. From Equation (7-6) it follows that

$$0 = 1 - \underbrace{\bar{P}_u(R^c)}_{=1} \leq P(R) \leq \underbrace{\bar{P}_u(R)}_{=1} = 1, \quad \text{for any } R \subseteq X,\ R \neq \emptyset \tag{7-8}$$

Because *every* probability distribution on X verifies Equation (7-8), the observer cannot describe in any way the system's aleatory profile. As expected, then, $\bar{P}_u$ represents a genuine lack of information about the system from the observer's point of view.

7.4.3 Possibility Functions

Similar to probability distributions, outer probability measures can assume many different forms depending on the estimation problem at hand. Appropriate formulations exist, for example, if the observer wishes to learn the aleatory profile of the system from a family of parameterized distributions [Houssineau, 2018b]. We focus here on the specific class of o.p.m.s of the form

$$\bar{P}(R) = \max_{x \in R} f(x), \tag{7-9}$$

for any region $R \subseteq X$, where f is a non-negative function with supremum one, or *possibility function*, on X.

We here point out the similar structure between the expression of a *probability distribution* through its density in Equation (7-2), and the single-possibility formulation Equation (7-9) above. Because p.d.f.s are functions on the state space X that are easy to approximate with a variety of approaches, such as Gaussian Mixtures (GMs) [Alspach and Sorenson, 1972] or Sequential Monte Carlo (SMC) implementations [Doucet et al., 2001], they are much easier to handle than probability measures in practical estimation problems. Similarly, the single-possibility formulation (Equation 7-9) is flexible enough to describe a wide range of uncertain components and is particularly convenient to handle in Bayesian inference mechanisms.

For the remainder of the chapter we switch our focus from probability distributions and o.p.m.s to p.d.f.s and possibility functions. When the information that an observer possesses is represented by a possibility function f on X, they infer that the system's aleatory profile must be such that

$$1 - \max_{x \in R^c} f(x) \leq \int_R p(x)\mathrm{d}x \leq \max_{x \in R} f(x), \quad \text{for any } R \subseteq \mathrm{X} \tag{7-10}$$

where we use the generic notation p to describe a potential candidate for the system's p.d.f. p_X.

A Few Simple Examples

A first example of a possibility function involves the indicator function 1_A on some region $A \subseteq \mathrm{X}$ of the state space, i.e., for any point $x \in \mathrm{X}$.

$$1_A(x) = \begin{cases} 1, & \text{if } x \in A, \\ 0, & \text{otherwise,} \end{cases} \tag{7-11}$$

Figure 7-5 illustrates the coarseness of the information represented by the possibility function 1_A.

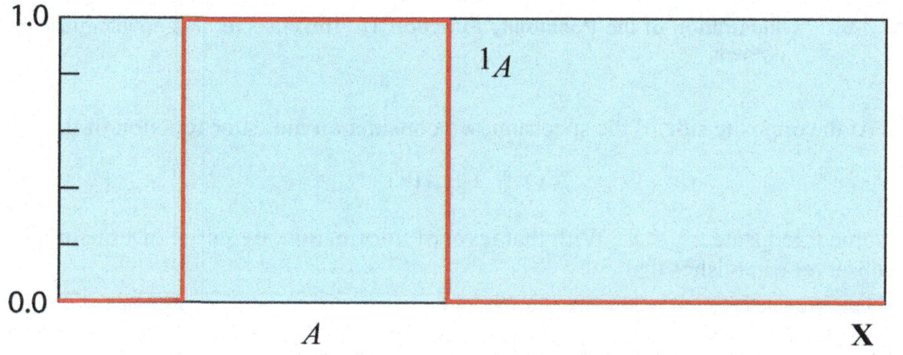

FIG. 7-5. Illustration of the Possibility Function 1_A. The observer knows the system lies within A, and nothing more.

The observer does not have information on the frequency pattern of the system's outcomes, except that it cannot reach states outside A; that is, they only know that the system lies somewhere in A. This information is not enough to determine the system's aleatory profile, since any p.d.f. p whose mass is concentrated within A is still a potential candidate. For example, the state of the system could be uniformly distributed within A, or uniformly distributed within a subset of A, or the system could be deterministic and assume some fixed state within A, etc. Thus, the possibility function Equation (7-11) provides a simple representation of the information given by *negative evidence*, i.e., when the observer has established where the system *cannot be*, such as in the initial orbit determination procedure [DeMars and Jah, 2013].

The least informative o.p.m. Equation (7-8) is easily expressed with the possibility function Equation (7-11) if we select $A = \mathrm{X}$, i.e.,

$$f(x) = 1_{\mathrm{X}}(x) = 1 \tag{7-12}$$

for any $x \in \mathrm{X}$, as depicted in Figure 7-6. By definition, every p.d.f. on X concentrates its mass in X and thus the possibility function (Equation 7-12) does not characterize in any way the system's aleatory profile (Figure 7-6).

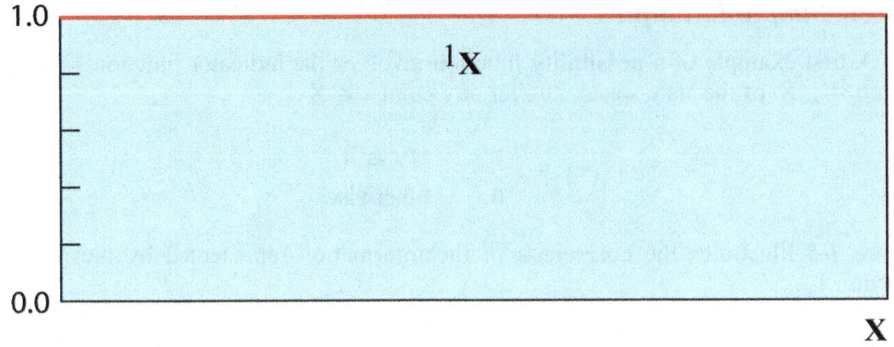

FIG. 7-6. Illustration of the Possibility Function 1_{X}. The observer has no information on the system.

At the opposite side of the spectrum, we consider an indicator function of the form

$$f(x) = 1_{\{x^*\}}(x) \tag{7-13}$$

for some fixed state $x^* \in \mathrm{X}$. With that level of information, by virtue of Equation (7-10), the observer establishes that

$$\int_R p(x)\mathrm{d}x = \begin{cases} 1, & \text{if } R \ni x^*, \\ 0, & \text{otherwise,} \end{cases} \tag{7-14}$$

for any region $R \subseteq \mathrm{X}$. The only p.d.f. that satisfies Equation (7-14) is the Dirac delta distribution δ_{x^*} concentrating its mass on x^*; thus, the observer has established that the system is deterministic and takes state x^* *almost surely* (Figure 7-7).

Gaussian Possibility Functions

Gaussian p.d.f.s are widely popular in the modeling of Bayesian filters, that is, sequential inference algorithms such as the Kalman filter, because they are convenient to handle in inference rules and they provide a simple description of a random phenomenon through its first two statistical moments (mean and covariance). For example, they often serve to model the precision and accuracy of physical sensors or the process noise in an orbital propagation model. The possibility function

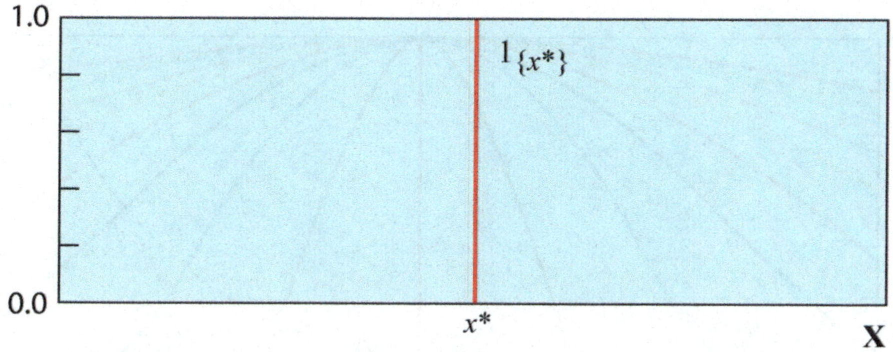

FIG. 7-7. **Illustration of the Possibility Function $1_{\{x^*\}}$.** The observer has full information on the system, and knows it takes state x^* with certainty.

$$\bar{N}(x;\mu, S) = \exp\left(-\frac{1}{2}(x-\mu)^t S^{-1}(x-\mu)\right) \tag{7-15}$$

for any $x \in \mathrm{X}$, is called the Gaussian possibility function on X with parameters μ and S and plays a similar role in the o.p.m. estimation framework [Houssineau et al., 2020]. It is convenient to refer to μ and S as the expected value and covariance, respectively, of the Gaussian possibility function $\bar{N}(\cdot;\mu, S)$. This is, however, an abuse of language since they do not have the same interpretation as their statistical namesakes: the Gaussian possibility is *not* a p.d.f., and does not characterize a random phenomenon.

Figure 7-8 illustrates a family of Gaussian possibility functions sharing the same mean but increasing covariances. We see that if the covariance tends to zero, the Gaussian possibility function tends to the indicator $1_{\{x^*\}}$ and represents the maximum level of information on the system. Conversely, if the covariance tends to infinity, the Gaussian possibility function tends to the indicator 1_X and represents the minimum level of information on the system.

7.4.4 Bayesian Filtering with Outer Probability Measures (o.p.m.s)

Because probability distributions and o.p.m.s are derived from the same fundamental tools in measure theory, we can develop rigorous Bayesian inference mechanisms to construct o.p.m.-based tracking algorithms as an alternative to the usual p.d.f.-based approaches. Analogues of probabilistic concepts such as conditional possibility functions, Markov transitions, likelihood functions, or Bayesian inference rules can be derived within the estimation framework with o.p.m.s [Houssineau and Bishop, 2018]. Practical implementations of o.p.m.-based tracking algorithms can be inspired by classical p.d.f.-based techniques; in particular, an extension of the SMC principles [Doucet et al., 2001] to possibility functions has been proposed in [Houssineau and Ristic, 2017], and the extension of the GM principles [Alspach and Sorenson, 1972] is explored in [Houssineau, 2018a].

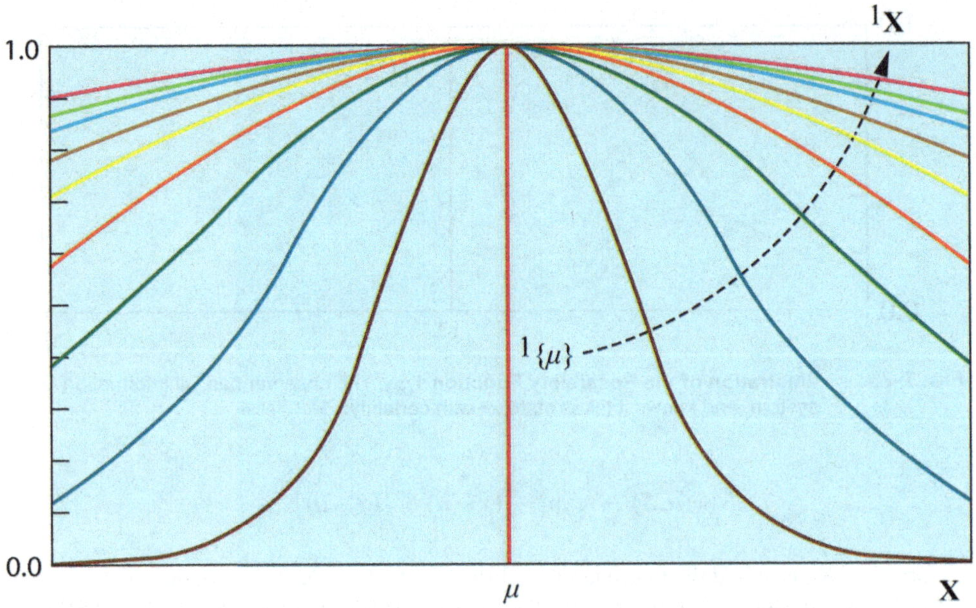

FIG. 7-8. An Illustration of a Family of Gaussian Possibility Functions $f_n(\cdot) = \bar{N}(\cdot;\mu, S_n)$, with Increasing Covariance S_n [Delande et al., 2018b]. When $S_n \to 0$, f_n converges to the indicator function $1_{\{\mu\}}$; when $S_n \to \infty$, f_n converges to the indicator function 1_X.

In the context of SDA, possibility functions are useful for tracking a single RSO initialized by an initial orbit determination procedure [Delande et al., 2018a]. Another tracking scenario exploits a Bayesian filter adopting a p.d.f.-based formulation for the object's state and the orbital propagation model, and possibility functions for the modeling of the error of the two-line elements (TLEs) extracted from the U.S. Strategic Command (USSTRATCOM)'s catalog [Delande et al., 2018b]. In particular, this study shows that random and uncertain components can be used concurrently within the o.p.m. framework, in order to design a Bayesian filtering solution according to the specifics of the scenario.

7.5 Revisiting the Collision Assessment Example

We now revisit the collision risks analysis introduced in Section 7.3, assuming that the information maintained by the catalog of objects is represented with possibility functions instead of p.d.f.s.

7.5.1 Conjunction Analysis with Possibility Functions

We make the same assumptions as in Section 7.3.1, except that

- The positions of the two RSOs are described by two independent uncertain variables, each of them characterized by a Gaussian possibility function with constant covariance.

Similar to the probabilistic assessment introduced earlier, the focus of the analysis then shifts to the two-dimensional conjunction plane in Figure 7-3. A collision is still defined by the event $\|p_*\| \leq r_c$, i.e., when the relative planar position p_* in the conjunction plane is smaller than the radius of collision r_c, whose probability is given by:

$$P_c = \int_{\|p_*\| \leq r_c} p_{p_*}(p_*) dp_* \tag{7-16}$$

where p_{p_*} is the p.d.f. characterizing the relative planar position p_*. However, the p.d.f. is unknown to the space analyst, in the general case, because the aleatory profile of the position p_* is not available from the catalog.

From the two six-dimensional Gaussian possibility functions describing the kinematic states of the objects at the time of closest approach, the analyst determines instead the three-dimensional Gaussian possibility functions f_p, describing the relative position p of the two RSOs, parameterized with some expected value μ and some covariance Σ. The change of variables on Gaussian possibility functions described in this paragraph has a structure similar to its counterpart with Gaussian p.d.f.s. [Delande et al., 2019].

From the possibility function f_p they obtain the two-dimensional Gaussian possibility function f_{p*}, describing the relative planar position p_*, parameterized with some mean value μ_* and some covariance Σ_*. Using Equation (7-10) we bound the probability of collision Equation (7-16) as

$$1 - \max_{\|p_*\| > r_c} \overline{N}(p_*; \mu_*, \Sigma_*) \leq P_c \leq \max_{\|p_*\| \leq r_c} \overline{N}(p_*; \mu_*, \Sigma_*) \tag{7-17}$$

7.5.2 Risk Assessment

As with the probabilistic assessment in Section 7.3, in order to ease the following discussion without loss of generality, we suppose that the space analyst knows the state of the target object. Thus, they only query the catalog for some information about the state of the risk object (green arrow in Figure 7-1), and they are provided with a six-dimensional Gaussian possibility function. Following the procedure described in the above section, they first produce the Gaussian possibility function f_p of the relative position of the two objects, then the Gaussian possibility function f_{p*} of the relative planar position in the conjunction plane, and finally they determine the upper and lower bounds on the probability of collision through Equation (7-17).

Figure 7-9 illustrates the upper and lower bounds on the computed probability of collision in conditions similar to those in Section 7.3. That is, we assume an expected relative position μ contained in the conjunction plane with a magnitude of 10 m, and a diagonal, isotropic covariance $\Sigma = \sigma^2 I_3$. Again, the discussion focuses on the influence of the standard deviation σ on the upper and lower probability bounds that the analyst estimated.

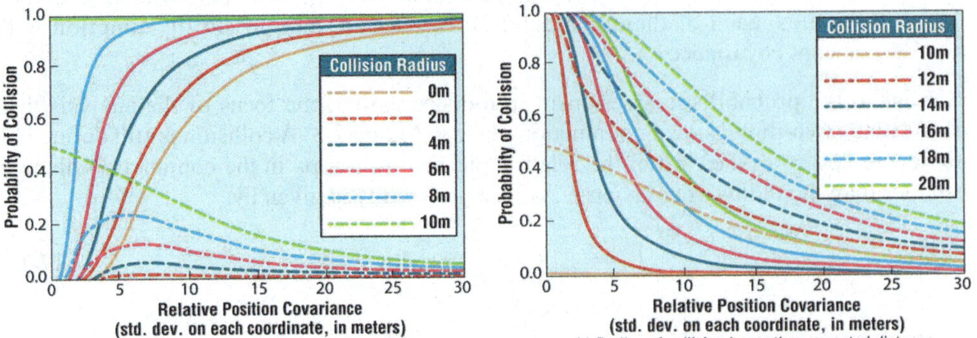

a) Radius of collision smaller than expected distance. b) Radius of collision larger than expected distance.

FIG. 7-9. Upper and Lower Bounds on the Probability of Collision with Respect to the Standard Deviation, on Each of the Three Coordinates, of the Relative Position Between Two RSOs [Delande et al., 2019]. The expected distance between the RSOs is fixed at 10 m. The left-hand plot depicts upper probability bounds (plain lines) for radii of collision ranging between 0 m and 10 m; the common, lower probability bound is the horizontal line with probability zero. The right-hand side depicts lower probability bounds (plain lines) for radius of collision ranging between 10 m and 20 m; the common, upper probability bound is the horizontal line with probability one. For the sake of comparison, the dotted lines illustrate the probabilities of collision computed in Section 7.3.

We focus first on the case where the radius of collision r_c is smaller than the expected planar distance of 10 m (Figure 7-9a). When the standard deviation σ is close to zero (left-hand side of the plots), the catalog gives enough confidence that the planar distance is within the vicinity of the mean estimated value of 10 m, which remains outside of the collision region. Thus, the upper probability (or credibility) of collision is close to zero, and a collision event is *known* to be unlikely. When the standard deviation grows, the catalog is less informative about the objects' respective position and does not exclude the planar distance to assume a value further and further away from the mean estimated value. Thus, the upper probability of collision grows and a collision event is increasingly possible. The standard deviation tending to infinity (right-hand side of the plots) indicates that the catalog possesses no information about the objects' respective state (e.g., the catalog does not identify this object). The upper probability of collision tends to one, while the lower probability of collision is still zero; that is, the space analyst cannot draw any conclusion regarding the risks of collision.

Similar conclusions apply in the case where the radius of collision r_c exceeds the expected planar distance of 10 m (Figure 7-9b). When the standard deviation σ tends to zero (left-hand side of the plots), the catalog gives enough confidence that the planar distance is within the collision region; the lower probability of collision is close to one, and a collision is *known* to occur. When the standard deviation grows, the catalog opens to the possibility that the planar distance might assume a larger value than the radius of collision. The lower probability of collision decreases accordingly, and the space analyst concludes that a non-collision event is increasingly possible. Again, when the standard deviation tends to infinity (right-hand side of the plots), the catalog possesses no information about the objects' relative position and the space analyst cannot conclude anything about the risks of collision.

We see that the conjunction analysis with possibility functions follows an approach similar to the usual conjunction analysis with p.d.f.s, and they both transcribe the initial information on the objects' states under similar assumptions. The credibilistic approach is premised upon the fact that realistic space catalogs are not informative enough to maintain an aleatory profile of the objects' states, and thus space analysts should not rely on p.d.f.s for their assessment of collision risks. Since the credibilistic approach does not evaluate directly the likelihood of a collision event, comparing its output to a threshold of acceptable collision risks does not lead the space analyst towards a straightforward, clear-cut decision regarding the necessity of an avoidance maneuver. The upper and lower probabilities of collision are nonetheless a principled quantification of the confidence of the estimation carried by the catalog and they provide natural grounds for the derivation of information-based collision avoidance policies, but this discussion is out of the scope of this chapter.

References

Alfano, S. Aug. 2003. *Relating Position Uncertainty To Maximum Conjunction Probability*. In AAS/AIAA Astrodynamics Specialist Conference, pages AA 03–548.

Alfano, S. and D. Oltrogge. 2018. *Probability of Collision: Valuation, variability, visualization, and validity*. Acta Astraunotica, 148:301–316.

Alfriend, K. T. et al. 1999. *Probability of Collision Error Analysis*. Space Debris, 1(1):21–25.

Alspach, D. L. and H. W. Sorenson. 1972. *Nonlinear Bayesian Estimation Using Gaussian Sum Approximations*. Automatic Control, IEEE Transactions on, 17(4):439–448.

Balch, M. S. et al. 2018. *Satellite Conjunction Analysis and the False Confidence Theorem*. ArXiv preprint, arXiv:1706.08565.

Carpenter, J. R. et al. 2017. *Relevance of the American Statistical Association's Warning on P-Values for Conjunction Assessment*. In AAS/AIAA Astrodynamics Specialist Conference 2017, pages AAS 17–614.

Delande, E. D. et al. 2018a. *A New Representation of Uncertainty for Data Fusion in SSA Detection and Tracking Problems*. In Information Fusion, Proceedings of the 21st International Conference on.

Delande, E. D. et al. 2018b. *Physics and Human-Based Information Fusion for Improved Resident Space Object Tracking*. Advances in Space Research. DOI: 10.1016/j.asr.2018.06.033 (preprint arXiv:1802.07408).

Delande, E. D. et al. January 2019. *A New Representation of Uncertainty For Collision Assessment*. In 2019 AAS/AIAA Spaceflight Mechanics Meeting.

DeMars, K. J. and M. K. Jah. 2013. *Probabilistic Initial Orbit Determination Using Gaussian Mixture Models*. Journal of Guidance, Control, and Dynamics, 36(5):1324–1335.

Dempster, A. P. 1967. *Upper and Lower Probability Inferences Based on a Sample From a finite Univariate Population*. Biometrika, 54(3-4):515–528.

Doucet, A. et al. 2001. *Sequential Monte Carlo Methods in Practice*. Statistics for Engineering and Information Science. Springer.

Dubois, D. and H. Prade. 1983. *Ranking Fuzzy Numbers in the Setting of Possibility Theory. Information Sciences*, 30(3):183–224.

Friedman, N. and J. Y. Halpern. 2001. *Plausibility Measures and Default Reasoning*. Journal of the ACM, 48(4):648–685.

Houssineau, J. 2018a. *Detection and Estimation of Partially-Observed Dynamical Systems: An Outer-Measure Approach.* ArXiv preprint, arXiv:1801:00571.

Houssineau, J. 2018b. *Parameter Estimation with a Class of Outer Probability Measures. ArXiv preprint, arXiv:1801.00569.*

Houssineau, J. and A. Bishop. 2018. *Smoothing and Filtering with a Class of Outer Measures.* SIAM/ASA Journal on Uncertainty Quantification, 6(2):845–866.

Houssineau, J. et al. 2020. *Elements of Asymptotic Theory with Outer Probability Measures.* ArXiv preprint, *arXiv:1908.04331*.

Houssineau, J. and B. Ristic. 2017. *Sequential Monte Carlo Algorithms for a Class of Outer Measures.* ArXiv preprint, arXiv:*1704.01233.*

Kalman, R. E. 1960. *A New Approach to Linear Filtering and Prediction Problems.* Journal of Basic Engineering, 82(1):32–45.

Klinkrad, H. et al. Apr. 2005. *Collision Avoidance for Operational ESA Satellites.* In Proceedings of the Fourth European Conference on Space Debris.

Shafer, G. 1976. *A Mathematical Theory of Evidence*, V*olume 1.* Princeton University Press, Princeton, RI.

Walley, P. 1991. *Statistical Reasoning with Imprecise Probabilities.* Taylor & Francis, London.

Worthy, J. L. III and M. J. Holzinger. 2017. *Use of Uninformative Priors to Initialize State Estimation for Dynamical Systems.* Advances in Space Research, 60(7):1373–1388.

Yen, J. 1990. *Generalizing the Dempster-Schafer Theory to Fuzzy Sets*. IEEE Transactions on Systems, Man and Cybernetics, 20(3):559–570.

Zadeh, L. A. 1965. *Fuzzy Sets*. Information and Control, 8:338–353.

Chapter 8

Space Domain Awareness and Decision-Making for Flight Safety, Environmental Sustainability, and Security

Marc J. Berkowitz
Lockheed Martin Corporation

Outer space is a complex domain. It is a unique operating environment with its own characteristics, natural hazards, and human-made threats. Space has no political boundaries and reflects the dynamic, uncertain, and dangerous international security environment on Earth. Awareness of conditions, activities, and events in space is essential to inform decisions about flight safety, environmental sustainability, and security that could impact individuals, nations, and even humankind.

Increasing numbers of nation-states, multinational consortia, and business enterprises are launching and operating spacecraft in pursuit of prestige, influence, knowledge, prosperity, and power. As space technology diffuses around the world, collaboration, commerce, and competition in the domain is increasing among nations, consortia, and enterprises. The network of relationships across the space community mirrors the multi-polarity and globalization of the international system.

While the volume of space is immense, a growing population of resident space objects and debris is increasing congestion in Earth orbits, electromagnetic and radiofrequency interference, and safety risks. The growing number of spacefaring entities and extent of space activity fuel competition for orbital slots and spectrum. Concurrently, there is increasing contention in space as threats to space assets have re-emerged because of mistrust and insecurity in international relations. Indeed, foreign powers pose threats to space systems and supporting infrastructure. Near-peers, regional powers, and even non-state actors have the ability to disrupt, deny, degrade, and destroy space assets or services.

The expanding scope and utility of space activities and attendant concerns about the growing risk of safety mishaps, environmental incidents, and security threats drive the demand for space domain awareness (SDA). Consequently, the U.S. Government (USG) issued a national policy statement to enhance space flight safety and ensure continued leadership, preeminence, and freedom of action in space [SPD-3, 2018]. The policy establishes goals, sets guidelines, and realigns roles and responsibilities for space situational awareness (SSA) data sharing and providing space traffic management (STM) services.

This chapter addresses the roles of SDA in space operational decision-making. First, we discuss the context for the evolution of the SDA mission and functions. Next, we examine the different authorities, levels, and information needs for SDA decision support. Finally, we address SDA for STM as well as space command and control (C2).

8.1 Context

The lexicon regarding awareness of conditions, activities, and events in outer space has changed since the beginning of the space age. It has evolved from space surveillance and SSA to SDA. These terms are not synonymous. Indeed, the terminology reflects changes in the scope of the mission and associated functions. These changes shifted the emphasis of SDA from cataloging space objects and providing early warning of ballistic missile attack to support for civil and commercial space operations as well as readiness for defensive and offensive space warfighting.

Space surveillance is "the observation of space and of the activities occurring in space [JSC, 2016]." SSA is "cognizance of the requisite current and predictive knowledge of the space environment and the operational environment upon which space operations depend (JSC, 2016)." For the purposes of national STM policy, SSA is defined as "the knowledge and characterization of space objects and their operational environment to support safe, stable, and sustainable space activities [SPD-3, 2018]." SDA is a significantly broader concept analogous to maritime domain awareness [CNO, 2007]. SDA is "the effective understanding of anything associated with the space domain that could impact the security, safety, economy, or environment of a nation."

During the Cold War, the U.S. national security establishment developed and fielded capabilities to prevent strategic surprise, provide indications and warning of attack, and conduct defense and intelligence space operations. While the U.S. intelligence community collected foundational, scientific, technical, and operational intelligence about foreign space activities, its primary focus was on observing activities on Earth to pierce the veil of secrecy cloaking Soviet leadership intentions and military capabilities.

Consequently, the U.S. Department of Defense (DoD) assumed responsibility for the surveillance of space mission. This primarily involved detecting, identifying, tracking, and cataloging objects launched into and operating in space. The driving imperative was to enable defensive preparations against nuclear-armed ballistic delivery vehicles and orbital bombardment weapons that could threaten national survival.

The space surveillance mission focused on metric tracking and cataloging to determine whether objects launched into space were weapons or benign spacecraft. It employed a network of dedicated, collateral, and contributing ground-based radar and optical sensors. The ability to determine if an object launched into space is a ballistic missile or anti-satellite weapon threatening U.S. national interests or a harmless satellite remains a core mission need. Secondary purposes of the mission that also remain basic needs are the ability to monitor compliance with international treaties and agreements as well as provide warning of overflight for operations security and force protection.

After the threat posed by the Soviet Union receded, space was no longer solely the province of the superpowers. Capabilities to launch and operate space systems proliferated around the world. Civil and commercial space activities increased significantly as numerous nations and enterprises began to launch, operate, and employ space assets. Similarly, the defense and intelligence uses of space expanded to involve other nations besides the United States and the former Soviet Union.

The need to adjust to changes in the international system and their implications for space operations prompted the transition from space surveillance to SSA. While the USSR's dissolution diminished concerns about threats to U.S. survival, the diffusion of political, economic, and military power in the international system was mirrored in space. This created a more complicated operating environment and the need for improved awareness of conditions, activities, and events in space.

SSA expanded the mission to include characterizing an object's capabilities as well as monitoring and assessing its operating behavior. The growth of space activities increased demand for flight safety support, including the ability to: analyze the risk of conjunctions; avoid collisions during launch, insertion, maneuvers, deorbit, and reentry; support anomaly resolution; deconflict electromagnetic and radiofrequency interference; and deconflict laser illuminations into space.

In the past decade, the further growth of civil, commercial, and national security space activities, as well as the reemergence of threats to space assets, prompted the evolution from SSA to SDA. The growing orbital debris population generated by increased launch and on-orbit activities, collisions between spacecraft and other objects (e.g., Iridium-33 and Cosmos-2251 in 2009), orbital breakups (e.g., Hitomi X-Ray Observatory in 2006), China's 2007 direct-ascent kinetic energy anti-satellite test, and the prospect of "mega" constellations with thousands of satellites in low Earth orbit, increased international concerns about flight safety, environmental sustainability, and security. This highlighted the need for enhanced awareness of congestion, competition, and contention in an increasingly complex operating domain.

In particular, the renewed threat demonstrated that foreign powers could hold space assets at risk and deny the United States freedom of action in the domain [DNI, 2019]. It encompasses a full spectrum of cyber, electronic warfare, radiofrequency, kinetic energy, directed energy, and nuclear weapons effects. Moreover, the threat involves terrestrial, cyber, and space weapons, including small orbital systems capable of rendezvous and proximity operations [Weeden, 2010; Weeden, 2015].

SDA thus continues to expand beyond such space surveillance and SSA functions as detecting, identifying, tracking, cataloging, characterizing, monitoring, and assessing. It now emphasizes predictive analytics to forecast anticipated patterns of behavior, provide indications and warning of hostile intentions and actions, and attribute hostile behaviors. Moreover, to support deterrence, control escalation, or, if necessary, defeat threats, SDA must support space control operations conducted to protect U.S. and friendly space assets, prevent or negate an adversary's use of space systems for hostile purposes, and assess battle damage or combat effects.

SDA capabilities are enhanced by the information technology revolution and attendant advances in tasking, collecting, processing, curating, accessing, discovering, fusing, analyzing, and disseminating data and information. In addition to intelligence, surveillance, and reconnaissance capabilities, we now have civil, commercial, academic, and international sources of SDA data. These include sensors that passively and actively track orbiting spacecraft (radar, optical, and beacon tracking), telemetry that enables the system to act as a sensor, data provided by space system owners and operators, sensors that passively and actively monitor the space environment, and in-situ monitoring, sampling, self-position, and anomaly reporting from onboard and offboard sensors across platforms and orbits. Modern capabilities for sensor orchestration, information management, fusion, artificial intelligence and machine learning, big data, and predictive analytics, among other things, are improving the performance of the system-of-systems executing the SDA mission.

SDA data and information may encompass origin, ownership, orbit, position, location, velocity, size, capability, and intention as well as historical orbital information. Such data and information feed decision support processes for STM to monitor responsible, safe, and orderly space flight behaviors across the operational lifetime of spacecraft. SDA support for STM involves monitoring, coordinating, facilitating, and deconflicting space activities to: assess environmental and operational risks; avoid or mitigate hazards; and preclude interference or collisions that endanger flight safety and environmental sustainability and adversely impact mission performance. SDA functions that are critical for STM include: space weather forecasting; conjunction analyses; launch and on-orbit collision avoidance; deorbit and reentry predictions; and space catalog maintenance.

SDA information sources also feed decision support for space C2 to plan, direct, monitor execution, and assess defense and intelligence space operations. This involves assuring unimpeded access to and operations in space to collect, generate, and relay information. It also involves countering an adversary's use of space for hostile purposes. SDA functions that are critical for space C2 include: treaty monitoring; overflight warning; threat characterization; indications and warning; attribution; intelligence preparation of the battlespace; operational preparation of the battlespace; combat identification; targeting; and battle damage or combat effects assessment.

8.2 Decision Levels, Authorities, and Information Needs

SDA capabilities must be designed to support three levels of decision-making: strategic, operational, and tactical. At the strategic level is the President of the United States with authority to make decisions in the roles of Chief Executive and Commander-In-Chief. The President may delegate authority to subordinate officials; for example, authority for the employment of weapons for national or collective self-defense may be delegated to the Secretary of Defense or further down the chain of command.

Authorities for public sector (civil, defense, and intelligence) space activities, as well as regulation of private sector or commercial space activities, are dispersed across the USG. The Administrator of NASA, the Secretary of Transportation, and the Secretary of Commerce are responsible for various aspects of civil space operations as well as regulating commercial space activities. The Secretary of Defense is responsible for defense space operations. The Secretary of Defense and Director of National Intelligence share responsibility for national reconnaissance space operations. The Secretary of State is responsible for registering space objects with the United Nations, and the Federal Communications Commission regulates orbital slots and use of the radiofrequency spectrum through the International Telecommunications Union (ITU).

At the operational level, the NASA Administrator is responsible for human spaceflight, space science, and space exploration. The Secretary of Commerce—through the Administrator of NOAA—is responsible for environmental monitoring, STM, and regulating the operation of private remote sensing space systems (e.g., commercial imagery). The Secretary of Transportation—through the Administrator of the FAA—is responsible for regulating commercial space transportation (i.e., launch and reentry operations). With respect to national security space missions, the Commander of U.S. Space Command (USSPACECOM) is responsible for defense space operations, and the Director of the National Reconnaissance Office is responsible for national reconnaissance space operations.

USSPACECOM was reestablished as a unified, geographic combatant command. It is assigned an area of responsibility (AOR) that begins 100 kilometers (62 miles) above Earth and extends beyond. An AOR is the geographical area associated with a combatant command within which a combatant commander has authority to plan and conduct operations. The Combined Forces Space Component Command and Joint Task Force Space Defense, subordinate to USSPACECOM, use two primary operations centers—the Combined Space Operations Center and National Space Defense Center [USSTRATCOM, 2017 a and b]. In the future, the USG may establish a combined joint task force, joint interagency task force, or other organizational structure to enable the use of all the authorities necessary to protect and defend U.S. national interests in space under a single operational commander. In addition to defense and intelligence, these will need to include authorities for homeland security (critical infrastructure protection, emergency mobilization, and management), law enforcement, civil space operations, and regulation of commercial space activities. The task force would enable the commander and staff to work with interagency partners, state, territorial, local, and tribal agencies, foreign military forces and government partners, non-governmental organizations, and the private sector.

At the tactical level of decision-making are USG (civil, defense, intelligence) and commercial space system owners and operators. This is the level at which responsibility rests for commanding operations of specific space platforms and associated infrastructure. The following are a few tactical level examples. The 2nd Space Operations Squadron of the 50th Space Wing commands the U.S. Global Positioning System. It operates the satellite constellation from the Global Positioning System Operations Center. NASA's Mission Control Center commands the International Space Station, while the Payload Operations and Integration Center controls experiments on board the station. The Satellite Operations Control Center under NOAA's National Environmental Satellite, Data, and Information Service's Office of Satellite and Product Operations commands the U.S. fleet of geosynchronous and polar orbiting weather satellites. Various commercial satellite operations centers in the U.S. and abroad command private constellations of telecommunications satellites.

Decision makers at each level have different authorities, requirements, and information needs for SDA support. This includes authorities for armed forces (Title 10 United States Code), war and national defense (Title 50), foreign relations (Title 22), national guard (Title 32), homeland security (Title 6), law enforcement (Title 28), telecommunications (Title 47), and national and commercial space programs (Title 51). Different levels of authority and decision support require different types of spatial, spectral, and temporal data and information. They also need different ways to visualize the data and information to facilitate STM, space C2, or other decisions. Consider, for example, differences in the SDA support needed between civil and national security missions or across the range of military operations: military engagement, security cooperation, and deterrence; crisis response and limited contingency operations; and large-scale combat operations.

A critical question for every decision-making level is whether conditions, activities, and events in space are consistent with prior experience, standard practices, and anticipated patterns of behavior or whether they are inconsistent and pose risks of a safety mishap, environmental incident, or threat. Decision makers base their decisions upon some degree of uncertainty, ambiguity, or confidence. Moreover, there is a difference between uncertainty based on randomness versus ignorance (see Chapter 7). SDA needs, of course, a plethora of information; Table 8-1 lists examples of such needs.

TABLE 8-1. Examples of Space Domain Awareness Information Needs. SDA needs go beyond those of space situational awareness.

What is the likelihood of a conjunction with another space object during launch and insertion into orbit during on-orbit operations and at end of life during disposal deorbit and reentry?
What is the likelihood of maneuvering in time to avoid collision with another space object?
What is the cause of a spacecraft anomaly?
What are the origin and source of electromagnetic or radiofrequency interference with transmissions or sensors?
What is a particular space object?
Who owns and operates the spacecraft?
What are its technical parameters and operational capabilities?

The international legal regime governing the conduct of space activities establishes parameters for certain operational behaviors (UNOOSA, 2017c). This regime includes: the 1967 Outer Space Treaty; the 1968 Agreement on the Rescue of Astronauts, the Return of Astronauts and the Return of Objects Launched into Outer Space; the 1972 Convention on International Liability for Damage Caused by Space Objects; the 1975 Convention on Registration of Objects Launched into Outer Space; and the 1979 Agreement Governing the Activities of States on the Moon and Other Celestial Bodies. In addition, the Constitution and Convention of the ITU govern the assignment of satellite positions and frequencies.

There are also bilateral and multilateral arms control treaties and agreements between the United States and the Soviet Union or Russian Federation that impose prohibitions or limitations on military activities in space [UNOOSA, 2017d, e, b]. These constraints include: nuclear or other weapons of mass destruction may not be placed into orbit, installed on other celestial bodies, or stationed in space; detonation of nuclear weapons in space is prohibited; testing any kind of weapons, establishing military bases, and conducting military maneuvers on other celestial bodies are prohibited; and interference with national technical means of verifying compliance with strategic arms reduction treaties is prohibited.

In addition, the United Nations General Assembly has adopted non-binding resolutions that the international community generally follows [UNOOSA, 2017a]. These resolutions include: the 1982 Principles Governing the Use by States of Artificial Earth Satellites for International Direct Television Broadcasting; the 1986 Principles Relating to Remote Sensing of the Earth from Outer Space; the 1992 Principles Relevant to the Use of Nuclear Power Sources In Outer Space; and the 1996 Declaration on International Cooperation in the Exploration and Use of Outer Space for the Benefit and in the Interest of All States, Taking into Particular Account the Needs of Developing Countries.

These treaties, agreements, and resolutions establish basic rule sets and operational guidelines. Nonetheless, there are no other international "traffic" laws, rules of the road, or codes of conduct for space operations that specify acceptable and unacceptable behaviors. Nor do we have any international mechanism for enforcing such rules. Norms are a standard or pattern of behavior considered typical or expected of a group. Norms for maritime operations evolved over centuries, while those for air operations developed relatively recently over the past century. Norms for safe, responsible, and orderly operations to, in, through, and from space, however, are only at a nascent stage of development.

The United Nations did not adopt the European Union's proposal for a formal Code of Conduct for Outer Space Activities. Transparency and confidence-building measures that establish rules of the road, codes of conduct, or procedures for dealing with incidents in space, however, would contribute to flight safety, environmental sustainability, and security. Such measures would help reduce the prospect of mishaps, misunderstandings, or misperceptions created by ambiguous or provocative operational behaviors.

Since the beginning of the space age, standard operating procedures have developed based on experience earned through trial and error. Procedures for range and launch safety, collision avoidance during orbit insertion and on-orbit operations, rendezvous and proximity operations with cooperative space objects, and spacecraft end of life, for example, have evolved from practical operational experience. Indeed, NASA and DoD have well established procedures and practices for human and robotic space flight safety. Such procedures provide the basis for the creation of norms and, over time, customary international law.

The United States, its allies, international partners, and friends should reinforce existing norms and create new international norms of safe and responsible spacefaring behavior. In particular, emphasis should be on measures for operational safety and environmental preservation. This might include, for example, collision avoidance, orbital debris mitigation, non-interference by radiofrequency emissions, and laser illumination deconfliction. Such measures would establish a baseline of normal, routine operational behavior patterns against which SDA capabilities could sense and evaluate deviations to discern risks of a safety mishap, environmental incident, or threat. Indeed, reducing mishaps, misperceptions, and misunderstandings by regulating the safe and orderly use of space will help draw clearer distinctions between acceptable, non-aggressive behaviors and unacceptable, aggressive behaviors that are hostile actions or demonstrations of hostile intentions.

8.3 Space Domain Awareness for Space Traffic Management

Space activities are integral to the global economy and information infrastructure. They enable the free flow of information, commerce, and trade around the world. Space services are embedded in critical national infrastructures, including telecommunications, finance, and transportation that enable governments to provide essential services to their citizens and allow private enterprise to conduct business. The global space industry annually generates hundreds of billions of dollars. In fact, the importance of space to U.S. scientific and technological development and economic well-being, as well as the continuing growth of commercial space activities, is one of the factors that prompted the USG to shift STM responsibilities from DoD to the Department of Commerce. The transition enabled an increased focus on civil and commercial needs, environmental preservation, enhanced transparency, and expanded international cooperation.

The International Academy of Astronautics defines STM as "the set of technical and regulatory provisions for promoting safe access into outer space, operations in outer space, and return from outer space to Earth free from physical or radio-frequency interference [Contant-Jorgenson, 2006]." For the purposes of national STM policy, STM is "the planning, coordination, and on-orbit synchronization of activities to enhance the safety, stability, and sustainability of operations in the space environment [SPD-3, 2018]." STM to support civil and commercial space operations, flight safety, and environmental sustainability encompasses the full spectrum of the operations life cycle. It begins with the initial launch phase and insertion into orbit. It continues from deployment and checkout throughout on-orbit operations. Finally, it concludes with end of life, disposal or deorbit, and (controlled or uncontrolled) reentry back into national air space. SDA, including space weather observations and forecasting, space object monitoring, and spectrum monitoring, is necessary to minimize safety risks and environmental incidents.

Civil and commercial owners and operators of space vehicles seek a safe, orderly, and predictable operating domain within which to conduct science, technology development, and commerce. Civil space activities include, for example: human spaceflight, space science, space exploration, environmental monitoring, weather forecasting, and natural resource management. Commercial space activities include: telecommunications; earth observation; manufacturing and materials processing; and tourism. In the future, they may also include: onorbit servicing, maintenance, and retrieval; debris removal; asteroid mining; and power generation.

The decision support priorities of civil and commercial space operators for STM fall into the following categories: human space flight safety; general space flight safety; conjunction analyses; launch and on-orbit collision avoidance; anomaly resolution; electromagnetic and radiofrequency interference resolution; end of life, disposal, deorbit, and reentry; and environmental sustainment and cleanup. Persistent monitoring, continuous catalog maintenance, and near- and real-time sensing are essential to perform effectively such STM functions.

In general, civil and commercial space system owners and operators want a continuous flow of high quality, accurate, and timely SDA data and information to keep informed about conditions, activities, and events in space bearing upon their missions, equities, and assets. Table 8-2 summarizes the fundamental questions they want to have answers for.

TABLE 8-2. Civil and Commercial Information Requests Examples. The burgeoning commercial space industry makes these questions especially urgent.

What are the expected operational conditions?
What is the current situation?
Are there any changes to expected conditions or the current situation?
If there are any changes then what are the anticipated future situation and mission impacts?

As they plan and execute civil and commercial space missions, owners and operators seek to optimize the operation and use of their assets to maximize performance or revenue generation. SDA data and information are essential to decisions about orbit determination, maneuver planning, propellant management, and predicting end of life. They are, moreover, essential to decisions about satellite health management, performance trending, payload management and reconfiguration, and anomaly resolution.

Owners and operators do not take lightly any decisions to change satellite operations in any way that impacts mission performance or lifetime in response to notification of a flight safety risk. The cost of developing and fielding most space assets remains high. Time is money in space operations, so satellite owners and operators want to maximize performance and endurance. Collision avoidance maneuvers impact performance and reduce operating life. Maneuvers in response to false alarms reduce precious resources and are expensive. Many of these maneuvers are now being automated based upon USSPACEOM products. The same applies, for example, to decisions about reconfiguring transponders because of electromagnetic and radiofrequency interference or reorienting sensors to avoid dazzling or damaging optics from laser illuminations.

Civil and commercial spacecraft owners and operators want to be aware of abnormal, unsafe, irresponsible, or disorderly behaviors that could affect their space operations, so they can make timely decisions about how to mitigate and manage the operational and financial risks. They are most interested in situations that could adversely impact their mission performance, business case, revenue generation, return on investment, and public services. Civil and commercial decision makers must be prepared to address a range of contingencies that require SDA support for STM. Environmental hazards such as solar storms, micrometeoroids, and other space weather phenomena can adversely affect spacecraft parts, components, and communications. Engineering anomalies may prevent solar panels from deploying properly, cause batteries to discharge, and produce other dysfunctions. Collisions with other satellites, spent rocket bodies, and debris, as well as breakups on-orbit, often cause catastrophic problems and loss of mission capability. Electromagnetic and radiofrequency interference can prevent transmission of commands and mission data.

Civil and commercial space assets may also be at risk because of security conditions in space. In the event of conflict, these assets could suffer collateral damage when defense and intelligence satellites are attacked. Conversely, civil and commercial space assets may be political, symbolic, or economic targets for an enemy who aims to undermine political will, societal cohesion, morale, or economic vitality. Indeed, such assets being employed to support hostilities may make them legitimate military targets. This may include, for example, commercial telecommunications satellites used for C2 of military forces, commercial imagery and civil weather satellites supporting targeting of weapons systems, and civil positioning, navigation, and timing satellites employed for weapons delivery.

The term STM tends to create the impression of a capability for actively managing space operations analogous to the federal air traffic control system within the broader context of the role performed by the Federal Aviation Administration and the International Civil Aviation Organization. The latter works with nearly 200 member states and industry groups to develop international civil aviation standards as well as recommended practices and policies to support safe, efficient, secure, economically sustainable, and environmentally responsible civil aviation. Member states use such practices and policies to ensure that their domestic civil aviation operations and regulations conform to global norms for safe and reliable aviation operations in every region of the world.

The USG, however, has no C2 capability for STM-like governance as it does for civil aviation. Indeed, orbital traffic monitoring may be a more appropriate term than STM given the physics and nature of flight beyond Earth's atmosphere. Consequently, the USG conducted a lengthy interagency process to examine alternative models and formulate national STM policy. The options included: private space traffic monitoring and coordination; civil space traffic monitoring and facilitation; and civil space traffic management [FAA; Nightingale, 2016; SAIC, 2016].

Space Policy Directive-3, "National Space Traffic Management Policy," issued on June 18, 2018, realigned roles and responsibilities for STM within the USG [SPD-3, 2018]. The Department of Commerce will provide basic SSA data and STM services to satellite operators, creating an open-source data repository of publicly releasable SSA data. It will also develop relationships with private entities to share SSA data, and create standards and best practices for pre-launch risk and on-orbit collision assessments. DoD will continue to maintain the authoritative catalog of space objects. NASA is responsible for updating orbital debris mitigation standard practices as well as developing satellite design and operations guidelines. The Department of Transportation, Department of Commerce, and Federal Communications Commission will incorporate standards into their respective licensing processes. The Departments of Transportation and Commerce will also develop STM best practices and guidelines. Finally, the Department of State will lead efforts on international transparency and space object registry for STM.

8.4 Space Domain Awareness for Space Command and Control

The United States has leveraged its position as the world's leading spacefaring nation for political prestige, international influence, scientific knowledge, technological development, economic well-being, and national security. America's society, economy, and way of life rely upon access to and use of space systems and services. Therefore, unimpeded access to and use of space is a vital national interest.*

Vital interests are conditions that are necessary to safeguard and enhance the United States' survival and well-being as a free and secure nation.

Realigning STM roles and functions to NOAA will enable DoD to focus on planning and executing national security space operations to deter and, if necessary, defeat threats to space assets and operations. The commander of the Combined Space Operations Center previously spent a significant amount of time on civil instead of national security missions. The shift from space surveillance and SSA metric tracking and cataloging for support of day-to-day STM will free up resources for space C2 decision support functions.

National security space system owners and operators, as well as military commanders, require an assured, secure, continuous flow of high quality, accurate, and timely SDA data and information. This helps keep them informed about conditions, activities, and events in space bearing upon their defense and intelligence missions, equities, and assets. Defense space activities include: support to space operations functions (e.g., command, control, and communications; missile warning; positioning, navigation, and timing; and environmental monitoring) to increase the effectiveness of military forces as well as space control functions to counter threats. Intelligence space activities include: treaty monitoring; intelligence-gathering; indications and warning; and support to military operations.

Space C2 to plan, direct, monitor execution, and assess national security space operations is fundamental to defense preparedness and prowess. C2 encompasses the exercise of authority and direction by a commander over assigned and attached forces to accomplish the mission [JCS, 2017]. Command includes both the authority and the responsibility to use resources to carry out assigned missions. Control is inherent in command. It entails managing and directing forces and functions consistent with a commander's authority. Control provides the means for commanders to maintain freedom of action, delegate authority, direct operations from any location, and integrate and synchronize actions throughout the operational area. Battle management is the

* President Donald J. Trump, National Security Strategy of the United States, (Washington, D.C.: The White House, December 2017), p. 31.

management of activities within the operational environment based on the commands, direction, and guidance from appropriate authority.

SDA data and information feed the decision support systems and processes for space operational decisions. National security space owners and operators need the same type of STM functions and services as their counterparts in the civil and commercial sectors to support flight safety and environmental sustainability for peacetime operations across the lifetime of their space assets. Similarly, they must have SDA to discern abnormal, unsafe, irresponsible, or disorderly behaviors that could affect their space operations and to enable timely decisions about mitigating risks.

National decision-makers, military commanders, and intelligence officials, of course, also have different decision support needs than their civil and commercial counterparts. They are concerned about having the quality and timeliness of SDA information to predict or warn of provocative, dangerous, or threatening behaviors. Provocative behaviors not only risk mishaps but are open to misinterpretation, could create misperceptions, or signify precursors to hostile actions or demonstrate hostile intentions. Moreover, national security officials need SDA to verify non-compliance with or breakout from treaty obligations, provide indications and warning of hostile intentions or acts, and support preparations for and execution of responses to threats. Table 8-3 shows the fundamental questions national security officials want to have answers for.

TABLE 8-3. Table 8-3. National Security Information Requests Examples. These needs exceed those of civilian owners and operators.

What are the expected operational conditions?
What is the current situation?
Are there any changes to expected conditions or the current situation?
If there are any changes, then what are their potential or actual impacts on their missions, equities, or assets?
Are the changes the result of natural phenomena or human activity?
If the changes are the result of human activity, are they benign or malign?
Does the activity constitute a hostile intention or hostile action warranting the use of armed force for national or collective self-defense?
What are the potential lines of operations and courses of action?
What are the potential adversary responses to the branches and sequels of a particular course of action?

In the event of a political crisis or outbreak of hostilities that begins in or extends into space, SDA data and information will be integral to providing decision support to political authorities, military commanders, and intelligence officials responsible for crisis management and national security. Indeed, if deterrence fails, the ability of SDA to feed the space C2 system and enable decision support for operational commanders and national decision-makers will be key to controlling escalation and defeating aggression. Among the many questions national decision-makers will want answered by operational commanders and intelligence officials are: Who demonstrated the hostile intention or committed the hostile action? What is the nature of the attack? What are the consequences of the attack? When will the degraded or lost capability be recovered or reconstituted? Where did the attack originate? How conclusive is the evidence of an attack? Why did the attack happen? What is the purpose of the attack?

National security decision support priorities for SDA will include: overflight warning; treaty monitoring; custody of suspicious space objects and identified threats; indications and warning; intelligence and operational preparation of the environment; combat identification; targeting; and combat effects assessment. The national security community must be prepared for a broad range of scenarios and dangerous behaviors across the conflict spectrum. These include addressing unknown or suspicious space objects to enable threat characterization and dynamic targeting. The challenge of dealing with such space objects is that adversaries are likely to conduct clandestine or covert operations in space designed to evade detection and attribution as well as achieve strategic or tactical surprise. We can expect aggressors to employ denial and deception to keep capabilities secret and gain an operational advantage.

In addition, SDA must support C2 of space control operations to ensure U.S. and friendly freedom of action in space and, if directed, deny such freedom of action to adversaries. In comparison to exoatmospheric kinetic impacts or nuclear detonations, it may be more challenging to provide indications and warning of cyber, electronic warfare, and directed energy attacks. (Credible and compelling evidence of such aggression and attribution of the source of the attack may also be difficult.) Supporting space C2 decision-making to evade, operate through, or defeat such threats is a fundamental SDA mission need. SDA support for space C2 is essential for effective operation of a dynamic, layered, defense-in-depth against threats to space security [Berkowitz, 2012]. Such a robust defense must be prepared, among other things, to deal with: the consequences of surprise and attrition in a deep crisis; a sudden threat surge against space-based command, control, communications, intelligence, surveillance, and reconnaissance systems during non-nuclear hostilities; and a multi-weapon space control campaign during regional or global conflict.

Another key scenario requiring SDA support for space C2-related decisions is assuring mission support to terrestrial military forces and other users under contested, degraded, or operationally limited conditions. SDA decision support must enable the operational commanders to dynamically shift and adapt to changing conditions and events during crises and conflicts. They will have to make operational decisions regarding how to synchronize space operations and mass or sequence force multiplication effects among multiple regional and functional combatant commands.

The space domain is a significantly more complicated operating environment for the national security community than it was during the previous century. The USG is in the process of shifting its space operational perspective from coordination and deconfliction of independent space operations to coordination, deconfliction, synchronization, and, in some cases, integration and interdependence of concurrent joint, interagency, and combined space operations. This includes evolving from joint to interagency and combined processes through the interaction of networked space operations centers, including the National Space Defense Center, Combined Space Operations Center, other USG space operations centers, and allied and commercial space operations centers. It is being facilitated by SSA Data Sharing Agreements with allies and commercial entities as well as a Combined Space Operations Memorandum of Agreement among the United States, the United Kingdom, Canada, and Australia [SSA Services and Info…., 2017; US DoD, 2014]. In the future, the Combined Space Operations agreement may expand to include Germany, France, Japan, South Korea, and other allies and friends.

Key issues to resolve to enable space C2 of interagency and combined operations include interoperability, data sharing, and security among DoD and other USG space sectors (i.e., intelligence and civil), the commercial sector, allies, friends, and other international partners. Intelligence, environmental, operational, and other types of SDA data and information will be necessary for awareness of conditions, activities, and events in space involving three categories of actors:

- Blue—USG defense, intelligence, and civil space assets, allied, coalition, and friendly government space assets, and domestic and international commercial space assets employed by the United States and its allies;
- Red—Adversary nation-state space assets as well as U.S. and third-party space assets employed by nation-state or non-state adversaries for hostile purposes; and
- Green—The rest of the world, non-governmental entity, and non-combatant space assets.

SDA drives the interaction of the intelligence/information and operations cycles for planning and conducting national security space missions. Space C2 must enable commanders to make decisions with speed, precision, and confidence. Smart, agile, and responsive space C2 is essential for: dynamic, adaptive planning; developing, assessing, and selecting alternative courses of action and lines of operations; predicting and assessing the effectiveness of assigned and attached forces and space asset employment; reacting to developing situations; generating space tasking orders for execution; assessing the results and consequences of intended and unintended effects; and learning, adapting, and replanning.

SDA decision support for space C2 is fundamentally about enabling and achieving information and decision superiority. This will be challenging given the nature of space warfare, particularly the reduced time for decision-making caused by the speed of effects ranging from milliseconds to minutes with cyber, electronic warfare, directed energy, radiofrequency, kinetic energy, and nuclear weapons as well as the risk of global and cascading effects. A comparative advantage will go to the side that operates within the sense-decision-action cycle of its adversary. This is why automated, machine-to-machine fusion, artificial intelligence and machine learning, predictive analysis, sense-making, and autonomy will be increasingly crucial for SDA and space C2 to enable humans and machines to do what each does best.

SDA data and information feeds for space C2 must support the formulation, analysis, and implementation of comprehensive, deliberate and contingency "whole of government" and "all of nations" plans. Such planning will depend upon SDA data and information to support operational space C2 about how to execute the military instrument of statecraft in concert with other elements of national power. It will also support the operational linkage to strategic-political intentions, war aims, political-military objectives, and desired end-state. In addition, such planning will involve decisions about establishing space postures, conditions, rules of engagement, and delegations of authority as well as triggers and thresholds for responding to hostile acts or demonstrated hostile intent. Planning will take into account different national policies, red lines or caveats, and rules of engagement among alliance or coalition members for effective shaping, deterrence, and warfighting options.

Moreover, SDA decision support for space C2 should enable the execution of seamless supported and supporting relationships among multiple geographic and functional combatant commands. Indeed, agility in moving back and forth from supported and supporting command relationships will be increasingly essential for complex operations. Consider, for example, the Asia-Pacific scenario involving a North Korean missile or space launch. Is it a space launch, an anti-satellite weapon, or a ballistic missile attack on Japan, other allies in the region, or the U.S. homeland? The C2 relationships among USSPACECOM, U.S. Indo-Pacific Command, U.S. Northern Command, and U.S. Strategic Command will change over time depending on the missile or space launch vehicle's trajectory and actual mission. Indeed, the following are key seams pertinent to space C2 that adversaries will seek to exploit and that the United States and its allies, friends, and international partners must actively manage to sustain unity of effort and unity of command: seams within the U.S. national security space sector between DoD and the Intelligence Community; seams between the U.S. national security and civil space sectors; seams between the U.S. public and private sectors; and seams between the United States, its allies, and international partners.

Finally, SDA for space C2 should enable decisions about multi-domain operations involving missions in other operating environments. Space systems operate simultaneously in multiple domains (space, cyber, and terrestrial). Space C2 must not only deal with that aspect of multi-domain operations, but also enable the proper interaction with terrestrial domains (land, maritime, air) and missions, particularly for nuclear deterrence, global strike, integrated air and missile defense, and special operations. Consequently, it must take into account different roles and needs across the chain of command including: the Director of Space Forces (Space Coordinating Authority) at geographic combatant commands; commanders of geographic and functional combatant commands; commanders of USSPACECOM joint or combined task forces; Commander of USSPACECOM; the Secretary of Defense; and the Commander-in-Chief.

8.5 Summary

Awareness of conditions, activities, and events in the space domain is essential to inform decisions about flight safety, environmental sustainability, and security that could impact individuals, nations, and even humankind. SDA is "the effective understanding of anything associated with the space domain that could impact the security, safety, economy, or environment of a nation." The evolution of terminology from surveillance of space, to SSA, to SDA reflects the growing scope of and demand for mission support.

Domestic and international changes prompted a realignment of SDA roles and responsibilities within the USG. The evolving scope and operational utility of space activities and related concerns about increased risks of safety mishaps, environmental incidents, and threats are driving demand for enhanced SDA support at the strategic, operational, and tactical levels of decision-making. Authorities for public sector space activities as well as regulation of private sector space activities are dispersed across the federal government. While we have similar needs for SDA decision-support across all space sectors, critical and fundamental differences exist in the data and information needed to support operational decisions involving STM and space C2.

In both STM and space C2, decision-makers need to be able to distinguish between routine, safe, orderly operational behaviors and deviations that pose hazards to space flight safety, risks to environmental preservation, or threats to space assets. Properly constructed international codes of conduct, rules of the road, or procedures for incidents in space that establish norms of responsible, non-aggressive spacefaring behaviors would be useful in this regard. Indeed, such norms might help facilitate indications and warning of hostile intentions or actions and associated changes to space conditions and postures based upon predetermined thresholds or triggers for operating a dynamic, layered, defense-in-depth against space threats.

National security, civil, and commercial space system owners and operators want a continuous flow of high quality, accurate, and timely SDA data and information to keep informed about conditions, activities, and events in the space domain to support their mission needs. SDA functions critical for STM-related decisions include: space weather forecasting; conjunction analyses; launch and on-orbit collision avoidance; deorbit and reentry predictions; and space catalog maintenance. In addition, SDA functions critical for space C2 decision support include: treaty monitoring; overflight warning; threat characterization; indications and warning; attribution; intelligence preparation of the battlespace; operational preparation of the battlespace; combat identification; targeting; and combat effects assessment. The future performance of SDA capabilities thus is essential to addressing natural hazards in space, avoiding flight safety mishaps, preserving the operating environment for the benefit of all humankind, and deterring or, if necessary, defeating threats to space assets.

References

Berkowitz, Marc J. "Shaping the Outer Space and Cyberspace Environments," in Scott Jasper, editor, Conflict and Cooperation in the Global Commons. (Washington, D.C. Georgetown University Press, 2012, pp. 155-170.

Contant-Jorgenson, Corrine, et al, editors. Cosmic Study on Space Traffic Management. (Paris, France: International Academy of Astronautics, 2006), p. 17.

Chief of Naval Operations. Maritime Domain Awareness Concept. (Washington, D.C.: U.S. Department of the Navy, 2007), p. 4.

Defense Intelligence Agency. "Challenges to Security in Space," January 2019.

Director of National Intelligence. "Worldwide Threat Assessment of the U.S. Intelligence Community," January 29, 2019, p. 16.

Federal Aviation Administration, Office of Commercial Space Transportation. "Towards a Civil Space Traffic Management System."

Joint Chiefs of Staff. Publication 1-03: Dictionary of Basic Military Terms. (Washington, D.C.: U.S. Department of Defense, 2010 (as amended through 15 February 2016), p. 220.

Joint Chiefs of Staff. Joint Publication 3-0: Joint Operations. (Washington, D.C.: U.S. Department of Defense, January 17, 2017), p. xiii.

Joint Chiefs of Staff. Joint Publication 3-14: Joint Space Operations Doctrine. (Washington, D.C.: U.S. Department of Defense, April 10, 2018).

National Air and Space Intelligence Center. "Competing in Space," December 2018, pp. 15-23.

National Space Policy of the United States of America. (Washington, D.C.: The White House, June 28, 2010), p. 3.

Nightingale, Emily S., et al. "Evaluation Options for Civil Space Situational Awareness (SSA)." (Washington, D.C.: Institute for Defense Analysis, August 2016).

Presidential memorandum. Space Policy Directive-3, "National Space Traffic Management Policy," The White House, June 18, 2018.

SAIC. "Orbital Traffic Management Study," Final Report. (McLean, VA.: SAIC, November 21, 2016).

"Space Situational Awareness Services and Information Agreement Between the United States of America and Japan, Effected by Exchange of Notes at Tokyo," May 28, 2013.

https://www.state.gov/documents/organization/219765.pdf, March 4, 2017.

Trump, President Donald J. National Security Strategy of the United States. (Washington, D.C.: The White House, December 2017), p. 31.

United Nations Office for Outer Space Affairs. "Documents and Resolutions Database." http://www.unoosa.org/oosa/documents-and-resolutions/search.jspx?&view=resolutions, March 4, 2017.

United Nations Office for Outer Space Affairs. "New Strategic Arms Reduction Treaty." http://www.unoosa.org/oosa/en/ourwork/spacelaw/treaties.html, March 4, 2017.

United Nations Office for Outer Space Affairs. "Space Law Treaties and Principles." http://www.unoosa.org/oosa/en/ourwork/spacelaw/treaties.html, March 4, 2017.

United Nations Office for Outer Space Affairs. "Treaty on Principles Governing the Activities of States in the Exploration and Use of Outer Space." http://www.unoosa.org/oosa/en/ourwork/spacelaw/treaties.html, March 4, 2017.

United Nations Office for Outer Space Affairs. "Treaty Banning Nuclear Weapons Tests in the Atmosphere, In Outer Space, and Underwater." http://www.unoosa.org/oosa/en/ourwork/spacelaw/treaties.html, March 4, 2017.

U.S. Department of Defense. "Stratcom, DoD Sign Space Operations Agreement With Allies." https://www.defense.gov/News/Article/Article/603303, September 23, 2014.

U.S. Strategic Command. "AFSPC Commander Becomes JFSCC, Joint Space Forces Restructure." http://www.stratcom.mil/Media/News/News-Article-View/Article/1386821/afspccommander-becomes-jfscc-joint-space-forces-restructure/, December 4, 2017.

U.S. Strategic Command. "Space Operations Center Gets New Name, USSTRATCOM begins Expanded Multinational Space Effort." http://www.stratcom.mil/Media/News/News-Article-View/Article/1141112/space-operations-center-gets-new-name-usstratcom-begins-expandedmultinational/, April 5, 2017.

Weeden, Brian. "Dancing in the Dark: The Orbital Rendezvous of SJ-12 and SJ-06F," The Space Review, August 30, 2010.

Weeden, Brian. "Dancing in the Dark: Recent Russian Rendezvous and Proximity Operations." The Space Review, October 25, 2015.

Chapter 9

End to End ArgusSat Example

Maarten Meerman, *Space Mission Designer*
Kathleen Conderre, *Lockheed Martin*
Pamela Magee, *PDM Aerospace*

9.1 Overview

The GEO environment is becoming busier, with more satellites planned for deployment and current systems coming to their end of life and maneuvering to a nearby graveyard orbit. A better understanding of the current precise location and orientation of a satellite in its GEO box allows for more accurate predictions of its future locations, which improves the ability to identify and prevent future collision hazards. More accurate tracking data also helps to identify a system whose performance is degrading. Satellite operators use this data to better plan satellite operations and reduce the likelihood of collisions causing a degradation of service or loss of mission. This chapter develops an example involving "inspector" satellites in near-GEO orbit as one means to address this challenge.

The mission of the hypothetical inspector satellite system, ArgusSat, is to image space assets in geostationary orbit to determine such parameters as size, aperture, power generation, operational status, etc. ArgusSat will use a set of small satellites at near-geosynchronous altitude, each carrying a particular sensor or combination of sensors. Prospective sensor types considered in this chapter are optical cameras, radar, and lidar.

The ArgusSat satellites will use orbital dynamics and on-board propulsion to navigate their way around space, and position themselves within safe imaging range of the various targets. The information gained by the first satellite will help optimize targeting and imaging by the subsequent satellites. The satellites will be delivered directly into a near-GEO orbit and will communicate with one or more user ground stations for target selection and data download. Each station will be able to see about a third of the orbital targeting sequences.

The ArgusSat System Design Process

A typical space mission design process follows a series of steps and iterations. Each step builds on the previous ones, developing more detailed information and deciding between alternatives, while iterations allow reassessment of earlier decisions, taking advantage of the latest information. Using hindsight, developers challenge assumptions and requirements to improve the overall mission.

Assuming a general mission requirement (and approved funding!), we used the following steps to guide the process, adapted from the Space Mission Analysis and Design (SMAD) FireSat project.

9.1.1 Current Methods of Identification

The Joint Space Operations Center (JSpOC) maintains the catalog of all artificial Earth-orbiting objects, and since the 1957 launch of Sputnik, has cataloged over 39,000 man-made space objects. Of the 16,000 objects that JSpOC is currently tracking, "about 5 percent…are functioning payloads or satellites, 8 percent are rocket bodies, and about 87 percent are debris and/or inactive satellites" [JSpOC Fact Sheet, 2018].

Four types of sensors to track and identify space objects. The JSpOC uses these sensors in a worldwide network of military and civilian assets called the Space Surveillance Network (SSN). The sensor data generates the Satellite Catalog, which contains the number, type, and orbits of all trackable space objects. The four types of sensors in the SSN are phased-array radars, conventional radars, electro-optical sensors, and space-based sensors, which use predictive techniques to track space objects. Phased-array radars comprise electronically steerable elements that generate pulses of radio waves, which then reflect back to a detection antenna, which in turn allows us to determine the object's location. Conventional radars use narrow beams, turning 360 degrees, to track targets and measure how the targets have moved since previous radar sweeps. Optical telescopes connect to computers and video cameras to form electro-optical sensors. This allows for real-time object tracking, recording, and analysis. Space-based sensors use infrared or optical instruments and operate without atmospheric interference [JSpOC Fact Sheet, 2018]. Figure 9-1 shows the locations of the SSN.

Figure 9-2 shows a sample of radar and phased-array antennas that track space objects. An example of a phased-array system is the FPS-85 radar at Eglin Air Force Base in Florida. A conventional radar system used by both NASA and the Department of Defense is the Haystack Radar, operated by the Massachusetts Institute of Technology (MIT)

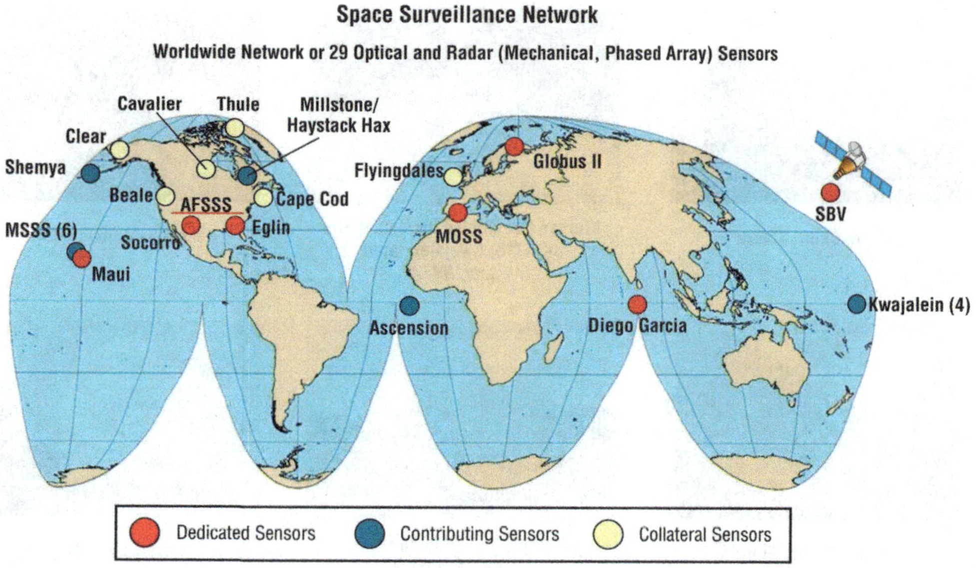

FIG. 9-1. Space Surveillance Network Locations. This network uses four types of sensors to track space objects [JSpOC Fact Sheet, 2018].

Lincoln Laboratory. Haystack can track objects between 1 and 30 cm that fly through its field of view, allowing the characterization of debris objects.

The space observation tracking and imaging radar (TIRA) in Germany comprises a tracking radar and an imaging radar. "The radar data of space objects and internally developed techniques are used to determine characteristic features of the objects, e.g., orbital elements, intrinsic motion parameters, orbital lifetime, target shape and size, ballistic coefficient, mass, and material properties" [FHR, 2018].

To characterize and identify a space object, we must identify such physical characteristics as material type. Material types of man-made objects are identifiable by the use of "low-resolution reflectance spectroscopy and comparing the absorption features and the shape of the spectra" [NASA Orbital Debris, 2018]. One ground-based system using this technique is the Air Force Maui Optical Supercomputing (AMOS) site, comprising a 3.67-m Advanced Electro-Optical System (AEOS) telescope, a 1.6-m telescope, two 1.2-m telescopes, and three 1-m Ground-Based Electro-Optical Deep Space Surveillance (GEODSS) telescopes. Figure 9-3 contains images of AMOS and AEOS. Optical measurements complement radar systems by identifying the physical characteristics of the space object using a different portion of the electromagnetic spectrum.

A space-based system eliminates atmospheric and weather problems in observations. The USAF's Geosynchronous Space Situational Awareness Program (GSSAP) satellites are located in near-geosynchronous orbit where they characterize other satellites in orbit around Earth. The satellites can rendezvous with a GEO satellite and use an electro-optical sensor that gathers precise data [GSSAP, 2018].

Cobra Dane, Alaska

Kiernan Reentry Measurement Site (KREMS) Kwajalein Atoll
Photo: NASA

Goldstone Antenna, CA

Eglin FPS-85, Florida

Tracking and Imaging Radar (TIRA), Germany
Photo: Fraunhofer FHR

Haystacks, MIT

FIG. 9-2. Radar Systems Used to Track Space Objects (NASA Orbital Debris and FHR). These systems use either radar or phased-array antennas.

Although all these methods provide a good amount of data for SDA, that an in-space asset with high resolution could fill.

9.1.2 Analogs for SDA

For understanding SDA, analogs and potential lessons learned could prove useful in developing a new system. One good analog is Maritime Domain Awareness (MDA). The text below is from a Security in Space White Paper, which summarizes how MDA relies on space access and how the data and information are invaluable to mission success. MDA has challenges (such as organizational ones) and practitioners understand the importance of balancing space and ground assets. The Patuxent Policy Group [2015] found the following:

> "MDA is all about space access and capability. The U.S. Navy (USN) has the primary responsibility for MDA, which dictates warnings and threats. The U.S. Navy has to rely on various national technological means, including assets in space, to maintain adequate MDA to support national security. We have greater threats and shorter response times. To win a war affordably in a market-based democracy, persistent Intelligence, Surveillance, and Reconnaissance (ISR) are invaluable. We do not coherently balance our ground assets with space. We need to correct this culturally, impose disrupters, and change the architecture of managing our ground and space assets. The United States has a huge variety of organizational entities, unique systems, ownership, and proprietary information, all of which contribute to creating

Air Force Maui Optical and Supercomputing (AMOS), Hawaii

Kirtland AFB Advance Electro-Optical System (AEOS), New Mexico

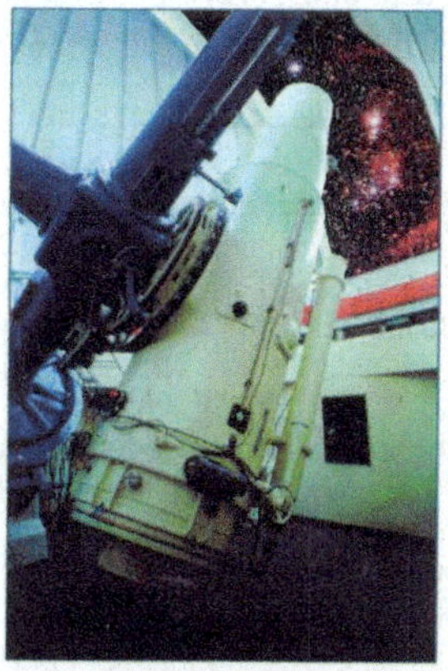

The Michigan Orbital Debris Survey Telescope (MODEST), Chile

FIG. 9-3. **Electro-Optical Systems Used to Track Space Objects.** One useful feature of such systems is that optical data are often immediately interpretable by humans [NASA Orbital Debris, 2018].

obstacles. The concept of embracing open-systems and open-source architecture brings with it increased security concerns and higher operating costs, along with the demand for more information and reconnaissance, not necessarily for maritime purposes." [Patuxent Policy Group, 2015]

9.2 Define Needs, Goals, and Objectives

In this section we identify stakeholders, both passive and active, associated with Space Domain Awareness (SDA) as defined in Chapter 1. In our interactions with SDA stakeholders we tried to understand their problems, concerns, and priorities, in order to capture and articulate their needs, goals, and objectives (NGOs). These NGOs form the basis of the problem or challenge that we aim to solve. Next, we define the top-level requirements so we can proceed to generate possible concepts of operation, as well as potential functional and physical architectures.

Chapter 1, Figure 1-5 provides a stakeholder context diagram showing the broad group of SDA stakeholders and their generic interfaces to our system of interest, the Distributed Space Situation Operations Center (DSSOC), which represents key stakeholders in the operational phase.

Stakeholders are the people, organizations, entities, systems, policies, and procedures which interact with or influence the system of interest, the DSSOC. Table 9-1 provides the list of stakeholders that we must engage to determine the true NGOs. Active stakeholders, those that touch the proposed systems or use the data generated by the systems during the operational phase, help us determine needed capabilities. Passive stakeholders do not actually use the systems but may add requirements and constraints that drive system characteristics.

TABLE 9-1. Active and Passive Stakeholders. Engaging in discussion with SDA stakeholders is crucial to understanding their problems and challenges well enough to propose credible solutions.

Active	Passive
Satellite providers	Insurance Companies
US Department of Defense – US Strategic Command (USSTRATCOM)	Space Lawyers (international space law, United Nations Office of Outer Space Affairs)
National Aeronautics and Space Administration (NASA)	Federal Aviation Administration (FAA) Commercial Space Office
National Oceanographic and Atmospheric Association (NOAA)	Federal Communications Commission (FCC)
National Reconnaissance Office (NRO)	Foreign governments
International space agencies (such as the European Space Agency (ESA))	International Telecommunications Union (ITU)
Satellite operators	
End-users of GEO satellite information	
Developers, manufacturers, trainers, end-user, testers, purchasers, handlers, logistics personnel, maintainers	
Air Force Satellite Control Network (AFSCN) Ground Stations	
Distributed Space Situation Operations Center (DSSOC)	

After identifying stakeholders, we must understand their needs and concerns. This can take place in the form of interviews or other research. For ArgusSat, the project team engaged with stakeholders and held interviews as well as research to determine their needs. Table 9-2 provides a sampling of responses.

TABLE 9-2. Space Domain Awareness Stakeholder Responses. These responses are very short but demonstrate the type of inputs that derive from stakeholder discussions.

Stakeholder	Needs
DSSOC	• Restrict to operational spacecraft—how to detect, track, and characterize GEO satellites • Standoff distance of 10 km (TBR) or less of the target satellite • There are multiple customers; would sell the service or data • Identify the debris. Can be done by determining the cross-sectional area, inertia, solar array size, power provided, thrusters, and type of stabilization • Resolution of 5–10 cm • A proof of concept mission could work • No specific requirement for operational life, but a proof of concept mission could be two years
Satellite providers	• Don't interfere with current satellite operations
US Strategic Command (USSTRATCOM), operates JSpOC [Barnes, 2009]	• Interfaces with the Space Surveillance Network • Data can be entered into the Space Catalog • Ensure space operations and spaceflight safety • Implement international treaties and agreements • Protect space capabilities • Protect military operations and national interests
NASA [National Space Policy, 2010]	• Preserve the space environment • Detect, identify, and attribute actions in space that are contrary to responsible use and the long-term sustainability of the space environment • Technologies and techniques to mitigate and remove on-orbit debris, reduce hazards, and increase understanding of the current and future debris environment • Maintain and improve space object databases; pursue common international data standards and data integrity measures; provide services and disseminate orbital tracking information to commercial and international entities, including predictions of space object conjunction
NOAA [WSSA, 2017]	• Reduce the risk of NOAA assets colliding with other spacecraft or orbital debris • When a close approach is identified and confirmed, appropriate action must be taken by one of the two objects in conflict in order to prevent a collision
ESA [ESA, 2018]	• Europe needs an autonomous SSA capability" • A system that gives the complete picture and is not "fragmented across national and institutional boundaries
Satellite Operators	• Protect current systems and investments • Determine characteristics of spacecraft • Collision avoidance • Better predictive analytics from more accurate and real-time data

Some specific stakeholder and potential customer statements are provided here as examples.

> Leopold stated that "*Pamela Melroy, a shuttle commander and former deputy director of the Defense Advanced Research Projects Agency's (DARPA's) Tactical Technology Office, also stressed the need for real-time space awareness. "We have to have more sensors to get more frequent updates, but these sensors don't have to be as precise or expensive as our current sensors," she told the space council. "The frequency of observations combined with our current sensor data makes up for the lack of precision." Ellis and Melroy also backed greater use of advances in distributed sensor networks and automation tools that can augment real-time situational awareness. "When you get a new piece of information, you need to be able to tell immediately whether something actually just happened on orbit or if one of your distributed sensors has been spoofed," Melroy explained. "It is possible to automate that."* [Leopold, 2017]

> According to the Patuxent Policy Group [2015], *"[T]here is a worldwide reliance on satellites for private, public, and defense activities. A loss of space assets would have an immediate, negative, and almost incalculably large impact to the U.S. and global economies."* (pg. i)

> *"Space situational awareness is not easy. We do not want everyone to know what is out there. We must accept the premise that our space assets will be attacked. We must maintain reliability and we must take risks, whether acting as a government agency or in the public/private partnership."* (pg. 11)

The project team received the inputs shown in Table 9-2, as well as comments from key stakeholders, and had to integrate and blend these comments to generate a top-level need statement. The overall mission need statement is:

> **Discover, locate, characterize, and identify space objects located in geostationary orbit to inform possible courses of action by the Distributed Space Situation Operations Center.**

The functions mentioned in the mission need statement (discover, locate, characterize, and identify) are described in various chapters of this volume. Chapter 2 describes discovery and change detection. Chapter 4 addresses orbital dynamics and tracking of objects. Characterization of objects is introduced in Chapter 1, and identification of objects is described in several chapters, notably Chapters 5, 7, and 10.

Further, these functions take place at some level in space situational awareness and space traffic management activities, but SDA stakeholders are interested in improving current capabilities. So when provided with options to provide and enhance detection, location, characterization, and identification, we assume, based on discussions with stakeholders, that the following attributes are relevant:

- Degree to which functions are improved
- Flexibility provided to the DSSOC

- Survivability of the systems
- Cost of the proposed solution

There are many different ways to meet stakeholder needs, so it takes quite a lot of analysis to decide on the best approach. In this chapter we *assume* that an integrated space-ground solution would be adequate to demonstrate the concepts described in this book, but we don't claim that this approach is necessarily the best.

We now state the goals and objectives.

9.2.1 Goals and Objectives

The goal of the ArgusSat System is to provide relevant and timely information on objects in GEO to decision-makers associated with the DSSOC to help them decide on the best courses of action. Examples include situations involving potential collisions, newly-found objects, inconsistent object behaviors, and predictive analysis of the future state of a spacecraft in GEO.

Primary Objective: Characterize and identify objects in geostationary orbit and provide data to operators in near-real time.

Secondary Objectives:

1. Significantly enhance optical and infrared sensing of objects in GEO
2. Improve real-time orbital location accuracy
3. Increase the type and number of space object characterization and identification sensing options

At this point, the stated needs, goals, and objectives are solution-independent, and they form the basis of the problem statement for this example. The challenge is to provide top-level requirements and key performance parameters so we can create options for providing the desired capabilities, analyzing the options, and selecting approaches for further study.

9.2.2 Develop Top-Level Qualitative Requirements

After many meetings and exhaustive discussions with stakeholders, the project team gleaned top-level qualitative requirements and key performance parameters (KPPs) (requirements so important that if not met, stakeholders would not support the project). For this example we divide the top-level requirements into functional, operational, and constraints, as shown below.

Functional:

Performance: 10-cm or less resolution optical (mid-scale), full-color images of GEO objects (KPP)

Coverage: Entire geostationary ring (360 degrees) at least once every 30 days (KPP)

Responsiveness: Near real-time object characterization data sent to Distributed Space Situation Operations Center within 30 minutes of acquisition (KPP)

Operational:

Duration: System operational for at least two years

Availability: System able to provide characterization data to DSSOC 98% of the time

Survivability: Natural environment only, with awareness of nearby vehicles

Data Distribution: Distributed Space Situation Operations Center to disseminate data

Data content, form, and format: object detection, location, characterization, and identification data shall be provided in a format TBD

Revisit capability: Updated GEO object characterization data accessible at least every 30 days with the possibility to characterize selected GEO object within 24 hours (KPP)

Constraints:

Cost: Less than $TBD million per year for operations plus development cost

Schedule: Operational within three years from authority to proceed

Regulations: DOD regulations and procurement constraints apply

Political: Responsive to Department of Defense need to protect space assets

Interfaces: Compatible with Distributed Space Situation Operations Center

9.3 Mission Characterization

9.3.1 Mission Interfacing to Its Stakeholders and Users

Context Diagram

A context diagram identifies system scope and interactions with active stakeholders. The stakeholders act as external systems that interact with the system in use, and the interactions take the form of inputs and outputs in the diagram. Figure 9-4 shows the context diagram for ArgusSat.

Sequence Diagram

Figure 9-5 depicts a sequence diagram for a command from the satellite operator through the ground station to the satellite. It shows a dialog between the system and external systems.

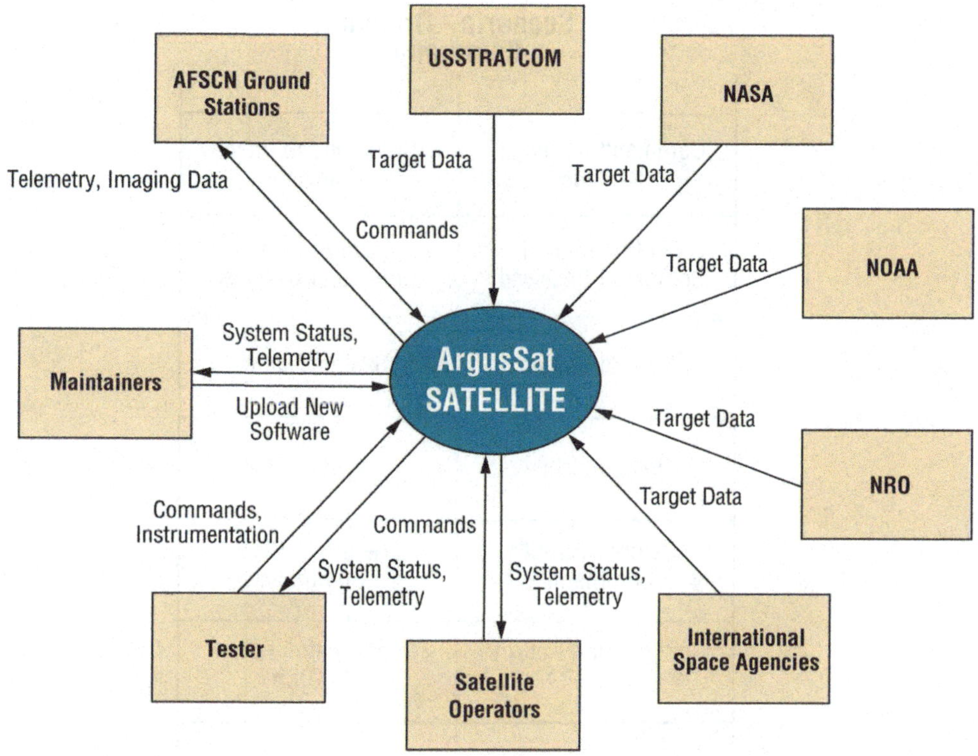

FIG. 9-4. **Context Diagram.** The context diagram shows interactions between the system and its stakeholders.

9.3.2 Characterize the Mission

Geostationary satellites are high above Earth, with an orbital altitude of 35,786 km. The resolution possible through optical imaging of such satellites depends on the diameter of the optics and the distance to the satellite; thus, any telescope trying to image these satellites from Earth must be extremely large. Even a 5-meter astronomical telescope can only resolve about 4 m of detail at geostationary distance, and on top of that, with most such satellites keeping one facet Earth-facing, only one side of the satellite is visible. One good aspect is that if the Earth-facing facet is sunlit, the telescope location underneath will be in darkness, but that highlights the fact that any particular telescope can only see one section of the geo-arc. For coverage around the arc, many telescopes would be necessary, all of them very large. The Rayleigh limit formula tells us that to achieve a resolving power of 10 cm, we would need a telescope at least 200 meters in diameter!

A wise person said long ago: "If the mountain does not come to me, I shall have to go to the mountain." The modern equivalent is to send an inspection satellite up close to the target, thus reducing the imaging distance, and allowing the telescope to be much smaller for the

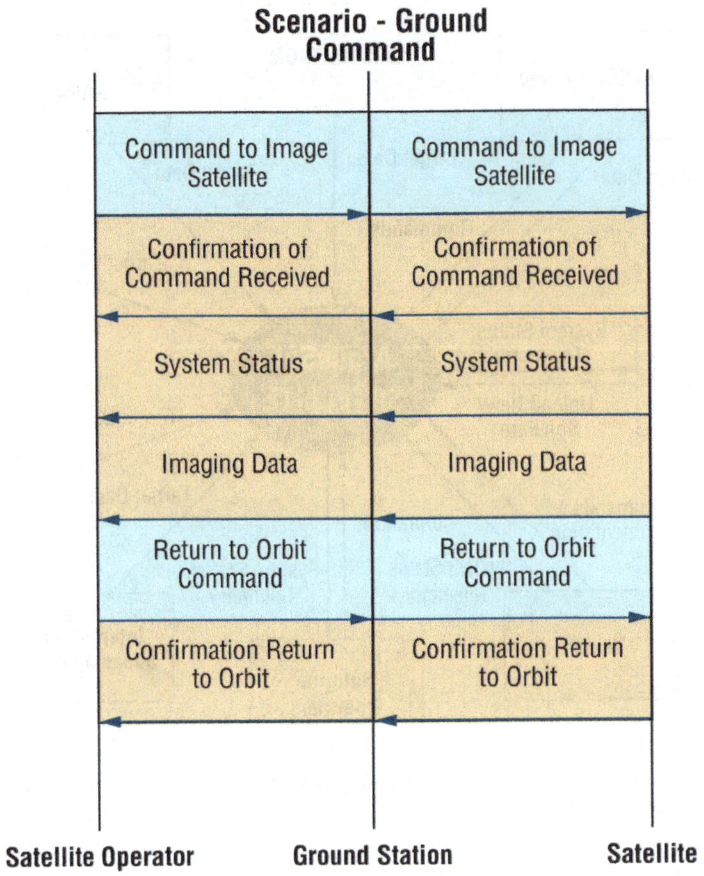

FIG. 9-5. **Sequence Diagram—Command From Ground.** This diagram shows the sequence of actions associated with a single event—a command to image a satellite.

same 10-cm resolution. Unlike a ground-based telescope, such a satellite can not only cover any position around the GEO belt, but it can also image the target from different directions. Orbital dynamics create a space path where the inspection satellite travels all around the target satellite automatically within one day, and at the same time keeps the path aligned with the Sun location. This has the effect that as the satellite travels around the target, the Sun seems to rotate with it, providing the same sunlight at all sides for good optical imaging.

The ArgusSat basic mission concept is therefore a proof of concept mission wherein an imaging satellite is placed in a sub-GEO orbit, which allows for ArgusSat to meet the customer requirement for a 30-day revisit. Figure 9-6 depicts this situation. When ArgusSat is within 10 km of the target satellite, a ground controller would task it to take images of the target satellite, and transmit the data back to an Air Force Satellite Control Network (AFSCN) ground station. From there it can be distributed to the ground controllers and any other end users who need the data. This mission would need to be operational for two years, and the payload would have to provide a resolution of 10 centimeters or less to be useful.

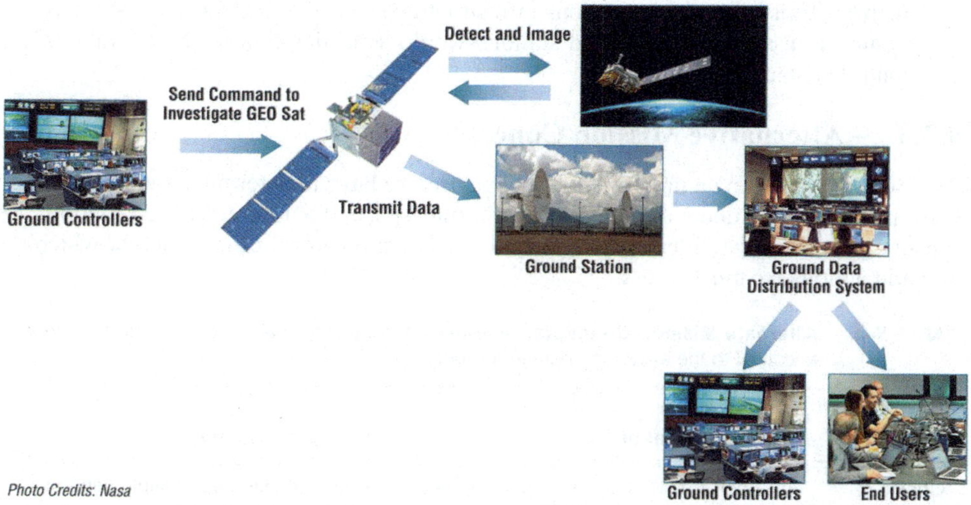

FIG. 9-6. **Basic Mission Concept.** This sequence of images depicts a nominal use of the ArgusSat system [NASA Images, 2018, 2017, 2014].

The primary components of the mission concept are data delivery; tasking, scheduling, and control; communications architecture; and mission timeline. The payload capability will be key to the system, as the customer needs 10-cm or less resolution, but ArgusSat cannot be too intrusive to the target satellite, so it must maintain a standoff distance of at least 10 km. This drives the entire mission concept. To attain this resolution and provide useful data to the end-user, we must conduct a payload trade to determine which configuration will best characterize the target satellite. We also need to determine the effect on the propellant and delta V. The customer wants to be able to revisit these satellites; revisit is useful and provide validation of ground predictive tools. It also gives historical data to potentially identify operational trends before a failure occurs. This data helps prevent collisions and allow operators to act if there appear to be issues with the GEO satellites. Below we define the components of the chosen mission concept.

Data Delivery:

The current plan is to use existing AFSCN ground stations. Since the satellite system will be in GEO, it should have a good view of ground stations as needed. The driver here will be the data rate, which will size the communications system.

Tasking, scheduling, and control:

The satellite will have multiple customers, and operations would be from one control center, linked to the ground station data network.

Communication architecture:

The initial mission concept is to use a satellite in or near GEO to image GEO satellites and downlink the information back to a ground control center for analysis. Existing communications infrastructure such as AFSCN ground stations would be used, and the ground stations would send the data to the operations control center via the internet.

9.3.3 Alternative Mission Concepts

Before arriving at an preliminary architecture, we have to determine alternate mission concepts and better understand the system's tradespace. There are different options for sensors, number of satellites involved in the system and control of the satellite which are all valid options we must assess.

TABLE 9-3. Alternate Mission Concepts. Determining architecture alternatives and their impacts is critical to the systems engineering process.

GEO Inspection	Single Sensor	Multiple Sensors
Choice of sensors	Each satellite has single sensor: either more satellites, or every month another sensor type is used. Likely makes satellites cheaper	Each satellite has multiple sensors: every month all measurements are taken. Satellites more expensive
360 target locations, each visited once per month	60 targets per month per satellite: 6 satellites for full monthly coverage, less time per measurement, and from one side only	30 targets per month per satellite: More time per target location, and full circle round the target possible
Satellite control	Person-in-the-loop operations: Must have groundstations round the world, must have high-rate downlink for video	Autonomous: Satellites visit locations, find targets, image / sense as needed, transmit data at leisure

9.3.4 System Drivers

Before defining the system architecture, we must identify the main systems. System drivers are the principal mission parameters that influence performance, cost, risk, or schedule and which the user or designer can control. These were determined by mission element or subsystem:

Target Satellite – no system drivers, the target satellite is operational, and the system we develop should not impact target satellite operations

Launch Element – on-orbit mass, altitude, and inclination

Ground Operations – scheduling, processing time, storage delay, user stations

Ground Systems – data rate, storage delay, latency

Orbit/Constellation – altitude, inclination, number of spacecraft

Spacecraft Element – mass, power, data rate, ADCS, power, delta V, propellant, payload resolution, number of spacecraft, encryption

Command, Control, and Communications Architecture – data rate, processing time, storage delay

We have to keep in mind that any requirement must be testable, both on the ground to verify that the design is suitable for build and launch, and in orbit to show that the mission is successfully implemented. The mission requirements here should not include any high-level customer aims, such as "Must limit illegal occupation of GEO slots". Whether such aims can be fulfilled is not under control of the contractor implementing the mission.

9.4 Mission Evaluation

9.4.1 Critical Requirements

Critical requirements are those that dominate the space mission's design and therefore most strongly affect performance and cost. The following are the critical requirements per mission element or subsystem:

Target Satellite – no critical requirements from the target satellite as it is operational, and the system being developed should not impact target satellite operations

Launch Element – coverage

Ground Operations – data distribution, commanding

Ground Systems – data distribution

Orbit/Constellation – distance to target satellite, resolution, delta V

Spacecraft Element – resolution, propellant

Command, Control, and Communications Architecture – data distribution

9.4.2 Mission Utility Evaluation

The purpose of mission utility analysis is to "quantify the system's performance and its ability to meet the ultimate mission objectives" [Wertz and Larson, 1999]. To fully provide this mission analysis, we must develop performance parameters and measures of effectiveness (MoEs) or figures of merit (FoMs). Performance parameters quantify how a system works through constraints such as power efficiency, instrument resolution, or coverage statistics. MoEs directly quantify how well the system meets mission objectives. Tables 9-4 and 9-5 list the performance parameters and MoEs for the GEO satellite system, using the process from SMAD [Wertz, 1999].

TABLE 9-4. Performance Parameters. Performance parameters quantify how the system works.

Performance Parameter	How Determined
Number of satellites that can be imaged	Analysis and simulation
Mean time between observations	Analysis
System response time	Simulation

TABLE 9-5. Measures of Effectiveness (MoEs). In contrast to performance parameters, MoEs quantify how well the system meets its objectives.

Goal	Measure of Effectiveness	How Estimated
Orbital accuracy	Distance from target to provide high-resolution image	Simulation
Response time	Time from ground command to the first image received of the target satellite	Analysis
Collision avoidance and predictive analysis	Accurate image and location data of target transmitted to the ground	Analysis
Time-based data collection	Revisit time for target satellite (gather historical data over time)	Simulation

9.4.3 Mission Concept Design Selection

Based on the performance parameters, MoEs, stakeholder analysis, and mission needs, the chosen initial concept is a single satellite in near-GEO orbit. This proposed system meets the overall mission objectives—it can provide timely data to ground controllers of a selected satellite and is technically feasible. The preliminary results show that high-resolution imagery is attainable through the in-space solution, and this resolution is better than Earth-based systems. A single satellite design also reduces the complexity of this initial proof of concept. It supports the political objectives identified during the stakeholder analysis phase, and the mission support infrastructure is already in place, as current facilities and ground systems are available.

After deciding on an initial concept, we conducted trades and analyses in order to further refine the design and ensure the system meets the requirements.

9.5 Results of Trades and Analyses

9.5.1 Subject Trade

A subject trade process helps to define a payload trade. The SMAD text outlines a process for the subject trade, which we followed for the satellite system [Wertz and Larson, 1999].

1. Determine mission objectives
   - The primary objective is to identify, track, and characterize objects in GEO
2. Determine possible subjects to meet those objectives
   - Cross-sectional area, inertia, solar arrays, thruster signature, stabilization, external materials (paint color and type, mylar, reflectance from Sun)
3. Determine a broad class of ways the spacecraft can detect or interact with the possible subjects
   - Physical features (cross-sectional area, array size, shape, etc.)—radar or visible
   - Inertia—radar
   - Solar arrays—visual or infrared (IR) (heat or light reflectance)
   - Thruster plume—IR
   - Stabilization—visible
   - External materials—visual or IR
4. Determine if the subject is passive or controllable
   - For controllable subjects, perform the trade of putting functionality on the subject, in the space system, or the ground system
     - This does not apply to this system as the subject is not controllable by the satellite system.
   - For passive subjects, determine general characteristics that the system can detect
     - The goal of the system is to characterize the subject. General detectable characteristics include heat, reflectance, and radar return from an object.
5. Determine whether to use multiple subjects and payloads
   - Multiple subjects are necessary to provide the complete data set to properly characterize the system. Multiple payloads could be beneficial as many characteristics can be defined and used either at distinct phases of the mission or to determine different subject characteristics. For example, a radar with a large sweep angle could initially "spot" a target from a larger distance, and visual and IR sensors could gather data on the major characteristics of the spacecraft, along with the radar signature to identify the spacecraft and its features.
6. Define and document initial subject selection
   - Optical
     - Visible RGB
     - Specialized bands
     - Near-infrared
     - Thermal infrared
   - Lidar
   - Radar
7. Review selection frequently for alternate methods and possible use of ancillary subjects

- Iteration of the design and payload selections will be performed and described in subsequent sections of this chapter.

9.5.2 Payload Design and Sizing Process

The payload capability is the key requirement from the customer and the main system design driver that will allow the system to complete its mission goals. To be able to characterize GEO satellites, we must be able to image them. We need to make several payload options and trades. The process is outlined in the SMAD textbook (pgs. 245–299) and we detail it in the remaining sections of this chapter.

9.5.3 Orbit

The first parameter to trade is the orbit. For this mission, we assume that the launch vehicle can inject the satellites into GEO. ArgusSat would be injected into GEO, and then two burns (~13 m/s delta V) would place it into an elliptical orbit with an apogee altitude of 35,991 km, and a perigee altitude of 35,286.4 km. This allows for 2 degrees motion per day (ArgusSat could get within 10 km of two GEO slots per day), at a reasonable speed, which allows plenty of time for measuring each satellite. However, one satellite in this orbit does not meet the requirement of a 30-day revisit. In fact, the revisit time is closer to 182.5 days, and we must therefore do a trade. Since the chosen imaging techniques are lidar, radar, and visual, several combinations of satellites and instrument configurations could work and will be discussed more later in this section.

Basic Orbit Calculations

ArgusSat will start is mission after being launched straight into GEO orbit. It transfers into an inspection trajectory using its on-board propulsion system. The inspection trajectory is eccentric, traversing below and above GEO altitude as needed. The satellite will require additional delta V for inclination correction (North-South Station Keeping), and along-track position correction (East-West Station Keeping), to counteract perturbations from the Sun, Moon etc. This takes about 53 m/s/year, and ultimately at End-of-Life the satellite will need about 10 m/s to raise itself to a suitable graveyard orbit. Any planned maneuver should be compared to any perturbation correction, to take advantage of any combined operation, or canceled maneuver delta V.

The first maneuvers, shown in Figure 9-7, set ArgusSat onto a transfer orbit, with perigee the required distance below the GEO belt. At perigee a second firing increases apogee altitude above the GEO belt. For a simple 1-degree per day GEO travel trajectory, visiting each 1-degree slot in turn, and approaching each slot to within 10 km at perigee, the following calculations apply:

First we need the semi-major axis (a_t) of the transfer orbit:

h = altitude

R_E = radius of the Earth (6378.14 km)

h_{GEO} = geosynchronous orbit altitude (35,786 km)

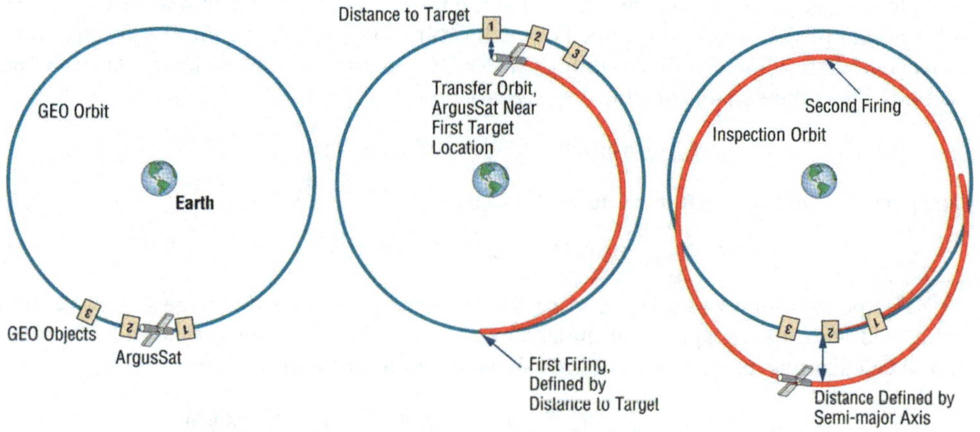

FIG. 9-7. **Basic Inspection Orbit Dynamics.** Transferring from the initial GEO orbit to an inspection orbit takes several steps, breaking up the calculations and maneuvers into separate operations that create an efficient Hohmann transfer.

$$a_t = (h_{apogee} + h_{perigee} + 2 * \mathrm{R_E}) / 2 \tag{9-1}$$

$$= (h_{GEO} + (h_{GEO} - 10\ \mathrm{km}) + 2 * \mathrm{R_E}) / 2$$

$$= (35{,}786 + 35{,}776 + 2 * 6378.14) / 2$$

$$= 42{,}159.14\ \mathrm{km}$$

The starting velocity at the first firing point in GEO (before any firing has taken place) is

$$V_{circ\_GEO} = \sqrt{(GM_E / a_{circ\_GEO})} \tag{9-2}$$

$$= 3074.66\ \mathrm{m/s}$$

where G is the universal constant of gravitation, M_E is the mass of the Earth and $a_{circ\_GEO} = \mathrm{R_E} + h_{GEO}$.

The velocity at apogee of the transfer orbit, Vt_a, is then used to determine the first ΔV

$$Vt_a = \sqrt{2 * (GM_E/(\mathrm{R_E} + h_{apogee}) - \mathrm{GM_E}/(2 * \mathrm{a_t}))} \tag{9-3}$$

$$= 3074.48\ \mathrm{m/s}$$

The required ΔV of the first firing is the difference in velocity:

$$\Delta V_{apogee} = |V_{circ\text{-}GEO} - Vt_a| \tag{9-4}$$

$$= 3074.66 - 3074.48 = 0.18\ \mathrm{m/s}$$

Next we need to establish the size (a) of the inspection orbit. We first to determine the orbit period of the inspection orbit. For each orbit of ArgusSat, the GEO satellites must move one degree more. GEO satellites move 360 degrees in one sidereal day (86,164 seconds); thus their angular motion, n_{GEO}, is:

$$n_{GEO} = 86{,}164 / 360 = 239.34 \text{ s/deg}$$

The period, T, of ArgusSat must be the GEO period plus 239.34 s

$$T_{ArgusSat} = 86{,}164 + 239.34 = 86{,}403.3 \text{ s}$$

The period allows us to determine the semi-major axis of the ArgusSat inspection orbit, and with that its apogee altitude, as we already have the perigee altitude from the initial calculations, as shown in the right-hand diagram of Figure 9-7.

$$a_{ArgusSat} = ((T_{ArgusSat} / (2 * \pi))^2 * GM_E)^{(1/3)} = 42{,}242.18 \text{ km}$$

$$h_{ArgusSat\_apogee} = 2 * a_{ArgusSat} - (2 * \mathrm{R_E}) - h_{perigee} = 35{,}952.1 \text{ km}$$

This defines the inspection orbit of ArgusSat as 35,952.1 km × 35,776 km to ensure a daily visit of each subsequent 1-deg GEO slot.

We now work out the velocity at perigee for both the transfer and inspection orbits to determine the ΔV for the second firing:

The velocity at perigee of the transfer, Vt_p, orbit is

$$Vt_p = \sqrt{(2 * (GM_E/(R_E + h_{tp}) - GM_E/(2 * at)))}$$

$$= 3075.21 \text{ m/s}$$

$$V_{perigee} = \sqrt{(2 * (GM_E/(R_E + h_{ArgusSat\_perigee}) - GM_E/(2 * a_{ArgusSat})))}$$

$$= 3078.23 \text{ m/s}$$

$$\Delta V_{perigee} = 3078.23 - 3075.21 = 3.02 \text{ m/s}$$

The total $\Delta V - (\Delta V_{apogee} + \Delta V_{perigee})$ for ArgusSat to transfer from GEO to a 1-degree per day inspection orbit is 3.2 m/s, using two firings about 12 hours apart. Figure 9-8 illustrates the inspection orbit and ArgusSAT's proximity to three GEO satellites over a period of several days. Figure 9-9 illustrates the same inspection orbit in Figure 9-8 with the view point fixed relative to Earth's surface.

If the inspection trajectory has to traverse the GEO belt in the opposite direction, the period must be shorter than for GEO by the same amount, and the same calculations apply. Similar adjustments can be made to change the 10-km approach to a different value, including a negative value indicating nearest access to be above GEO altitude.

As shown in Figure 9-8, ArgusSat now travels in its elliptical inspection orbit, approaching each subsequent 1-deg GEO slot when it is near its perigee, and crossing the GEO altitude between apogee and perigee.

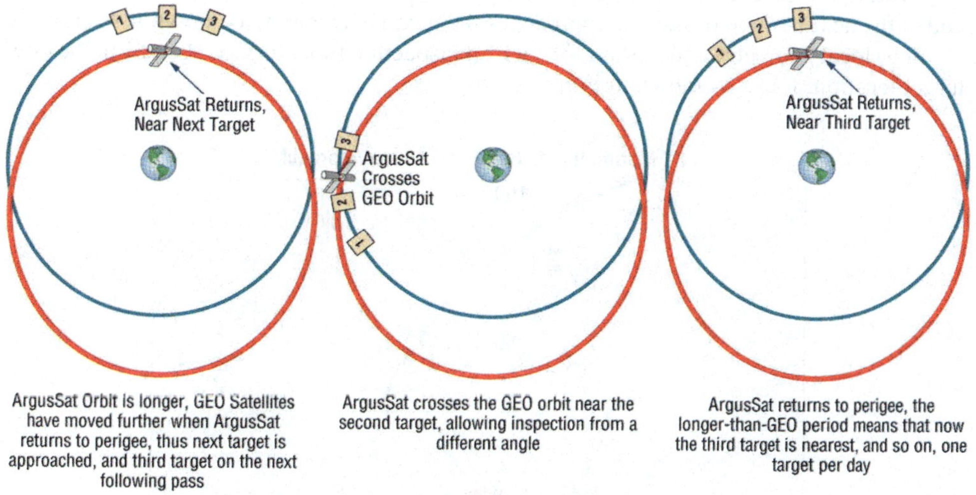

FIG. 9-8. **Orbit Motion Showing the Inspection Orbit After the Initial Transfer from GEO.** After the two firings the inspection satellite is in a slightly elliptical orbit, and every orbit its position has shifted relative to the satellite in the GEO belt.

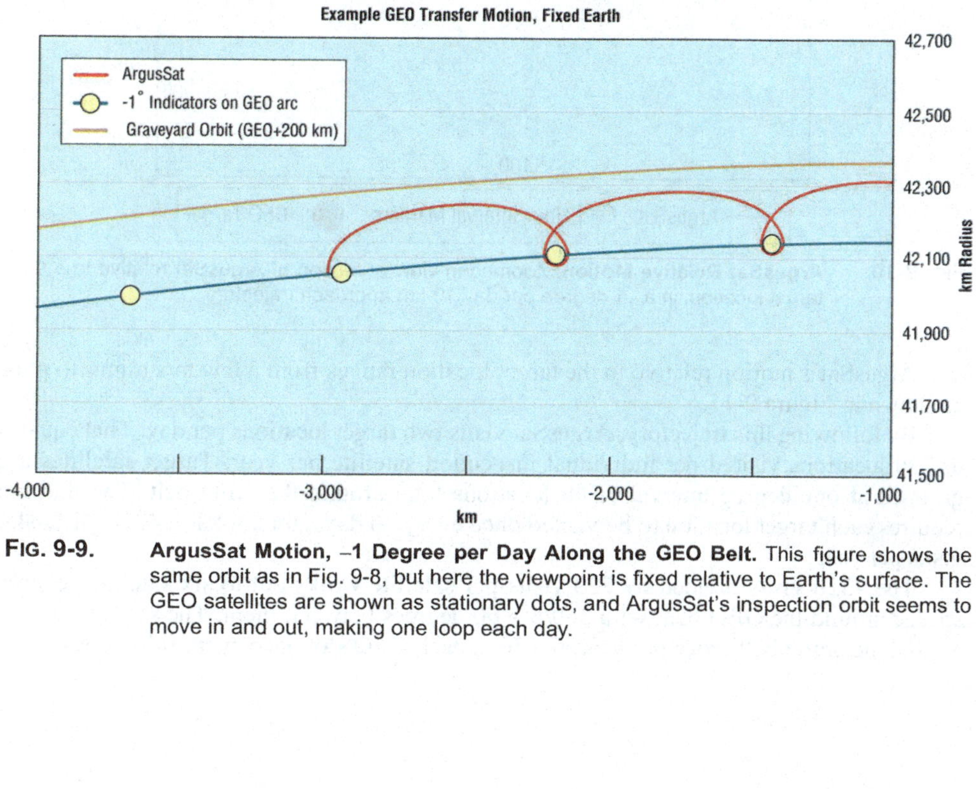

FIG. 9-9. **ArgusSat Motion, –1 Degree per Day Along the GEO Belt.** This figure shows the same orbit as in Fig. 9-8, but here the viewpoint is fixed relative to Earth's surface. The GEO satellites are shown as stationary dots, and ArgusSat's inspection orbit seems to move in and out, making one loop each day.

During the GEO crossings our satellite can inspect the target from different directions, and at the next perigee passage it approaches the next GEO target, as shown in Figure 9-9.

The local geometry and motion for such an inspection trajectory, as seen from the GEO target location, looks as shown in Figure 9-10.

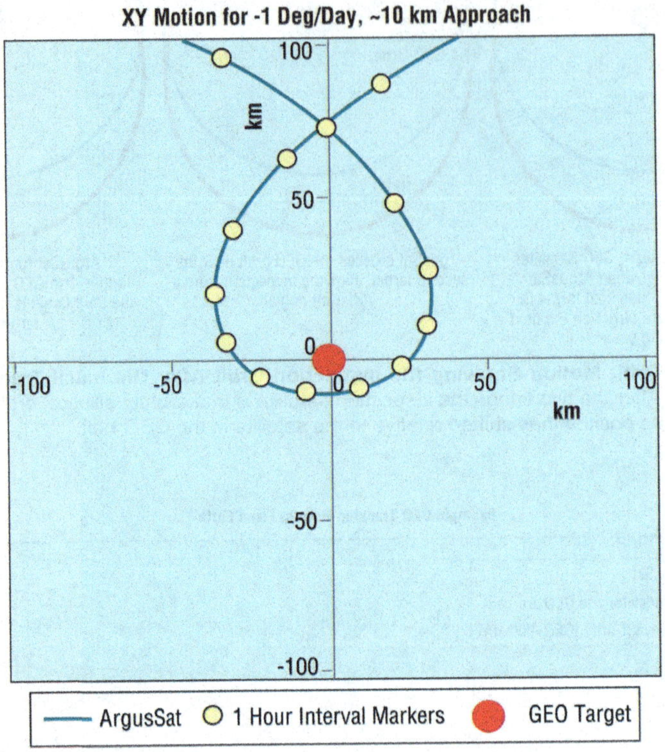

FIG. 9-10. **ArgusSat Relative Motion.** Zoomed-in view of motion of ArgusSat relative to a GEO target location, in a −1 degree per day 10-km approach trajectory.

ArgusSat's motion relative to the target location ranges from a few m/s minus to plus, as shown in Figure 9-11.

By following this trajectory, ArgusSat visits two target locations per day. That equates to 730 locations visited per individual inspection satellite per year. Target satellites are grouped at one-degree intervals: 360 locations total around the GEO belt. The mission requires each target location to be visited once every 30 days, for a total of 4320 GEO-slot visits per year.

The 4320 visits divided by 720 visits per satellite works out to six satellites evenly spaced around the GEO belt, with 360/6 = 60 degrees between them. The satellites travel around the entire belt twice per year, reaching each GEO slot once every thirty days.

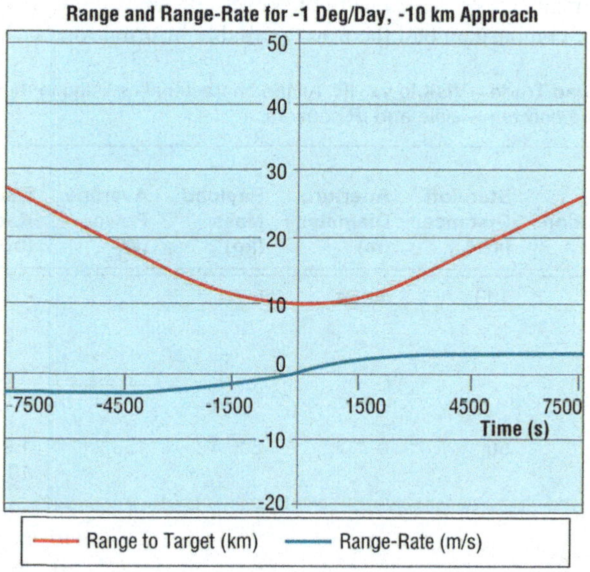

FIG. 9-11. **ArgusSat Relative Range and Range-Rate Over Time.** This chart shows a time-axis of the approach speed and distance for ArgusSat as seen from the target location. The vertical axis numbers indicate km and m/s for the respective curves.

9.5.4 Payload Trade

Now that we've selected an orbit, we need to do a trade on the payload type and determine a preliminary payload size. The typical mission would entail every satellite having all sensing payloads on board, giving a full suite of measurements every thirty days for each GEO slot. A cost reduction option could be to include a subset of instruments per satellite. That way the individual satellites would become smaller and cheaper, with simpler downlink and other design requirements. Each GEO slot would still be visited once every thirty days, but each time a different selection of measurements would be taken.

The first example walks through a system design of multiple satellites, each with one sensor, using the optical sensor trade as an example. Then the system sizing and cost will be shown for an example using multiple sensors on each satellite (optical, lidar, and radar).

9.5.5 Payload Trade—One Sensor per Satellite

Starting with the optical example, and using SMAD (pg. 255), we used a visual wavelength of 6×10^{-7} m and an infrared wavelength of 4.2×10^{-6} m to get a first estimate of the aperture size for the 10-cm resolution at a standoff distance of 10 km. Our research yielded a sample set of current visible and IR sensors to perform the sizing against; Table 9-7 and Figure 9-12 give details. Table 9-6 shows the sizing data and resolution for the visible and IR sensors. To properly compare, we needed a few key pieces of information: the aperture, dimensions, mass, and power. Not all the information was available for each of the sensors. In many cases the aperture or power consumption was not available, so the sensors chosen for comparison had the most available information.

TABLE 9-6. Payload Trade—Visible vs. IR. All the characteristics listed in this table are relevant in trading between visible and IR sensors.

Payload Type	Resolution (m)	Standoff Distance (km)	Aperture Diameter (m)	Payload Mass (kg)	Average Power (W)	Data Rate (bps)	Existing System Used for Sizing
Visible	0.10	100	1.464	-	-	-	Note – quite large aperture, do not continue
		50	0.732	306.4	153.2	1.918E+10	High Resolution Advanced Imaging System
		25	0.366	25	6.3	9.631E+09	Iguana
		10	0.146	1.8	1.5	3.862E+09	NanoCam (with aperture ratio of 2.0)
		8	0.117	0.9	0.7	3.135E+09	NanoCam
		7	0.102	0.6	0.5	2.744E+09	NanoCam
		6	0.088	0.4	0.3	2.352E+09	NanoCam
		5	0.073	0.2	0.2	1.933E+09	NanoCam
		3	0.044	0.1	0.1	1.160E+09	NanoCam
Visible	0.05	100	2.928	-	-	-	Note – quite large aperture, do not continue

TABLE 9-6. Payload Trade—Visible vs. IR. (Cont.) All the characteristics listed in this table are relevant in trading between visible and IR sensors.

Payload Type	Resolution (m)	Standoff Distance (km)	Aperture Diameter (m)	Payload Mass (kg)	Average Power (W)	Data Rate (bps)	Existing System Used for Sizing
		50	1.464	-	-	-	Note – quite large aperture, do not continue
		25	0.732	306.4	153.2	3.852E+10	High Resolution Advanced Imaging System
		10	0.293	25.7	6.4	1.545E+10	Iguana
		5	0.146	1.8	1.5	7.731E+09	NanoCam (with aperture ratio of 2.0)
		3	0.088	0.4	0.4	4.640E+09	NanoCam
Infrared	0.10	10	1.025	-	-	-	Note – quite large aperture, do not continue
		5	0.512	105.1	52.6	1.960E+09	High Resolution Advanced Imaging System
		3	0.307	22.7	11.4	1.177E+09	High Resolution Advanced Imaging System
		1	0.102	0.6	0.2	3.924E+08	Iguana

TABLE 9-6. Payload Trade—Visible vs. IR. (Cont.) All the characteristics listed in this table are relevant in trading between visible and IR sensors.

Payload Type	Resolution (m)	Standoff Distance (km)	Aperture Diameter (m)	Payload Mass (kg)	Average Power (W)	Data Rate (bps)	Existing System Used for Sizing
Infrared	0.05	10	2.050	-	-	-	Note – quite large aperture, do not continue
		5	1.025	-	-	-	Note – quite large aperture, do not continuc
		3	0.615	840.8	420.4	7.842E+09	High Resolution Advanced Imaging System
		1	0.205	4.4	1.1	1.570E+09	Iguana

Initial observations on the visible vs. IR trade show that the IR instrument is much larger than the visual instrument at the same aperture and resolution. Since aperture will drive the mass as well as power and propellant use, we chose the visible spectrum instrument to proceed with further analysis. The standoff distance determines the aperture diameter; a larger distance increases the size of the payload. Initially, the closer standoff distances appeared to produce a very small aperture payload, and research confirmed that such small aperture payloads do exist as commercial products. We used these in the preliminary sizing of the system. The data rates are quite high, in the Gbps range. This was still apparent even after adjusting the duty cycle per orbit period to 20%, since the sensor will not be running all the time, but only when imaging a target satellite. This data rate will drive the communications system, and we discuss trades on this in later sections of this chapter.

A close standoff distance of 5 km was the first trade, but seemed too intrusive to the target satellite. Therefore, we selected a 10-km standoff distance for the initial spacecraft sizing. The aperture size is small, but we found systems such as the NanoCam, which has an aperture diameter of 0.146 m and an estimated resolution of 0.1 m at a 10-km standoff distance. The mass and power are also quite small. See Table 9-6 for additional details.

TABLE 9-7. Off-the-Shelf Visible and IR Sensors. Using appropriate off-the-shelf equipment saves much of the cost of developing it from scratch. The resolution values in this table are for Earth Surface from the specified altitude. For comparison, correlate the ratio of 10 km/altitude, and then the resolution times that ratio. For example, Crystal Space has a resolution of 600 m from an altitude of 600 km. This system at a 10 km standoff would have a resolution of 10 m.

Name	Aperture (m)	Resolution (m)	Dimensions (mm)	Altitude at Resolution (km)	Mass (kg)	Power (W)	Wavelengths
Chameleon (CubeSatShop)	unavailable	9.6	200 × 94 × 94	500	1.35	3.5	Visible and Near Infrared
Gecko (CubeSatShop)	unavailable	39	97 × 96 × 60	500	0.48	unavailable	Visible
NanoCam (GOMSpace)	0.070	30	86.0 × 91.7 × 97.2	650	0.2	0.8	Visible
Crystal Space (Crystal Space)	unavailable	600	45 × 25 × 45	600	0.05	0.075	Visible
Planet Doves (Safran 2015 and Planet 2016)	0.099	3	10 × 10 × 30	475	5	unavailable	Visible
Iguana 12U (SCS)	0.25	2	22cm × 22cm × 33cm	500	8	8	Visible and Near Infrared
High Resolution Advanced Imaging System (Dubaisat-2, Aldhafri, and ESA 2015)	0.400	1	200cm × 75cm (diameter)	600	50	100	Visible and Near Infrared

9.5.6 Revisit on Exact Delta V (ΔV)

To continue with sizing ArgusSat, we need to determine the required ΔV. From GEO to the targeted elliptical orbit is about 3.2 m/s ΔV accomplished by two burns. If the system has six spacecraft, then rephasing maneuvers will be necessary to separate each spacecraft by 60 degrees within their orbit. For rephasing, the longer the timeframe, the less ΔV, but ArgusSat should be operational quickly, so we chose a period of two weeks between launch and operation of the spacecraft. In this case, the ΔV (worse case for sizing of 180 deg) total for one spacecraft is 208 m/s, broken down as follows (total includes the start and stop maneuvers):

	SPACEVIEW™ 50 SV-50	SPACEVIEW™ 65 SV-65	SPACEVIEW™ 70 SV-70
Satellite Class	SmallSat	SmallSat	SmallSat
Aperture	0.5 m	0.65 m	0.7 m
Payload Mass	90 kg - 130 kg	110 kg - 150 kg	120 kg - 160 kg
Imaging Power	200 W - 275 W	245 W - 305 W	250 W - 320 W
Resolution @500 km (GSD)	0.35 m - 0.5 m	0.3 m - 0.45 m	0.25 m - 0.4 m
Sensor Capabilities	• Staring • Scanning • Motion Imagery/Video • Low Light	• Staring • Scanning • Motion Imagery/Video • Low Light	• Staring • Scanning • Motion Imagery/Video • Low Light
Bands	• Visible Pan/Color • Pan, 4- or 8-Band MS • SWIR/MWIR	• Visible Pan/Color • Pan, 4- or 8-Band MS • SWIR/MWIR	• Visible Pan/Color • Pan, 4- or 8-Band MS • SWIR/MWIR

	SPACEVIEW™ 24 SV-24	SPACEVIEW™ 35 SV-35	SPACEVIEW™ 42 SV-42
Satellite Class	NanoSat	MicroSat	MicroSat
Aperture	0.24 m	0.35 m	0.42 m
Payload Mass	< 10 kg	20 kg - 35 kg	25 kg - 40 kg
Imaging Power	10 W	70 W - 170 W	70 W - 170 W
Resolution @500 km (GSD)	0.9 m - 1.1 m	0.7 m - 1.0 m	0.5 m - 0.75 m
Sensor Capabilities	• Staring • Motion Imagery/Video • Low Light	• Staring • Scanning • Motion Imagery/Video • Low Light	• Staring • Scanning • Motion Imagery/Video • Low Light
Bands	• Visible Pan/Color	• Visible Pan/Color • Pan, 4- or 8-Band MS	• Visible Pan/Color • Pan, 4- or 8-Band MS

FIG. 9-12. SpaceView Series of Payloads. System tradeoffs include considerations of payload characteristics [Harris, 2018].

- ΔV (highest) is 146 m/s at 180 deg phase at 14 days (total includes start and stop maneuvers)
- Add deorbit into graveyard orbit ~11 m/s
- Add ΔV for getting from GEO to elliptical ~3.2 m/s in two firings

- Subtotal: 160 m/s
- Add 25% margin: 40 m/s = 200 m/s
- Add in 5% for attitude control = 8 m/s
- 208 m/s total ΔV

9.5.7 Payload Characteristics—Radar, Optical, and Lidar

The section above outlines a simple trade between two optical sensors, visual and IR. This section discusses more details and design characteristics of radar, optical and lidar sensors. With the relatively short distance (still many km) between it and the target, the inspector satellite can take detailed images and measurements using various suitable sensors. Typical inspection instruments include optical cameras in the visible and near-infrared wavelengths, which can produce photographic still and video images. Radar systems can locate the subject, possibly identify it through its received radar signature, and use the signals for relative navigation in space. Radar is also good for ranging, sizing, and motion indication of the target. A lidar can detect internal vibrations of the target, (determining whether the satellite is still active), and can produce precise dimensions and volumetric measurements, e.g., solar panel and antenna size, instrument aperture diameter, and thruster size. A thermal infra-red imager can determine the temperature and temperature variations across the target. All these measurements combine to give a good overview of the capabilities and status of the target satellites.

9.5.8 Radar Basics

The target must be at a fairly great distance, and the trajectory of the inspector must be carefully controlled to remain outside the particular range to prevent any inadvertent interaction with the target.

Radar works by transmitting a pulsed signal with an antenna sweeping the direction that the target is expected to be in, or all the way around in case the location is initially unknown. The radar pulse reflects off the target in many directions, and some of the signal returns to the inspector. The returned signal pulses are received and analyzed to yield information about the target. Space radar is easier than Earth-based radar, as it involves no signal delay, attenuation, ground reflections, or atmospheric scattering and refraction.

The time from sending to receiving of the pulse depends on the distance. The signal travels at the speed of light (c_0); thus the distance to the target R (range) is:

$$R = c_0 * t / 2 \tag{9-5}$$

t = time in s

c_0 = speed of light (3E8 m/s)

R = range in m

Example:

t is 110 microseconds: $t = 0.00011$ s, $R = 3E8 * 0.00011 / 2 = 16500$ m (16.5 km)

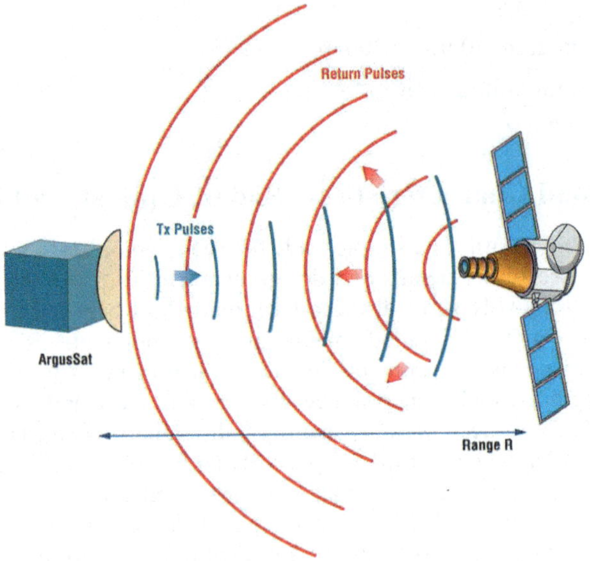

FIG. 9-13. **Radar Ranging to the Target.** Here the inspector uses radar pulses to determine the distance to the target by measuring the time it takes for pulses to return. By analyzing the frequency shift of the return signal, it determines the relative motion rate, and by analyzing returned signal strength it determines the relative size of the target

The time from transmission to reception is divided by two, as the signal travels there and back, and the range R is thus associated with half the time.

The resolution of the measurement depends on the length of the pulse: the shorter the pulse, the more accurate the distance, but the target has many facets that might be meters apart from each other. The range-resolution, S_r, is:

$$S_r = c_0 * \tau / 2 \tag{9-6}$$

S_r = range-resolution in m

τ = length of the pulse in s

Example:

If the pulse is 1 μs long: $\tau = 0.000001$ s, $S_r = 3E8 * 0.000001 / 2 = 150$ m.

Any target sections closer together than 150 m will appear as one irregular item, with an extended return pulse. In addition, the length of the return pulse determines the distance between the nearest and furthest part of the target, as measured radially out from the inspector. If the return pulse is the same length as the transmitted pulse, the reflection must have been from a single surface. If the return pulse is spread 10% longer than the original length, then the target must have surfaces that are apart by at least 10% of the range resolution—15 meters for the above example.

As mentioned, the distance is measured radially, so the measured value between points on, for example, a target satellite with large wings depends on its orientation towards the inspector. We can measure the relative velocity between the target and the inspector by two different methods. In the first, two subsequent measurements taken with a known time between them give the radial velocity, V_r:

$$V_r = (R_2 - R_1) / t_m \quad (9\text{-}7)$$

V_r = radial velocity in m/s

$R_2 - R_1$ = difference between the measured distances in m

t_m = time between measurements in s

Example:

The time between two selected pulses is R_2 s. The first distance, R_1, is 12.406 km, and the second distance is 12.401 km.

$$V_r = (12{,}406 - 12{,}401) / 2 = -2.5 \text{ m/s}$$

A second method of determining relative motion is to measure the change in frequency of the received signal:

$$V_r = \Delta f * c_0 / (2 * f_{tx}) \quad (9\text{-}8)$$

Δf = difference between transmitted and received frequency (Hz)

V_r = radial velocity (m/s)

f_{tx} = radar transmit frequency (Hz)

The difference in frequency is measured in the payload, and if the transmitted radar frequency is, for example, 12 GHz, with a received signal of 1500 Hz higher, the relative velocity is

$$V_r = 1500 * 3E8 / (2 * 12E9) = -18.75 \text{ m/s}$$

The negative value indicates that the distance between target and ArgusSat is getting smaller at the time of measurement.

Accuracy

The measurement accuracy should be better than the resolution of the distance value. The main disturbance in space, in the absence of, e.g., atmospheric and ground scatter, is random noise from background radiation. A higher signal bandwidth and stronger signal both improve the quality, and thus the range error δR becomes smaller. A perfectly accurate measurement would require a perfect square-wave pulse, to which the time measurement is related, which would require infinite bandwidth. Thus, the rise-time of the pulse has a non-zero value, and the measurement is defined at a particular point of the rising edge, typically √(2)/2, about 0.707 of the maximum value. The actual value will have noise on it, and hence the precise time will vary from pulse to pulse.

We calculate the range error as:

$$\delta R = c_0 / (2 * B * \sqrt{(2 * \text{SNR})}) \quad (9\text{-}9)$$

δR = random error in m
B = bandwidth in Hz
SNR = signal to noise ratio (must be >> 1 to yield a good measurement)

Pulse Repetition Frequency

The pulse repetition frequency, or PRF, is the number of pulses per second that the radar sends out. If the return pulse is received after several pulses have gone out, then the signal becomes ambiguous, because it will be impossible to know which initial pulse the return pulse relates to, and thus the range cannot be calculated. The time between pulses thus defines the maximum distance over which the radar can measure distance without employing additional measures like coding, frequency variation, etc.

With a PRF of 1 kHz, the signal travels around 300 km between pulses, so the maximum distance, based on half the overall travel time, is about 150 km.

The measurement accuracies here relate to the relative distances between the target and the inspector. The actual location of any target will require knowledge of the inspector's location by onboard GPS, ground ranging, or triangulation between Earth, Sun, and Moon, requiring appropriate navigation equipment on board.

Angular Measurement

We can determine the direction of the target coarsely by knowing the beamwidth of the radar antenna. If the inspector receives a reflected radar pulse, then the target must be located in the beamwidth.

The beamwidth of the antenna (the angular size of the main antenna lobe), depends on the frequency of the radar signal and the diameter of the dish:

$$BW = 70 * (c_0 / f_{tx}) / D \tag{9-10}$$

BW = beamwidth in degrees
f_{tx} = transmit frequency in Hz
D = dish diameter in m

Thus, for an X-band radar using a 0.3-m dish, the beamwidth will be about 7.5 degrees. When rotating the antenna or the entire inspector while taking measurements, the variation in the strength of the reflected pulses could aid in interpolating the results to achieve better angular resolution.

The gain of the antenna is calculated similarly:

$$G = 0.6 * \pi^2 * D^2 / (c_0 / f_{tx})^2 \tag{9-11}$$

f_{tx} = transmit frequency in Hz
D = dish diameter in m

We give 0.6 as a typical efficiency factor, accounting for the fact that some of the aperture is blocked by the feed, and that not all signal transmitted by the feed is captured by the reflector.

The antenna gain is also a factor in determining the radar cross-section.

Radar Cross-section

The measured strength of the reflected signal depends on the apparent size and reflectivity of the target, and the orientation of the facets towards the inspector. This is called the radar cross-section, denoted σ:

$$\sigma = 4 * \pi * r^2 * S_r / S_t \tag{9-12}$$

σ = radar cross-section in m^2

r = effective radius of the target in m

S_r = the scattered power density at the target (W / m^2)

S_t = the transmitted power density at the target (W / m^2)

These values are defined at the target, as σ is intrinsic to the target. A metal sphere with a projected area of 1 m^2 has a σ of 1 m^2, and this sphere has the following diameter:

$$D_{sphere} = 2 * \sqrt{(1/\pi)} = 1.1284 \text{ m} \tag{9-13}$$

Thus a metal sphere of 1.13 m diameter shows up as a radar size of 1 m^2

A corner reflector directs all received energy straight back to the transmitter, and hence its σ is very large. Its reflection works as an aperture, and is thus frequency-dependent. The shape of the three panels making up the corner reflector drive the overall size. For a reflector of three squares with sides of length L = 0.762 meter each, and mounted at 90 deg from each other, the formula for effective area of an X-band (12 GHz) radar is:

$$\sigma = 12 * \pi * L^4 / \lambda^2 \tag{9-14}$$

$$= 12 * \pi * 0.337 / 0.025^2 = 20336 \text{ m}^2$$

This is similar to the optical reflecting paint used on street signs that light up far in the distance and stand out from the surrounding area. Satellites sometimes have radar or optical corner reflectors to aid accurate position determination from the ground or to aid laser ranging.

The Radar Equation

The received signal strength is related to the transmitted power, the wavelength, the antenna gain, the distance to the target, and the radar cross-section of the target. We can put all those parameters into one equation and then solve for R, the range to the target. This is known as the radar equation.

$$R = \sqrt[4]{(P_{tx} * G^2 * \lambda^2 * \sigma/(P_{rx} * (4 * \pi)^3)} \tag{9-15}$$

R = range, distance between the radar imaging satellite and the target in m

P_{tx} = transmit power in W

P_{rx} = received power in W

G = antenna gain (dimensionless factor)

λ = wavelength of the signal in m
σ = radar cross-section in m^2

Rewriting for the radar cross-section, we obtain:

$$\sigma = P_{rx} * (4 * \pi)^3 * R^4 / (P_{tx} * G^2 * \lambda^2) \quad (9\text{-}16)$$

Given that we know the transmit power, the wavelength, and the gain of the antenna, and that we can separately determine the range R to the target from measuring the time it takes to receive the return pulse, we can calculate the radar cross-section (σ) of the target by measuring the received signal strength. This value of σ helps identify the target and gives other information. If, for example, σ varies over time, with each return signal being different, the target might be rotating. If σ varies in the same way as that from another target, the two targets might be similar satellites.

Summary

Radar determines the distance, location, and radial velocity of targets by transmitting radiofrequency pulses towards them and measuring the time it takes to receive the reflected signals. The strength and distortion of the shape of the reflected signals, as well as any Doppler shift in the received frequency, give additional information about the maximum radial size, the radar cross-section, and the radial velocity of the targets.

9.5.9 Optical Imagers

Sunlight illuminating the target is reflected by its various surfaces, and a telescope on the inspector can receive this reflected light. The light is then focused on a light-sensitive sensor. The sensor, typically either a CCD (charge-coupled device) or a CMOS (complementary metal-oxide-semiconductor) array, absorbs the photons as electron displacement energy, and this electrical charge appears as a voltage on each pixel. These pixels are read out and digitized sequentially by the inspector's camera electronics, thus turning the image into a computer data file that can be stored, processed, transmitted to the ground, and displayed for analysis. This is much the same as the camera in modern smartphones, but these would not show much detail at many miles distance, and they would not deal with the scene of a very bright target against a very dark background, creating a high dynamic range.

Telescope

The telescope includes the zoom lens that receives the light in its aperture and projects it onto the sensor. Its focal length must be large enough to magnify the target image for the desired image resolution, and its aperture wide enough to ensure that the image is as sharp on the sensor as the pixels can sample. As opposed to Earth imaging from space, where the satellite motion in orbit means that there is little time available for light gathering for each image, the lens aperture does not have to be any larger than required for the resolving quality: the imager can use a relatively long exposure time to collect as much light as needed to achieve the desired signal-to-noise ratio (SNR). This means that each pixel receives enough photons to allow low-noise conversion into digital information.

The telescope consists of lenses, mirrors, or a combination of the two, coupled with beam splitters, filters, and baffles as needed. The whole optical grouping is held in place by the metering tube. This is the main structure holding the lenses and mirrors at the right distance from the sensor, over the expected temperature range, and typically contains the adjustment mechanisms used before launch to fine-tune the setup. The adjustment between the mirrors is called alignment, and this typically has to be accurate to less than one micrometer in all three axes. The alignment creates a focused image of all the light coming into the telescope. Focusing, on the other hand, is the positioning of the sensor at the exact place where the focused image appears, so that the pixels of the sensor can sample the image. The rear depth of focus is the allowed range of the sensor position, and this depends on the size of the pixels and focal length of the telescope. The longer the focal length, the longer the rear depth of focus, so focusing a telescope with a long focal length is easier than focusing a camera with a short focal length.

The projection size of the image on the sensor depends on the focal length and on the distance between the telescope and the target. The ratio between the overall size and a single pixel is the same for the actual target and the projected image, so we can relate the projection size of a single pixel on the sensor to the target sampling distance (TSD) on the target, together with the range (R) to the target and the focal length (f):

$$\text{TSD} / R = \text{Pixel} / f \tag{9-17}$$

TSD = distance between pixel positions at the target in m
R = distance between the inspector and the target in m
Pixel = linear size of the actual pixel on the sensor in m
f = focal length of the telescope in m

Similarly, the sensor size defines the image size at the range using the same formula, but replacing the sensor pixel dimension with the sensor size, and the projected pixel dimension with the image size at the target distance:

$$\text{Field-of-View} / R = \text{Sensor} / f \tag{9-18}$$

Field-of-View = image length in m
Sensor = sensor size in meters

Figure 9-14 depicts this relationship.

Example:

Telescope with a focal length of 500 mm ($f = 0.5$ m), a sensor with 5 micrometer pixels (1 pixel = 5E-6 m), at 10 km from the target ($R = 10{,}000$ m):

$$\text{TSD} = R * \text{Pixel} / f \tag{9-19}$$

$$= 10{,}000 * 5\text{E-}6 / 0.5 = 0.1 \text{ m}$$

Assuming that the lens is wide enough and that the target is suitably illuminated, the image produces a 10-cm pixel distance on the target, measured on the sensor's 5-micrometer pixels. Figure 9-15 gives an example of such a target.

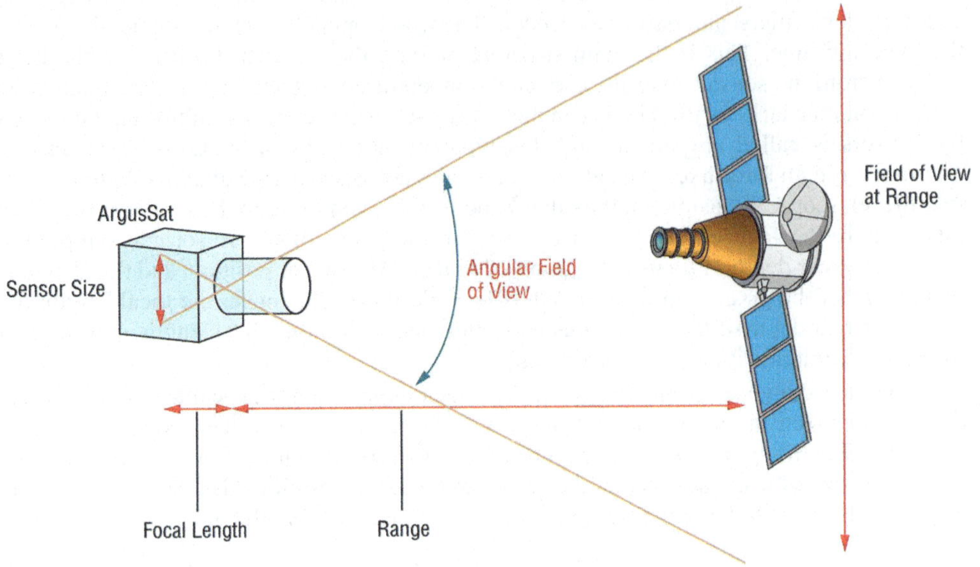

FIG. 9-14. Optical Imaging of Target Satellite. The image size depends on the range, the target size, the sensor size, and the focal length of the imager.

Image Credit: Nasa

FIG. 9-15. Target Sampling Distance. This is a simulated satellite image with a target sampling distance of about 10 cm. [Photo courtesy NASA]

Rayleigh Limit

Nature fixes a minimum physical size for a lens to be able to resolve the detail on the target and project it clearly on the sensor. Sampling the sensor with pixels that are smaller than the detail visible in the projection does not improve the information. The Rayleigh Limit is related to the wavelength of the light and the diameter of the optical system (typically the first mirror or lens).

$$\text{Rayleigh Limit} = 1.22 * \lambda * R / D \quad (9\text{-}20)$$

1.22 = factor pointing to the first minimum of the interference pattern of two resolvable pixels
λ = wavelength of the light in m
R = range from the inspector to the target in m
D = diameter of the lens or mirror in m

Example:

For light in the middle of the visible band, around 600 nm, and a lens of 7.5 cm diameter, at a distance of 10 km, the resolvable detail (TSD) in the image is

$$\text{TSD} = 1.22 * 0.0000006 * 10000 / 0.075$$

TSD = 0.0976 m

This is slightly better than what the sensor can detect; thus, the projected image takes full advantage of the capability of the sensor. The TSD is very close to the TSD calculated from the focal length and pixel size, giving us a balanced system.

The target will appear very bright against a very dark background, with sometimes direct sunlight mirror reflection directly to the inspector. Hence, any terrestrial automatic exposure control system or algorithm will not work very well or at all. Similarly, if the Sun is shining from a less favorable direction onto the target, the light level from the target will be very different again. With the available time to image, the best practice is to take many images at different exposure times in quick succession for each scene, and from them select those sub-scenes that have the best SNR: enough light to allow good analog to digital sampling and conversion, without saturation.

Image Size

With the requirement to take images at a 10-cm sampling distance at the target, and a typical target no larger than 30 meters end-to-end, the maximum number of pixels across the target is about 300. Thus a sensor of 1000 × 1000 pixels (1 MP) would easily suffice to see any target, and allow some pointing inaccuracy as well. The angular aperture of the telescope to support such a 1-MP sensor is determined as follows: The diagonal size of the image at the target distance is $\sqrt{2}$ * 1000 pixels * 0.1 m/pixel = 141.4 meters. An inverse tangent function can be used to determine the telescope viewing angle for a range of 10 km and half the diagonal size of the image.

$$\text{Telescope viewing angle} = 2 * \tan^{-1}(0.5 * 141.4 / 10{,}000) = 0.81 \text{ degree}$$

This means that the telescope can only see a very small area of space, and it will need to rely on the radar or very detailed orbital information about the target to be able to find it with the camera. A larger sensor would need a wider angle telescope, which makes it much more expensive, especially for a mirror system, but it would still only see a very small region of space. A second, wide-angle, camera, which would only detect the target as a few bright pixels, could serve as a targeting sensor, helping to locate the target and point the larger telescope.

Image Compression

The images, when converted to a digital image file, can be compressed to a much smaller size than the initial file: most of the image is black, those pixels need not be transmitted, and an RLC (run-length compression) algorithm is simple to implement. The rest of the file with the image of the target in it can usually be compressed to about 25% of the initial size with little or no loss of information. A typical image file for transmission, using a 1-MP bayer-encoded color filter, could be very small indeed.

Example file size: the target makes up 10% of the image (the rest is black), those 100,000 image pixels are compressed 4:1, and if initially sampled at 10 bits per pixel, the total file size is 250 kbits. This can be transmitted over any data link in a short time, even in near real-time if needed. For video-type imagery, the rate should be about 30 frames per second. The transmission will typically involve around 20% overhead, so the transmission rate, TX Rate, should be at least:

$$\text{TX Rate} = 250{,}000 * 30 * 1.2 = 9 \text{ Mbps}$$

The target might be larger in the image, perhaps at closer range, and so for actual implementation, a rate of 20-30 Mbps might be appropriate. This is too high for S-band implementation, but for X-band many technical solutions are available that provide this and much higher data rates.

LIDAR

In its simplest form, a LIDAR (light detection and ranging) consists of a laser shining a dot onto the target and then using a camera to detect that dot. If the laser is pulsed, then the time from transmitting the pulse to receiving the reflected light in the camera defines the distance (range) R:

$$R = c_0 * \mathrm{t} / 2 \tag{9-21}$$

R = range in m

c_0 = speed of light (3E8 m/s)

t = time between sending and receiving the pulse in s

Similar to radar, the absence of an atmosphere removes from the measurement any problems with scattering, absorption, and refraction. Figure 9-16 illustrates this situation.

The laser can scan across the target in many short angular steps by using rotating mirrors, or by using chip-level micromachined mirrors. The laser is pulsed while scanning the target region, and the optical camera detects the return time and signal strength. With

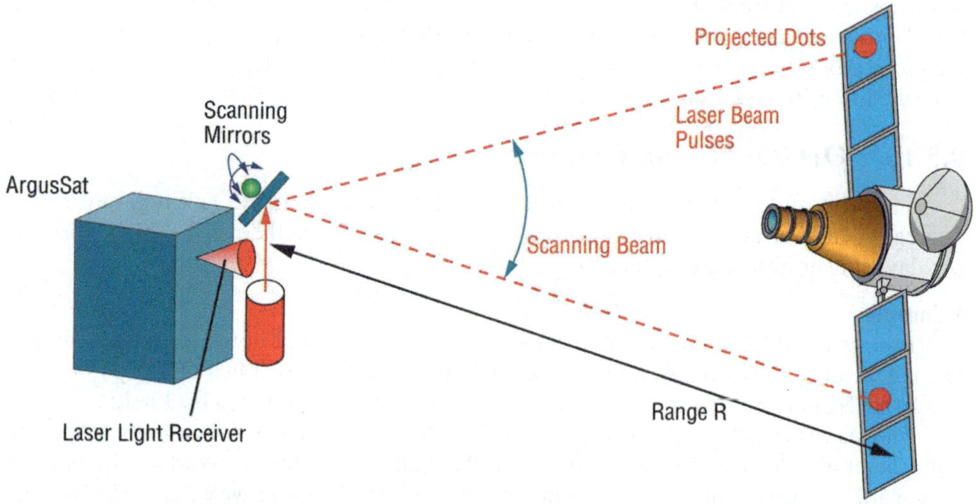

FIG. 9-16. Lidar Ranging. The sensor scans the tager with a series of light pulses, and analyzes the information captured in the return light much the same as for radar.

appropriate filters in the camera blocking all other signals, the reflections from, say, sunlight on the target should not unduly affect the detection of the laser dot, even at large distances. The knowledge of the instantaneous scan angle of every pulse allows the creation of a 3D-image file of the target as seen from the inspector. Repetitive scanning allows the detection of any motion of the target and any relative motion of parts of the target. Scanning from different angles allows us to build a full 3D model. The sensor need not perform any directional measurement: a single photodiode or photomultiplier to measure the time for the reflected light pulses to return to the inspector is all that is needed, possibly with a concentrator lens to improve sensitivity, and filters to reject any frequencies other than the transmitted laser light. The location of the projected dot is known from the original direction of the laser beam, and the distance: together they provide full 3D information.

Continuous pointing to a single surface spot on the target allows the detection of any vibrations of that surface, thus revealing, e.g., the rotation frequency of internal scan mirrors, the operation of cooling systems, etc. Lidars can operate at a high pulse repetition rate, over 100 kHz, and thus can detect vibration frequencies up to at least 50 kHz (the sampling frequency is at least twice the detectable frequency), which covers pretty much all expected mechanical vibration and rotation on a target satellite.

The measurement does not depend on sunlight illuminating the target, so the measurements can be performed from any direction, and also with the target in Earth eclipse.

Calibration of the Mission Sensors and Measurements

Each ArgusSat inspection craft can travel around a variety of targets, including well-characterized friendly satellites. Such targets can be used to calibrate the sensors, both against known characteristics of the target satellites, and possibly simultaneous Earth-based measurements, such as tone-ranging. As there are friendly targets in many GEO locations around the world, we would have many opportunities to recalibrate the sensors over the life of the mission.

9.5.10 Orbital Motion Examples

The orbit of ArgusSat can be adjusted so that the position at GEO altitude shifts by -2 degrees per day on average. With the right settings, ArgusSat will cross the GEO belt twice per day, at near one-degree intervals.

Example:

From a fixed GEO position it will take about 13.3 m/s total in two separate firings to change the orbit to an apogee radius of 42,369 km and a perigee radius of 41,664 km. This causes a period of 85,685 s, and a 2-degree per day motion along the GEO belt.

The maximum height above the GEO belt is 205 km, which crosses into the graveyard belt; therefore, the trajectory must be carefully monitored. Most expired satellites in the graveyard belt are no longer in 0-deg inclination and are therefore well out of the way, but they still cross the equatorial plane.

Figures 9-17 and 9-18 depict this trajectory. Figure 9-17 shows that the satellites are seen alternately from the east and the west, and that ArgusSat curves closer to the anti-Earth facet than to the Earth facet.

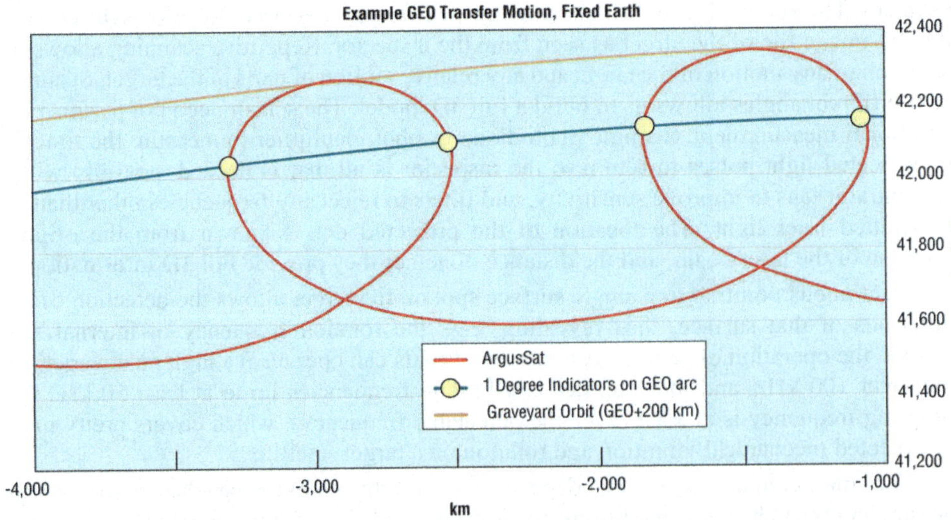

FIG. 9-17. Two-Degree Eastward Drift per day Inspection Trajectory, East-West Alternating Crossing. ArgusSat's trajectory relative to the GEO belt for a two-degrees, two targets per day, inspection path.

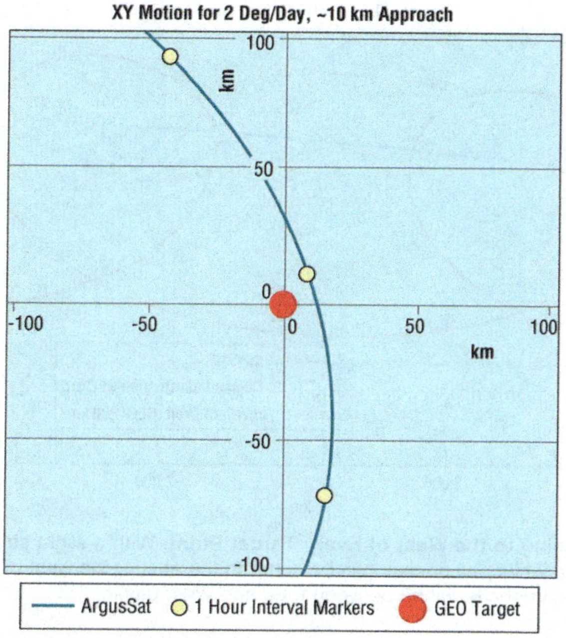

FIG. 9-18. Detail of Two-Degree per day Eastward Drift. The path of ArgusSat passes neat the anti-Earth side of the satellite at much closer range than the Earth-facet passage. It is an artifact of the orbit dynamics of this double visit per day trajectory.

The GEO-belt crossing happens at true anomaly angles of about 115 and 245 degrees; thus, the Sun will be in a substantially different position, shifted about 130 degrees between the two visits. With positioning of perigee towards the Sun, ArgusSat will see each target with good Sun illumination.

A slight adjustment of the trajectory will cause all crossings to occur at either the east side or the west side of the target slots, as shown in Figure 9-19. This could be useful for a survey by the radar and lidar imagers, where Sun angle is less important.

If we reverse the motion of the satellite, we can have ArgusSat cross the Earth-facing facets of the targets at a better viewing angle, see Figure 9-20. This orbit should have apogee towards the Sun for best illumination of the satellite Earth facet. This is an orbit of about 36,293 km × 35,590 km (radius 42,671 km × 41,986 km). From GEO this takes two propulsive maneuvers, one of about 9.2 m/s and one of about 3.6 m/s.

Since this orbit reaches more than 500 km above the GEO belt, shown in Figure 9-21, it crosses the graveyard altitude and therefore the trajectory must be carefully timed and adjusted to prevent any close approaches.

Reverse GEO Motion

For a very rapid inspection of all targets, we have the option of putting a satellite in reverse near-GEO orbit. Assuming that the orbit is circular, and 10 km smaller than GEO,

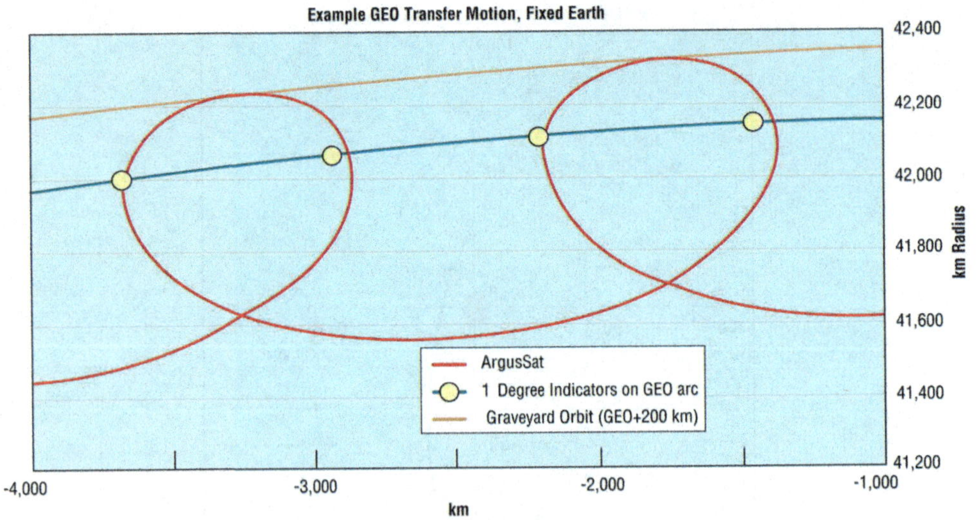

FIG. 9-19. Crossing to the West of Every Target Point. With a slight shift in the two target per day trajectory all passes can be either to the east or the west of the targets, instead of each target alternating between East and West passes.

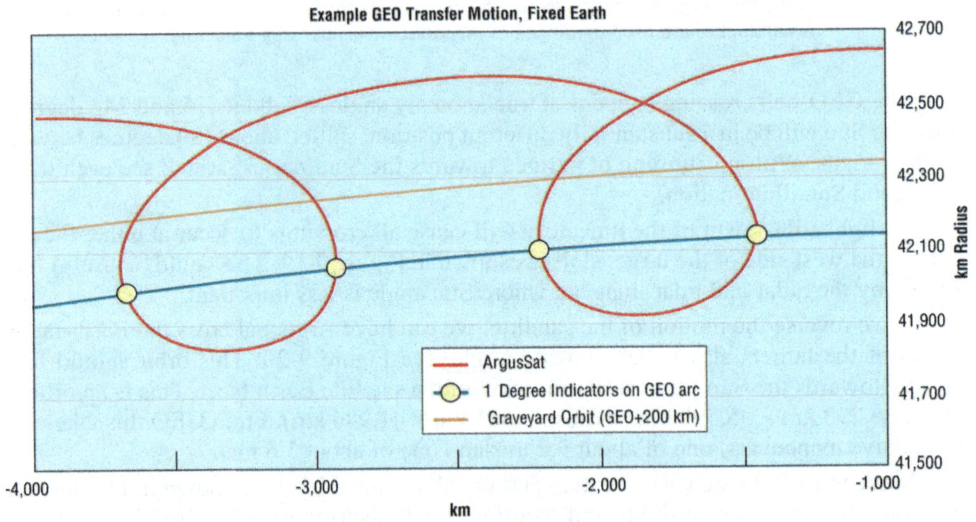

FIG. 9-20. Perigee Closer to GEO Altitude. Compared to the trajectory in Fig, 9-19, here the apogee is far above GEO, and perigee just below GEO. This creates a better viewing geometry for the Earth-facing side of the targets.

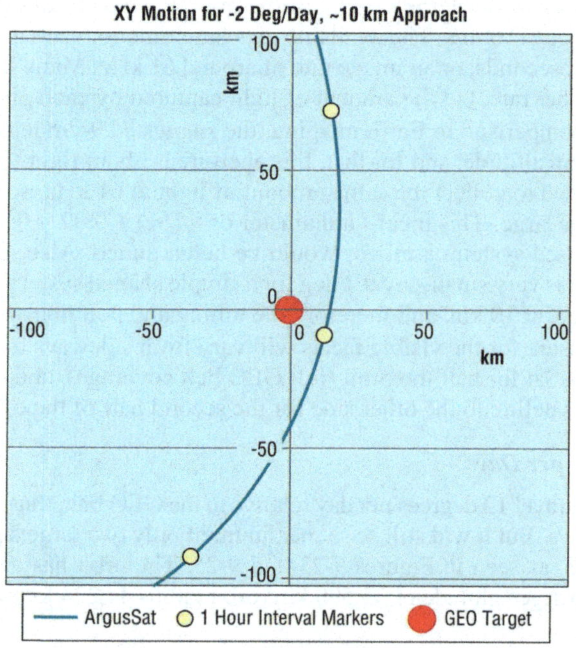

FIG. 9-21. Detail of Target Approach with Better Earth-facet Visibility. With apogee extending far beyond the graveyard orbit, ArgusSat curves tightly around the Earth-facing facet of the target for better imaging opportunities.

as shown in Figure 9-22, ArgusSat will see all GEO slots twice in one day. It makes a full orbit one way, and the GEOsats make a full orbit in the opposite direction; hence the relative motion is 360 degrees in about 12 hours.

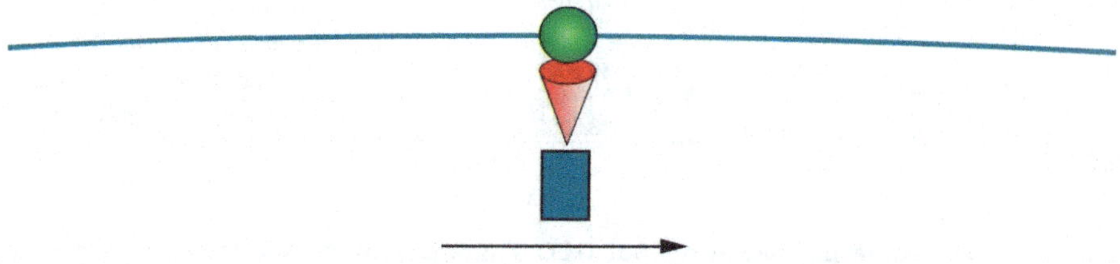

FIG. 9-22. Retrograde Near-GEO Orbit. The high relative speed of the target and inspector places a limit on exposure time for the sensor.

The crossing speed twice the GEO altitude in-orbit velocity of 3075 m/s, that is, an approach velocity of 6150 m/s. The available exposure time for each image is therefore no more than 16 microseconds, or an image rate of around 61 kHz. Many imaging sensors can read at a much higher rate, but the amount of light captured by each pixel will be limited.

As a simple comparison to Earth imaging, the suggested 9-cm lens would produce 6-m images at 600 km altitude, and for that, this aperture is about right to collect the light at 1 kHz. Here we need to collect the same amount of light at 61 kHz, so the lens must have an area 61 times as large. This means a diameter of √ (61) * 0.09 = 0.7 meter. This is too large for a lens-based system; a mirror would be better suited. Also, the field of view of such a large mirror is very small, say 0.1 deg for a simple shaped system, which is just about the size of the target at 10 km, and therefore the timing and pointing will be very difficult.

The Sun exposure for the visible facets will vary from sideways to full-on to the other side towards ArgusSat for half the orbit (full GEO belt coverage), and then from sideways to the back of the satellite to the other side for the second half of the orbit.

12-degree Motion per Day

ArgusSat can travel 12 degrees per day relative to the GEO belt, thus traversing the entire belt in about 30 days. But it will still see a maximum of only two targets per orbit. These can be adjacent targets, as seen in Figures 9-23 and 9-25. The orbit has an apogee altitude of 37,846 km and a perigee altitude of 35,590 km (orbit radius 44,224 km × 41,986 km).

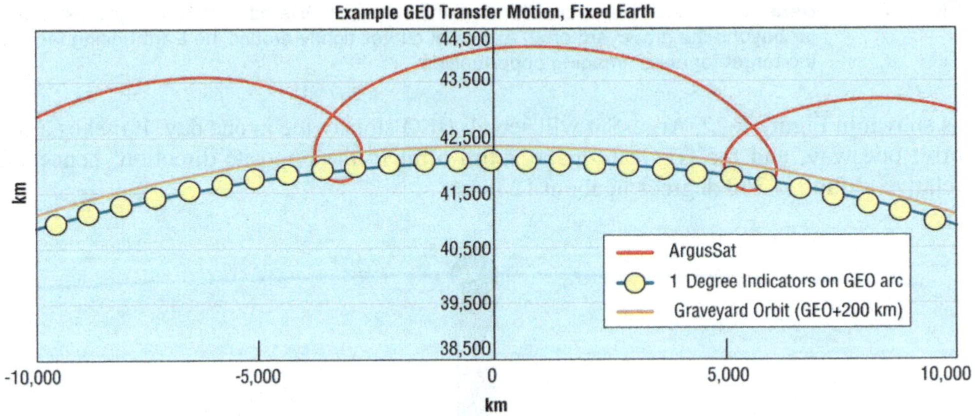

FIG. 9-23. 12-degree per day ArgusSat GEO Transfer Motion Above Graveyard Orbit. By raising apogee even further, a larger position shift per day can be achieved. Here we give an example 12-degree per day relative drift along the GEO belt shown, visiting two adjacent targets each day.

Other combinations of two targets are possible, with different settings of apogee and perigee, but the other ten target locations of each 12-location group will not be seen.

A similar trajectory moving in the opposite direction is shown in Figures 9-24 and 9-26, with an apogee of 35,982 km and a perigee altitude of 33,705 km.

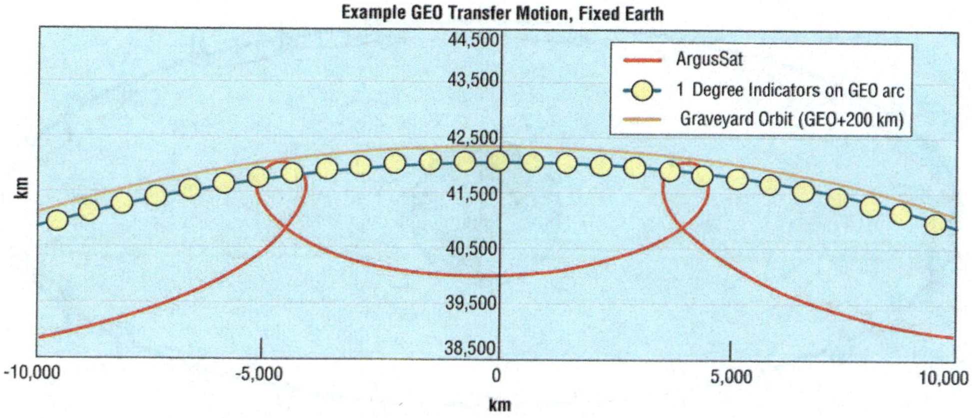

FIG. 9-24. Reverse Twelve-degree-per-day Drift. By making the semimajor axis (SMA) of ArgusSat smaller than GEO, the trajectory motion is reversed, still a 12-degree per day relative drift along the GEO belt, visiting two adjacent targets each day.

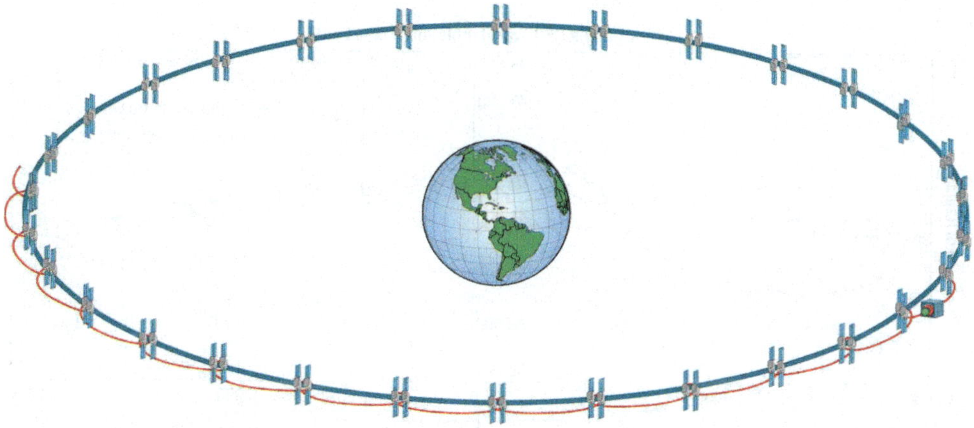

FIG. 9-25. 12-degree per day Inspection Trajectory. A 3D view of the 12-deg/day inspection trajectory, shown relative to the GEO belt; 15 days of the trajectory shown.

Sample GEO satellite pairs are shown 1 degree apart, every 12 degrees longitude. Larger or smaller steps are possible with adjustment of the orbit parameters.

Target of Opportunity

At any time a target of opportunity could require that ArgusSat change its trajectory to visit a slot not on its original path, but shifted a few degrees, and then after the diversion return to its original scheduled path. The example illustrated in Figure 9-27 shows a 6-degree per day trajectory (6 deg/day instead of 12 shown for clarity), visiting two slots each day. One day the target location is shifted by two degrees from nominal.

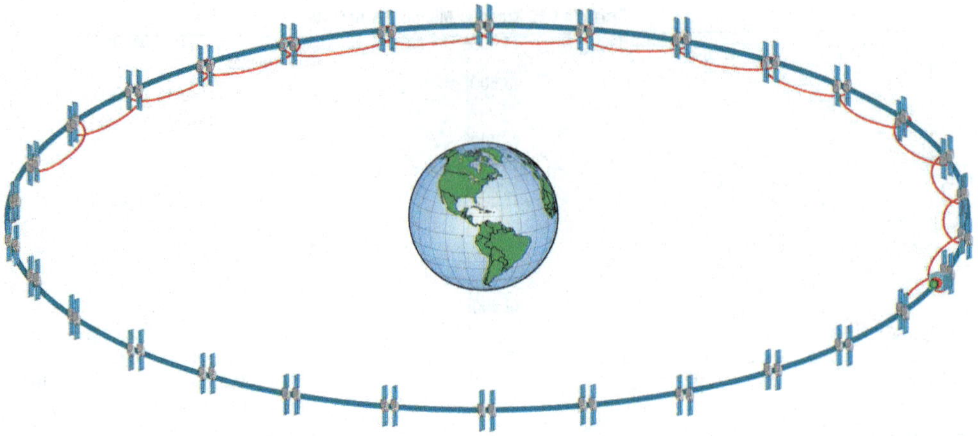

FIG. 9-26. **12-degree-per-day Reverse Drift.** Similar to above, 12 deg drift per day, two locations visited per orbit, but traveling in the opposite direction, so ArgusSat spends most of its orbit inside the GEO belt.

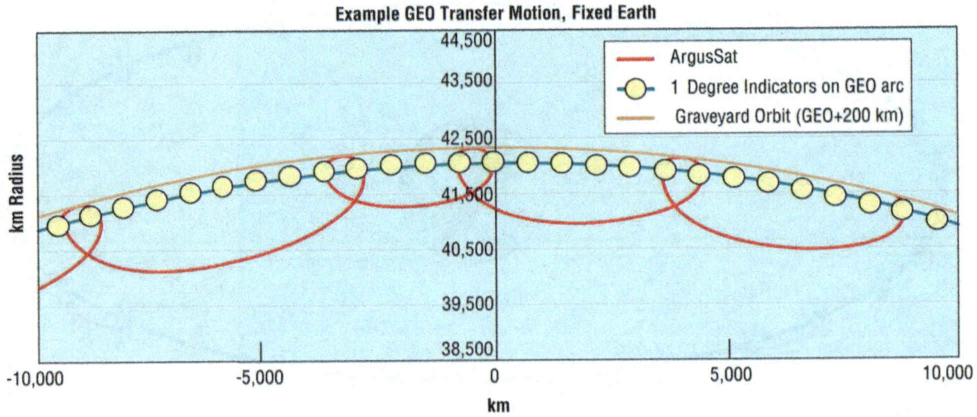

FIG. 9-27. **Target of Opportunity.** ArgusSat is shown traveling in a regular six degrees per day inspection trajectory. A slight shift in its orbit, with a simple in-line firing at apogee near the previous targets allows it to visit two targets of opportunity, and then with another two burns return to the original inspection path.

The transfer firing to shift from the basic inspection trajectory to the shifted trajectory is 5.89 m/s, performed at apogee of the inspection orbit. During the shifted trajectory, again at apogee between the two new target location, another single firing, this time 11.8 m/s, sets ArgusSat back toward its original next set of targets, and at apogee between those two scheduled targets a third and final single firing of 5.95 m/s restores the original trajectory rate. The required delta V depends on the size of the angular shift: the trajectory shift shown in the example diagram above requires 23.67 m/s in total. The same trajectory for a 1-degree shift requires only 11.83 m/s, and 5 degrees takes 59.2 m/s. This shows that the delta V per maneuver is linearly related to the shift. For this 6-degree per day trajectory each intermediate shift takes 11.83 (m/s) degree. Similarly, for a 12-degree per day trajectory an intermediate shift takes 12.25 (m/s) degree. The advantage of the twelve-degree-per-day trajectory with target of opportunity adjustments is that even a single satellite can get to any required target within 30 days maximum, traveling by a series of other targets on the way.

9.5.11 Preliminary Spacecraft Sizing—One Satellite, Optical Payload

In an ideal mission, every satellite would have all sensing payloads on board, giving a full suite of measurements every thirty days for each GEO slot. But to reduce cost and complexity of each individual spacecraft, we assume each satellite only has one sensor. This way, the individual satellites become smaller and cheaper, with simpler downlink and other design requirements for each. Each GEO slot would be visited once every thirty days, but each time a different selection of measurements would be taken.

After choosing the orbit and the payloads, we completed the preliminary spacecraft sizing using the SMAD design worksheet "Preliminary Spacecraft Sizing." See Table 9-8 gives the information for the one satellite, optical payload example.

TABLE 9-8. Preliminary Spacecraft Sizing—Single Optical. This table shows mass and power for a single spacecraft with an optical payload.

	Mass (kg)	Average Power (W)
Payload	1.8	1.5
Subsystems	4.8	6
Margin	1.7	1.9
Propellant	0.6	N/A
Loaded Mass	8.9	9.4

9.6 Detailed Subsystem Design

After initially sizing the system, we proceed with the detailed subsystem design.

9.6.1 Propulsion

The preliminary system sizing indicates a small satellite, such as a cubesat, with a required ΔV of 208 m/s. Using this delta V and an initial specific impulse of 300 s for a chemical propulsion system, we obtain the values in Table 9-9.

To move forward with the subsystem design, we've chosen the chemical propulsion system, since multiple burns will be necessary. After some market research we find several propulsion systems available in this size, mass, and power range.

TABLE 9-9. Initial Propulsion Subsystem Sizing. The values in this table are for a cubesat or other small satellite.

Specific impulse (s)	300
Initial stage mass (kg)	8.5
Final stage mass (kg)	8.1
Propellant mass (kg)	0.5
Inert mass (kg)	0.1
Thrust (N)	125.4

9.6.2 Communications

The communications subsystem needs to the right size for both uplink and downlink. Mission objectives and requirements must be definite enough to be able to trade alternate architectures before we determine the data rates. We then design the links, bearing in mind that the major cost drivers in this subsystem are antenna size and transmitter power. We can then determine the size and mass of the antennas, as well as the power and mass of the satellite transmitters.

There are three types of communications architectures: data collection, data relay, and tracking, telemetry and command. This is not a data relay system, so we don't include this as part of the architecture; ArgusSat is not a communications satellite. The link design is key to designing the communications subsystem. It shows the relationship between data rate, antenna size, propagation path length, and transmitter power and is defined in the link budget equation.

The satellite system will be imaging two targets per day with high-resolution imagery. The communications architecture needs to close for both the uplink and downlink cases. The satellite is in GEO, so has good coverage of ground stations. The drivers for the downlink communications system sizing are power, spacecraft antenna size, ground antenna size, and data rate.

For the communications downlink design, several types of antenna can handle the downlink needed to the ground. The transmitter needs to be an agile system, as the ArgusSat will be swinging around the target satellite to take images, and the antenna will need to hinge to a point on Earth to transmit the data. If we opt for an omnidirectional antenna, a much larger dish will be needed on the ground to make up the gain. For this design we chose a 10-cm X-band transmitter; a summary of the characteristics for the downlink design is in Table 9-10.

For the uplink communication system design, the receiver antennas on the spacecraft should be omnidirectional. We chose an S-band system, and a smaller 4-meter antenna on the ground; the operational costs are lower than for a larger 12-m ground system. Table 9-11 provides details on the uplink design.

TABLE 9-10. Communication System Sizing—Downlink. The downlink must be a good match for the satellite's mission.

Data Rate (bps)	1.0E+07
Spacecraft Transmitter Output Power (W)	20
Spacecraft Antenna Diameter (m)	0.11
Ground Antenna Diameter (m)	3.5

TABLE 9-11. Communication System Sizing—Uplink. These specifications reflect the size and functions of the ArgusSat system.

Data Rate (bps)	5.0E+04
Spacecraft Transmitter Output Power (W)	10
Spacecraft Antenna Diameter (m)	0.1
Ground Antenna Diameter (m)	4

9.6.3 Attitude Control

For the spacecraft system, we base requirements for attitude control on system-level and mission requirements; we then use an iterative process to size the attitude determination and control system (ADCS).

The first design solution to consider is the orbital insertion of the system. We have two choices: either the system performs the orbital insertion itself, or a launch vehicle injects the system into its operational orbit. We chose to have a launch vehicle perform the orbital insertion and found launch vehicles available that could launch the system directly into GEO.

For pointing, the system is not a typical Earth- or nadir-pointing satellite; rather, the pointing mode is inertial, which allows for payload targets of opportunity. Since reorientations are required due to multiple maneuvers and viewing of multiple satellites, 3-axis control is the most versatile. However, it is also a complicated system that requires more equipment and may require propellant, unless it uses magnetic torquers. This control mode does provide great accuracy, in the range of +/– 0.001 degrees to +/– 1 degree. Spin stabilization is fine for a medium accuracy system with few attitude maneuvers, but this system will be performing multiple maneuvers. A 3-axis ADCS typically requires sun sensors, star trackers, and inertial measurement units. Reaction wheels and thrusters are the usual actuators for 3-axis systems. This ADCS will also drive up the cost, as the accurate pointing requirements require better and higher cost sensors and actuators.

After we've chosen this baseline, the initial system and subsystem requirements for the ADCS are as follows:

1. The spacecraft orbital insertion shall be provided by the launch vehicle.
2. The spacecraft shall provide inertial pointing to within 0.1 degree accuracy.

3. The ADCS shall be sized for a maximum slew maneuver time of 450 seconds.
4. The ADCS shall be sized for a maximum slew angle of 30 degrees.

As seen in Table 9-12, we used the slew requirements to estimate the environmental and slewing torques.

TABLE 9-12. Attitude Control—Torque Estimates. Environment, orbit, and configuration all contribute to the torque.

Torque	(N-m)
Gravity gradient	3.781E-10
Solar radiation	4.822E-07
Magnetic	2.201E-07
Aerodynamic	0.000E+00
Total (RSS)	5.301E-07
Slewing torque	3.953E-06

After determining the torques, we need to size the different attitude control systems. The components for the ADCS are all small, as the satellite system itself is in the cubesat range. To determine feasibility and ensure there are off-the-shelf products in this range, we compared the reaction wheel sizing to existing reaction wheels sold by Blue Canyon technologies [Blue Canyon, 2018]. The reaction wheel sizing calculated for the satellite system shows a 0.03 m radius reaction wheel of mass 0.85 kg and angular momentum of 0.01 Nms. As Figure 9-28 shows, the Blue Canyon RWP050 is a 0.029 m radius reaction wheel with a mass of 0.24 kg and angular momentum of 0.050 Nms.

9.6.4 Power Subsystem

After sizing the ADCS, we sized the power system. For this, we had to determine a solar array efficiency. We chose an efficiency of 25% as a starting point, as most silicon solar arrays are in that range. For this small single-sensor satellite, 0.06 m^2 of solar arrays of mass 0.5 kg are needed to power the system. With a depth of discharge of 50% for the secondary batteries, small secondary batteries are necessary: 23.9 W-hr and a mass of 0.5 kg.

9.7 Final Spacecraft Sizing

After sizing all the subsystems, we did the final subsystem sizing, the results of which are in Table 9-13. The design decisions are as follows:

- ADCS: 3-axis, zero momentum
- Communications: separate uplink and downlink antennas and a traveling wave tube amplifier (TWTA) design

	(MicroWheel) RWP015	RWP050	RWP100
Momentum	0.015 Nms	0.050 Nms	0.10 Nms
Max Torque	0.004 Nm	0.007 Nm	0.007 Nm
Mass	0.130 kg	0.24 kg	0.33 kg
Volume	42 x 42 x 19 mm	58 x 58 x 25 mm	70 x 70 x 25 mm
Voltage	10 - 14 VDC	10 - 14 VDC	10 - 14 VDC
Power @ Full Momentum	< 1.0 W	< 1.0 W	< 1.0 W
Design Life	> 5 years	> 5 years	> 5 years

	RWP500	RW1	RW4	RW8
Momentum	0.50 Nms	1.0 Nms	4.0 Nms	8.0 Nms
Max Torque	0.025 Nm	0.07 Nm	0.25 Nm	0.25 Nm
Mass	0.75 kg	.95 kg	3.2 kg	4.4 kg
Volume	11 x 11 x 3.8 cm	11 x 11 x 5.4 cm	17 x 17 x 7 cm	19 x 19 x 9 cm
Voltage	22 - 34 VDC	22 - 34 VDC	22 - 34 VDC	22 - 34 VDC
Power @ Full Momentum	< 6.0 W	< 10.0 W	< 10.0 W	< 10.0 W
Design Life	> 10 years	> 10 years	> 10 years	> 10 years
Integrated Electronics	Yes	Yes	Yes	Yes
Typical Data Interface	RS-422			

FIG. 9-28. Reaction Wheels on the Market. The RWP050 is a good choice for the ArgusSat ADCS [Blue Canyon, 2018].

- Primary power source: deployable solar arrays
- Propulsion system: chemical
- Structure: monocoque

The 3-axis, zero momentum ADCS is required due to multiple maneuvers and viewing of multiple satellites, and it provides greater accuracy, in the range of +/– 0.001 degrees to +/– 1 degrees. This 3-axis ADCS also drives up the cost of the system, as the accurate pointing requirements require better and higher cost sensors and actuators. Due to the high data rates, we selected separate uplink and downlink antennas for the communications subsystem. The TWTA design type has greater mass but consumes less power than a Solid State Amplifier, so we chose this. Due to the relatively low power draw of the satellite system, there is little difference between the different types of power methods. The solar arrays are small and readily available on the market. The chemical propulsion approach was chosen over the electric propulsion due to the short duration of the mission and the multiple maneuvers. We chose the monocoque structure because it has a lower mass impact. Table 9-13 shows the final system sizing for the single visual sensor.

TABLE 9-13. Final System Sizing. This consolidates the sizing decisions made for the individual subsystems.

Loaded Mass (kg)	20.6
Peak Power (W)	61.2
Expected Volume (m^3)	0.21
Expected Body Area (m^2)	0.47
Expected Linear Dimension (m)	0.69
Solar Array Area (m^2)	0.06
Solar Array Mass (kg)	0.5

9.8 Operations Planning

The design work performed in previous sections of this chapter informs the estimates for the design cost. A key element, and typically most costly, is the full life cycle cost, much of which can be attributed to the operations cost. This satellite system will be downlinking a lot of imagery after imaging a target satellite every three days. The operations complexity for the design, flight systems, risk, and ground systems needs to be understood so we can characterize the impacts of the operations.

For the mission design and planning, we designated many of the events as medium or low complexity. Higher complexity elements were in the data rate section for real time vs. playback and a high timing accuracy for the system. For the tracking events, the peak data return rate is high, and the remaining items were low or medium since the concept of operations calls for one satellite rendezvous every three days.

For the flight systems, most of the operations complexity is medium, with several lower complexity attributes such as a low link margin. Higher complexity items include attitude control, which is 3-axis stabilized.

For operational risk avoidance, the risk posture selected was medium for all the items. There is nothing of high complexity for risk. The ground system is already designed, and most of the complexity is medium. The higher complexity items were the number of commands and activities generated.

9.9 Cost Modeling

With all the sizing completed, we carried out the cost modeling. We want to show a comparison of the end cost of six satellites with one sensor type each to six satellites with all sensors on board. The basis for the estimate assumes existing designs for all the components with a moderately modified design for structures and thermal. Table 9-14 details the results for a single sensor type on each of six satellites for the four cost models with a range of $21 million to $74 million.

TABLE 9-14. Single Sensor Cost Estimates. These costs do not include life cycle costs.

	USCM 7th Edition (SMAD, Tables 20-4 & 20-5) (FY00 - $M)	SSCM (SMAD, Table 20-6) (FY00 - $M)	SSCM (Ver 7.4) (FY00 - $M) (RSMC, Table 8-4)	SSCM (Ver 8.0) (FY00 - $M) (RSMC, Table 8-5)
Total cost (using S/C bus—total estimate)	$50	$21	$64	$74
Total cost (using S/C bus—sum of components estimate)	$62	$56	$64	$74
Total cost (using weighted average of S/C bus estimates)	$50	$34	$64	$74

The total lifecycle cost estimate from Table 9-15 includes the space segment cost and operational costs for the two-year mission for a single sensor type on each of six satellites. The life cycle cost ranges from $53 million to $106 million. A system-level requirement was for the cost not to exceed $50 million per year, and the operations costs are estimated at $32 million for the two-year period, which meets this requirement.

9.10 Multiple Sensor Satellite Design

The ideal satellite would have multiple payloads onboard, but this increases cost and complexity. In an effort to quantify the cost impact of multiple sensors, a simple scaling was performed to the single payload SMAD process, assuming three payloads (one optical, one lidar and one radar). Of course, a complete analysis would require more detailed information from sensor vendors and would best be performed when more mission cost drivers such as launch mass and volume are available. But for the purpose here, this gives us an idea of the impact on the system; about 27 kg more mass and considerably more volume and power than a single sensor spacecraft (Tables 9-16 and 9-17). And the cost impact, depending on the cost model used, varies from an increase in life cycle cost ranging from $36M up to $86M (Tables 9-18 and 9-19).

TABLE 9-15. Single Sensor Life Cycle Cost Estimates. The four cost models yield significantly different estimates.

	USCM 7th Edition (SMAD, Tables 20-4 & 20-5) (FY00 - $M)	SSCM (SMAD, Table 20-6) (FY00 - $M)	SSCM (Ver 7.4) (FY00 - $M) (RSMC, Table 8-4)	SSCM (Ver 8.0) (FY00 - $M) (RSMC, Table 8-5)
Lifecycle cost using "S/C bus - total" estimate for space segment costs	$82	$53	$96	$104
Lifecycle cost using "S/C bus - sum of components" estimate for space segment costs	$94	$88	$96	$104
Lifecycle cost using weighted average of S/C bus estimates for space segment costs	$82	$66	$96	$104

TABLE 9-16. Preliminary Spacecraft Sizing—Multiple Sensor. The advantage of having three different sensors comes at a cost of about 27 more kg of mass.

	Mass (kg)	Average Power (W)
Payload	10	100
Subsystems	27	160.3
Margin	9.3	65.1
Propellant	2.4	N/A
Loaded Mass	48.7	325.4

9.11 Summary

Space Domain Awareness is achievable with a small, cubesat-size satellite using a visible payload. This chapter proposed a concept and then sized a system for a 2-year proof of concept mission using as much existing hardware as possible to size a system that can image GEO satellites and send the data back to Earth. The system is launchable on existing launch systems and will have a development cost between $6 and $20 million depending on multiple or single sensor payloads and the cost model followed. The yearly operations cost was within the initial requirement set target, and the mission operations are of low or medium complexity with a few outliers, which were chosen due to design considerations.

TABLE 9-17. Final System Sizing, Multiple Sensors. The multi-sensor satellite requires considerably more volume and power than one with a single sensor.

Loaded Mass (kg)	48.7
Peak Power (W)	242.0
Expected Volume (m^3)	1.02
Expected Body Area (m^2)	1.36
Expected Linear Dimension (m)	1.17
Solar Array Area (m^2)	1.94
Solar Array Mass (kg)	17.37

TABLE 9-18. Multiple Sensor Cost Estimates, Without Launch Vehicle. The differing assumptions that drive each model yield vastly different estimates.

	USCM 7th Edition (SMAD, Tables 20-4 & 20-5) (FY00 - $M)	SSCM (SMAD, Table 20-6) (FY00 - $M)	SSCM (Ver 7.4) (FY00 - $M) (RSMC, Table 8-4)	SSCM (Ver 8.0) (FY00 - $M) (RSMC, Table 8-5)
Total cost (using S/C bus - total estimate)	$70	$57	$91	$160
Total cost (using S/C bus - sum of components estimate)	$95	$92	$91	$160
Total cost (using weighted average of S/C bus estimates)	$80	$71	$91	$160

This project met the initial goal of characterizing and identifying objects in GEO and providing data to operators in real time to predict future behavior. The secondary goals should be achievable with more ground processing and sharing of data between different end-users.

The design meets all the key performance parameters: the constellation and orbit design allow revisit of any point every 30 days, the sensors provide 10-cm optical resolution, and, assuming suitable ground stations available when needed, the data can be provided to the ground infrastructure within 30 minutes, even as live-streamed data if desired.

TABLE 9-19. Multiple Sensor Life Cycle Cost, Without Launch Vehicle Cost. Operational costs add about $32M to each estimate.

	USCM 7th Edition (SMAD, Tables 20-4 & 20-5) (FY00 - $M)	**SSCM (SMAD, Table 20-6) (FY00 - $M)**	**SSCM (Ver 7.4) (FY00 - $M) (RSMC, Table 8-4)**	**SSCM (Ver 8.0) (FY00 - $M) (RSMC, Table 8-5)**
Lifecycle cost using "S/C bus - total" estimate for space segment costs	$112	$89	$124	$192
Lifecycle cost using "S/C bus - sum of components" estimate for space segment costs	$127	$124	$124	$192
Lifecycle cost using weighted average of S/C bus estimates for space segment costs	$113	$103	$124	$192

References

Al Dhafri S, et al. 2012. *Dubaisat-2 High Resolution Advanced Imaging System (Hirais).* International Astronautical Congress.

Anz-Meador, P. and D. Shoots. 2018. NASA Orbital Debris Quarterly Newsletter. 22(1): 10-12. Retrieved from https://orbitaldebris.jsc.nasa.gov/quarterly-news/pdfs/odqnv22i1.pdf

Barnes, S. *Space Situational Awareness Users' Perspective.* 29 April 2009. Air Force Space Command. Retrieved from: https://www.swpc.noaa.gov/sites/default/files/images/u33/BARNES%20SWW%202009.pdf

Blue Canyon Technologies. c2017. *Reaction Wheels.* Retrieved from http://bluecanyontech.com/reaction-wheels/Blue Canyon

Chameleon Imager. CubeSatShop (c2009-2018). Retrieved from https://www.cubesatshop.com/product/chameleon-imager/

Crystal Space. 2015. *Crystal Space Micro Camera System.* Retrieved from http://crystalspace.eu/wp-content/uploads/2015/12/Camera-summary.pdf

Dubaisat-2. c 2018. Retrieved from http://spaceflight101.com/spacecraft/dubaisat-2/

European Space Agency (ESA). 8 April 2015. *Deimos-2 Imagery User Guide* Retrieved from https:/ /earth.esa.int/documents/10174/2605491/Deimos-2-Imagery-User-Guide

ESA. *About SSA* (14 March 2018). Retrieved from http://www.esa.int/Our_Activities/Operations/Space_Situational_Awareness/About_SSA

Force Support--Air Force Satellite Control Network (2018). Retrieved from http://www.au.af.mil/au/awc/awcgate/au-18/au18004a.htm

Fraunhofer Institute for High Frequency Physics and Radar Techniques (FHR). 2018. *Space observation radar TIRA*. Retrieved from https://www.fhr.fraunhofer.de/en/the-institute/technical-equipment/Space-observation-radar-TIRA.html

Gecko Imager. c 2009-2018. CubeSatShop. Retrieved from https://www.cubesatshop.com/product/scs-gecko-imager/

GeoHack. Retrieved from *https://tools.wmflabs.org/geohack*

GOM Space. *High performing Camera-system for Earth Observation Projects*. Retrieved from https://gomspace.com/Shop/payloads/earth-observation.aspx

GSSAP Satellite Overview (c 2018). Spaceflight 101. Retrieved from http://spaceflight101.com/spacecraft/gssap/

Harris. c 2018. *Spaceview*. Retrieved from https://www.harris.com/solution/spaceview

Holzinger, M. J. and M. K. Jah. 1 January 2018. *Challenges and Potential in Space Domain Awareness*. Journal of Guidance, Control and Dynamics. 41(1):15 18.

Gen Dempsey, M. E., Chairman of the Joint Chiefs of Staff. 2013. Joint *Publication 3-14 Space Operations*. May 2013: II-1. Retrieved from http://www.jcs.mil/Portals/36/Documents/Doctrine/pubs/jp3_14.pdf

JSpOC. Fact Sheet. 2018. Retrieved from http://www.stratcom.mil/Media/Factsheets/Factsheet-View/Article/976414/usstratcom-space-control-and-space-surveillance

Larson, W. J. and M. Jah. 2017. November 3. *Space Domain Awareness Workshop.*

Leopold, G. (11 Oct 2017). *Space awareness, persistence emerge as priorities*. Retrieved from https://defensesystems.com/articles/2017/10/11/milspace.aspx

NASA. 24 August 2014. *Antenna Gallery*. Retrieved from https://www.nasa.gov/content/antenna-gallery/#lowerAccordion-set1-slide1

NASA. 8 November 2017. *NASA Sets Media Coverage of Rescheduled NOAA Weather Satellite Launch*. Retrieved from https://www.nasa.gov/press-release/nasa-sets-media-coverage-of-rescheduled-noaa-weather-satellite-launch

NASA. 27 March 2018. *Be a Flight Director: NASA Accepting Applications for Mission Control Leaders*. Retrieved from https://www.nasa.gov/press-release/be-a-flight-director-nasa-accepting-applications-for-mission-control-leaders

NASA Orbital Debris Office. 2018. *Photo Gallery*. Retrieved from https://orbitaldebris.jsc.nasa.gov/

National Space Policy of the United States of America. 28 June 2010. Retrieved from https://www.nasa.gov/sites/default/files/national_space_policy_6-28-10.pdf

Patuxent Policy Group. 2015. *National Security in Space the Ultimate "High Ground": Deterrent or Provocation? What are the Implications for the United States Naval Services?* Volume V, Number I, May 2015. Page 11, Retrieved from http://www.smcm.edu/democracy/wp-content/uploads/sites/29/2014/03/National-Security-In-Space-2014-White-Paper.pdf

Planet Satellite Imagery Products. 2016 November. Retrieved from https://www.planet.com/docs/spec-sheets/sat-imagery/

Safyan, M. March 2015. *Overview of the Planet Labs Constellation of Earth Imaging Satellites.* Retrieved from https://www.itu.int/en/ITU-R/space/workshops/2015-prague-small-sat/Presentations/Planet-Labs-Safyan.pdf

SCS. 2017. *Iguana Imager.* Retrieved from http://scs-space.com/wp-content/uploads/2017/10/Iguana-Brochure.pdf

Web-based Space Situational Awareness (WSSA). 23 June 2017. Retrieved from https://www.fbo.gov/index?s=opportunity&mode=form&id=652c99d2f28c21f8c738254a0410c710&tab=core&tabmode=list&=

Wertz J. R. and W.J. Larson. 1999. *Space Mission Analysis and Design (SMAD)*. Third Edition. Microcosm Press (CA) and Springer (NY).

Chapter 10

Space Object Identification: Fuse, Use, Analyze, and Inform

Dr. Shawana P. Johnson, GISP
Founder, CEO, Global Marketing Insights, Inc.
Timothy Stratman, *Chief Master Sergeant, USAF (Ret.), President, Kestrel Corporation**
Dr. Michael S. Warren, *Co-Founder, Descartes Labs, Inc.*

Near-future Space Object Identification has to be more than a catalog number, identification of components, physical orientation, or parameters of its orbit when it comes to national defense. It will do little good to just state these kinds of characteristics if space-based assets have a nefarious purpose. The question is, "How would we know?" Present satellites of most interest are highly dynamic as illustrated in Figure 10-1. In order to understand probable future intentions, we have to quantify their behavior in the most recent past and project that forward. Following a philosophy similar to the International Rules of Engagement for Positive ID, it's less about an equivalent biometric identification (the present state-of-the-art), and more about whether the subject is functioning presently as a

* With contribution and support from Eugene W. Butler, Founder, CEO, Kestrel Corporation.

legal combatant or non-combatant...and in the moment, whether or not a subject is truly portraying Hostile Intent of an immediate threat, e.g., detection of a chemical or electrical thrust "burn" followed by a secondary object moving away from the primary on a potential intercept course, or just witnessing a failed satellite service module separation or micro-meteor accident with debris. The latter would likely not call for an active response, while the former might, heavily dependent upon the behavior that a satellite took leading up the event. The required behavior has to be observed over a meaningful period of time (temporal), on a global scale. The myriad of possibilities with thousands of interacting satellites makes conventional smart human analyst-based solutions not able to scale to the problem at hand. This capability calls for a new process as illustrated in Figure 10-2.

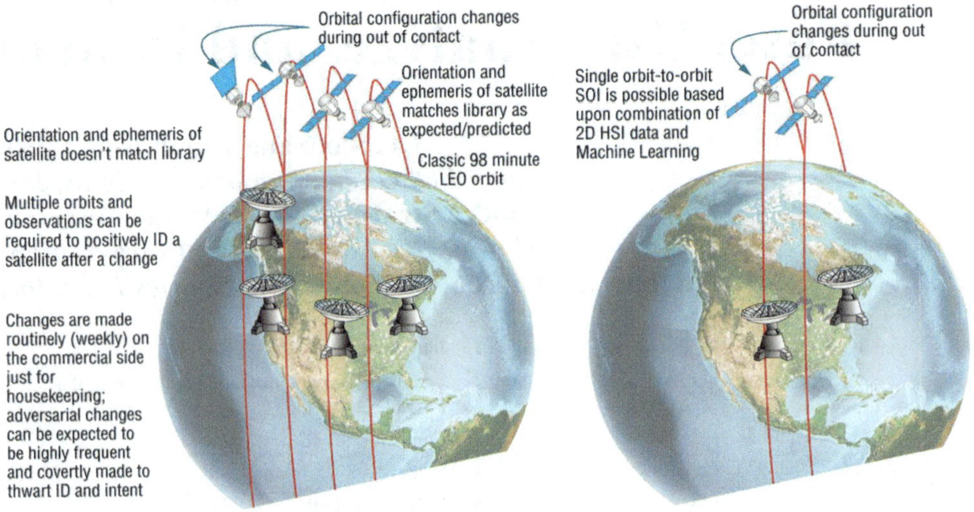

FIG. 10-1. **Chapter Vision.** Left image: State-of-the-art. If a satellite is not where expected, several observations and orbits may be necessary for space object identification (SOI). Right image: Art-of-the-possible (and operational need). Combination of 2D hyperspectral imaging (HSI) observation, fusion of other path data, and machine learning/artificial intelligence (ML/AI) enable single orbit SOI. Benefit of "the possible": "Chain of custody" is reinforced on every observable orbit, and decision making time regarding SOI and intent decreases by at least an order of magnitude. System/Analysis Components: In addition, a powerful and robust data ingest and analysis solution is required to manage all the data captured through these observations, as Figure 10-2 demonstrates.

Problem Statement: Space-enabled technologies provide support to many societal daily activities, both military and commercial. Applications range from positioning, navigation, and timing (GPS, timestamps, autonomous transportation); communications (TV, Internet, phone); remote sensing (weather forecasting, environmental monitoring, intelligence, surveillance, and reconnaissance); to science (research and exploration). Thus, losing access to space-based services would have far-reaching effects on both our economy and our military. Having seen our reliance on space-enabled operations, rival foreign

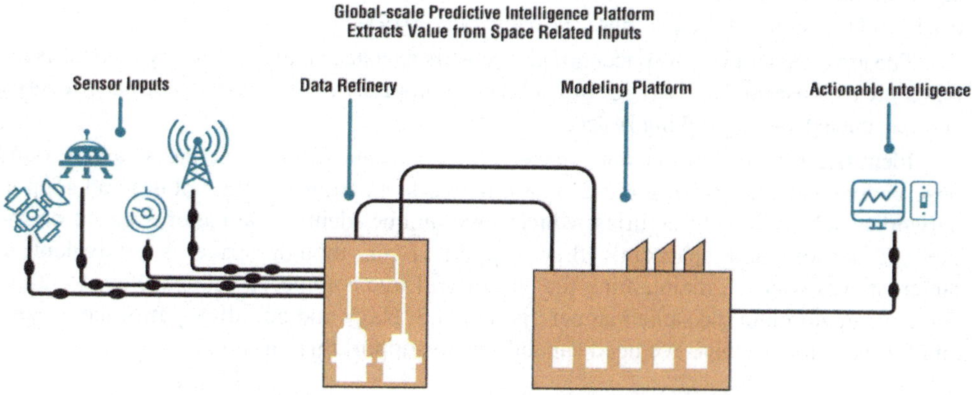

FIG. 10-2. Expeditious Advancement Chapter Vision. This machine learning/artificial intelligence (ML/AI) solution is currently available and provides a fuse, utilize, analyze and inform (FUAI) system for space object identification (SOI) operating in a commercial cloud environment with space and terrestrial characterizations.

governments, in attempts at space control or superiority, are actively developing capabilities that threaten and deny others' ability to use space [DIA, 2019], [CSIS, 2019]. "New threats to commercial and military uses of space are emerging, while increasing digital connectivity of all aspects of life, business, government, and military creates significant vulnerabilities. During conflict, attacks against our critical defense, government, and economic infrastructure must be anticipated" [US DoD, 2018], [NASIC, 2019]. Our adversaries will continue to improve their ability to access, operate, and exploit their positions in and through space [Coats, 2019], [Lambakis, 2017].

The Department of Defense's August 2018 Report on Organizational and Management Structure for the National Security Space Components of the Department of Defense focused on eight essential capabilities in the area of space. This chapter deals with two of these elements:

1) The need for global and near-real time space situational awareness, and
2) Highly-scaled, low-latency, persistent, artificial-intelligence-enabled global surveillance.

These requirements go to the heart of the vision for this chapter as Figures 10-1 and 10-2 illustrate.

While various sensor systems can support tracking and custody of various space objects as they orbit Earth, the objects of most interest from a national security perspective are active; what an object did, is doing, or looks like it's going to do is most important. Former NRO Director Marty Faga recently highlighted this level of need and urgency: "A big problem at least currently for all of our systems, military, and IC [intelligence community], is that many forms of attack will come with no warning, like cyber, jamming and laser. Moreover, kinetic ASAT [anti-satellite] attack in LEO (low-Earth orbit) takes eight minutes from launch to strike. It will be quite a while before our ability to detect, track, and respond will be that fast." [Adamczyk, 2019]

The growing number of spacecraft and sensors in space creates difficulty in identifying type, intent of use, and activity of these objects, complicating the associated space security and response decision-making process.

Identification of sensors and spacecraft can occur from space-based and ground observations such as global ground station networks (stationary and mobile) and support networks such as launch facilities, which have unique identifiable patterns based on the type of sensors and spacecraft they support. Even though space is equivalent to "international waters", maintaining the "peace and free commerce" of space has to fall to some organization in the same manner that the U.S. Navy and our allies patrol the oceans, and for the same reasons. We need capabilities to support this mission.

Smart Doesn't Scale

The tremendous volume and diversity of data provided by new sensor deployments simply overwhelm traditional forms of analysis. This is a recurrent theme in many scientific and commercial realms; the exponential decrease in the cost per transistor in a processing unit or sensor (roughly an order of magnitude every five years according to Moore's Law) leads to an exponential increase in data volume over time [Wikipedia, 2018]. The lurking analysis difficulties may lie unrecognized while the data remains within the grasp of a smart analyst, but as the exponential growth crosses the threshold of human cognition, producing actionable information from the data becomes intractable using existing methods. As an analogy from astronomy, identifying the type of a supernova from spectral measurement of its electromagnetic emissions (collected from a small set of exquisite instruments) was tractable using human experts when the rate of discovery was a few per month. Now, however, multi-gigapixel cameras and kilo-fiber spectrographs can observe hundreds of supernovae in a single night. The job of the human expert must be passed to the realm of artificial intelligence where it can scale. The most promising current approach is deep learning with a convolutional neural network (ConvNN) (Figure 10-3).

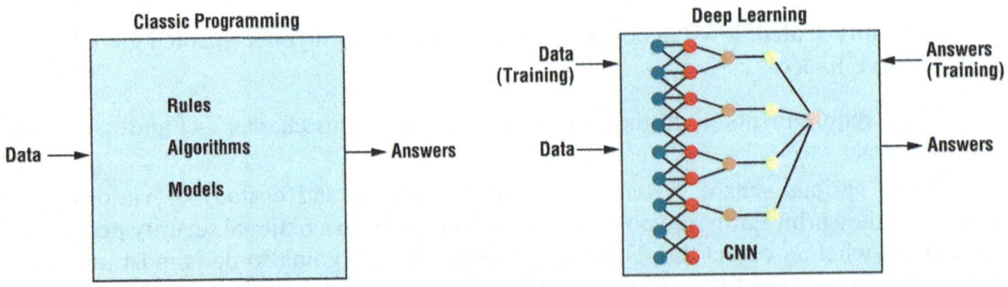

FIG. 10-3. Smart Doesn't Scale—From Classic Programming to Deep Learning. Utilizing convolutional neural networks (ConvNN) allows the analyst or data scientist to focus on solving their problem or challenge while the deep learning model does the heavy lifting of data discovery and development.

Deep learning takes advantage of the discovery that given ample training data and corresponding answers (labeled data, or ground truth), a computing system using a ConvNN can learn to produce the correct answers from input data it has not seen before. A ConvNN is a type of artificial neural network used in image recognition and processing that is specifically designed to process pixel data. Traditional neural networks are not ideal for image processing and must be fed images in reduced-resolution pieces. A ConvNN has its "neurons" arranged more like those of the frontal lobe of the brain, the area responsible for processing visual stimuli in humans and other animals. The layers of neurons are arranged in such a way as to cover the entire visual field, avoiding the piecemeal image processing problem of traditional neural networks [SearchEnterpriseAI, 2018]. This obviates the need for labor-intensive and time-consuming development using domain expertise about appropriate rules and classical programming incorporating algorithms and models.

A deep learning model has many parameters embedded in it, just as the weights (or the strength of connections between units) do in the neural network. Typically, the number of these parameters is in the tens of millions, so training such a system requires a sufficiently large training dataset to avoid overfitting. To be successful, a training dataset generally contains hundreds of thousands of examples. The best known of these is Google's efforts to use machine learning/artificial intelligence to identify cats [Gershgorn, 2017]. In terms of training data for satellite imagery, there are multiple expert domains online (through space agencies globally, such as NASA, ESA, JAXA and others) and many research groups (GitHub, etc.). These share the labeled image datasets they have collected with the rest of the learning community to further machine learning research on satellite imagery.

Using a deep learning framework for understanding crucial signal outputs of the sensors and observed spacecraft type characteristics (pixel information by semantic segmentation) encompasses deep fully convolutional neural network (ConvNN) and other machine learning processes. These include finding the data needed, pre-processing all the required data, and developing algorithms to apply to the data to create accurate models. As Figure 10-4 shows, the final result is the best or "golden" model to create applications to solve the end-users' problem.

With the coming onslaught of space satellites and the resultant data loads (Figure 10-5), employing a deep learning model to understand and identify what is in space is one of the few approaches that can encompass and begin to parse the data to let us understand what we are facing.

This chapter discusses the potential data availability of space objects (e.g., spacecraft) and how to manage that data to make valuable connections and observations while using convolutional neural networks (ConvNN). The ConvNN mines this data to monitor for current and new sensor and spacecraft presence in space and to provide predictive data outputs for alerts as well as for general spacecraft object identification (SOI) information. In addition to SOI, ground-based measurements also provide predictive intelligence concerning SOI and are part of the SOI global ground station network (fixed and mobile) and the associated launch facilities. Figure 10-6 shows an example of the largest and one of a dozen commercial and quasi-government ground station networks providing global land and sea downlink capabilities [KSAT, 2020].

Once we collect space-based data, we must analyze it. The basics of the data analysis architecture have been vetted and utilized in current practices with MatLab [Spoorthi et al.,

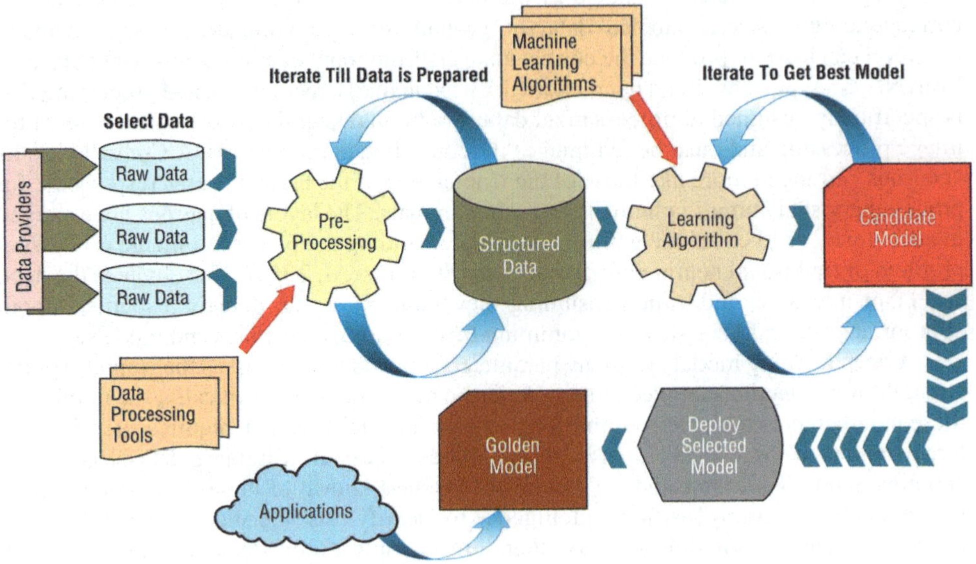

FIG. 10-4. **Machine Learning Processes.** These machine learning processes must be applied to understanding the signal outputs in order to identify spacecraft. [Mittal, 2017]

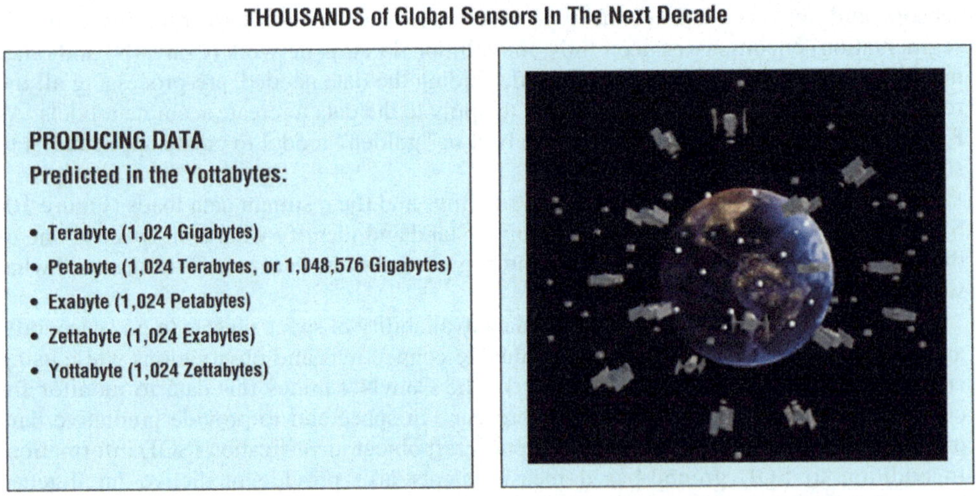

FIG. 10-5. **Space-Based Sensors.** Ubiquitous visibility of Earth through space-based sensors over the next decade will update our picture of the planet in high resolution potentially every 20 minutes, and along with other data sets, will produce yottabytes of data.

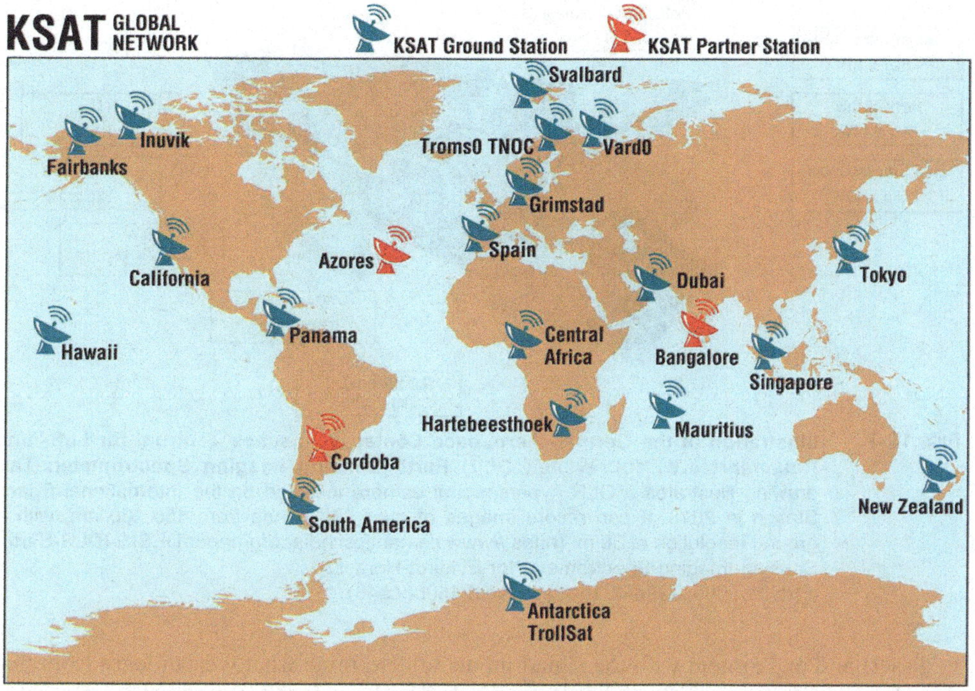

FIG. 10-6. Kongsberg Satellite Services (KSAT) Global Ground Station Network. KSAT is the world's largest ground station network, founded in 1967 as the Tromso Telemetry Station. It now provides more than 140 antennas at 21 sites worldwide.

2018] and asteroid mining [Berthier et al., 2016] and consist of encoder networks, with corresponding decoder networks followed by pixel classification layers. A pixel classification layer provides a categorical label for each image pixel, allowing the system to mathematically determine what the pixel represents (for instance, spacecraft characteristics such as an antenna, solar panel, etc.). We can apply this same framework to demonstrate the art of the possible in space object identification with automated detection and analysis systems.

Among other things, this chapter presents a hyperspectral-based sensor (described later), providing signal inputs into this system allowing us to classify spacecraft characteristics. A hyperspectral sensor operates in narrow bands (10–20 nanometers) and could have hundreds or thousands of bands collected from one sensor, such as the one illustrated in Figure 10-7.

Traditional monochromatic or multispectral imagery generally refers to 3 to 10 discrete "broader" bands. A deep learning framework is necessary to manage this data abundance, and such a framework currently exists to manage these signal inputs. Figure 10-8 highlights its basic architecture. This chapter focuses on the "art of the possible." By applying deep learning frameworks we can capture the data about space objects and begin

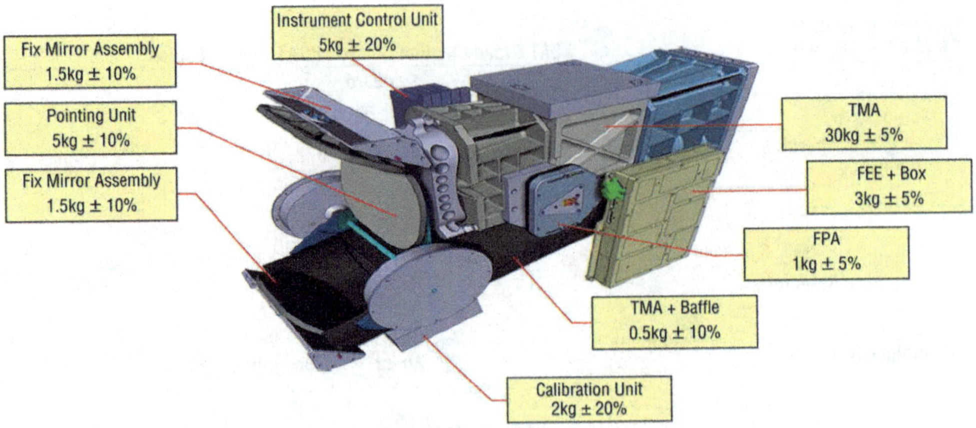

FIG. 10-7. Illustration of the German Aerospace Center (Deutsches Zentrum für Luft- und Raumfahrt e.V., abbreviated DLR) Earth Sensing Imaging Spectrometer. This drawing illustrates a DLR hyperspectral camera installed on the International Space Station in 2018. It can record images of over 235 bands from 450-100 nm, with a ground resolution of 30 m. (https://www.semanticscholar.org/paper/DESIS-(DLR-Earth-Sensing-Imaging-Spectrometer-for-Eckardt-Horack/41d9f8e500daa99ef4216c141559cd2450b8ca48)

to dissect and experiment with the signal inputs to determine what we can learn from that data. For a practical example and analogy we discuss how deep learning processes enable us to utilize Landsat 8 and Sentinel 1 Multispectral and Sentinel 2 Radar satellites.

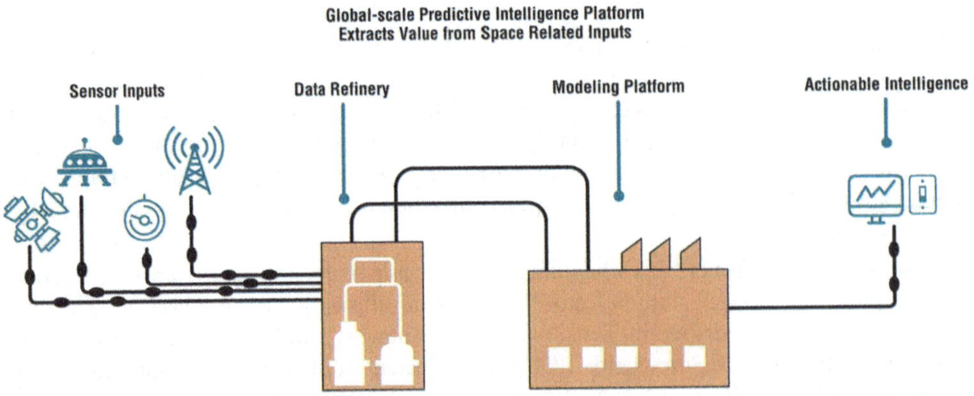

FIG. 10-8. Demonstrated and Currently Available Characteristics of FUAI System for Space Object Identification (SOI). This system operates in a commercial cloud environment with space and terrestrial characterizations. [Courtesy of Descartes Labs, Inc.]

10.1 Introduction

Positive identification of a space object relies on developing a unique fingerprint. This fingerprint has traditionally consisted of five elements: physical, functional, measurable, kinematic, and operational. Many space objects of greatest interest are physically small and have low electromagnetic (whether reflected or absorbed) signatures, making them not only difficult to detect but also hard to positively identify. All space observing sensors exploit the energy scattered from or emitted by an object as a primary way to characterize it. But these sensors have to be properly calibrated to produce precise measurements that match to an object in a database, and the right sensors need to be available. Ideally, a single sensor would have sufficient fidelity to sub-identify elements such as antennas, solar panel arrays, spacecraft bus, etc., that taken together would result in consistent positive identification of the object. But given the thousands of objects in orbit and the need to reduce the identification false alarm rate, we presently need observations from more than one sensor (though the desire to have single orbit SOI remains). In addition, we have to deal with the challenge of having sufficient observation time per object (given their typically low energy signatures).

This concern is compounded when one considers the anticipated increase in the number of low-Earth orbit (LEO) satellites and the fast passage of each object through a sensor's field of regard. This latter aspect is particularly difficult in places where operations are taking place and the number of sensors available is limited, negatively affecting the ability to have as tight a "chain-of-custody" on each satellite as desired. Therefore, there is a clear and urgent need to be able to rapidly fuse, use, analyze, and inform (FUAI), especially as augmented by machine-learning to support sparse sensor networks. An example is speed cross-cueing and hand-off from low-resolution sensors that might be over-the-horizon to high-resolution sensors, and processing the collected data to quickly and confidently perform SOI.

We can contrast the feasibility for developing a robust machine-learning (ML)-based SOI system with similar requirements already demonstrated by current commercial satellite systems. Persistent surveillance by satellite systems will continue to evolve and expand as additional satellite capabilities are added to the constellations and as ground support systems increase. For example, the entire 40-year global collections of Landsat 4-8 and Sentinel 1 and Sentinel 2 are automated in terms of near real-time data ingest and cleaning and are always available for data analysis purposes (Figure 10-9). This involves over 15 petabytes of data in a commercial organization's Google cloud, which currently ingests a petabyte of electro optical/infra-red (EO/IR) data in less than 16 hours at sustained speeds of 23 gigabytes per second. This enables ingestion of nearly 2 petabytes of data every 24 hours.*

This capability is more than sufficient to ingest the full volume of earth observation data collected in a single day as well as the data from the 3,300 satellites predicted to launch between 2018-2027 [Businesswire, 2019].The learning which took place in data refinement and data training in these types of currently available machine learning environments has

* Data courtesy of Descartes Labs, Inc.

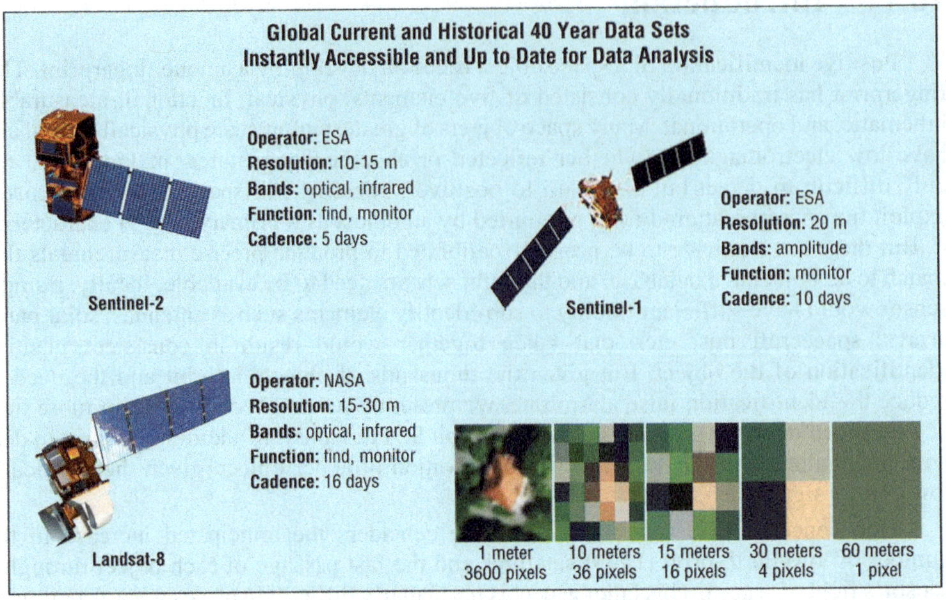

FIG. 10-9. Example of Global Data Sets Constantly On and Available in ConvNN. With over 15 petabytes of data at the ready and a terabyte of data added daily, this architecture stands ready to allow analysts and data scientists to spend the majority of their time focused on the problems they need to solve—not on data wrangling (searching and cleaning for scientific use). (Sentinel-2: https://sentinel.esa.int/web/sentinel/missions/sentinel-2; Landsat-8: https://www.usgs.gov/land-resources/nli/landsat/landsat-8; Sentinel-1: https://sentinel.esa.int/web/sentinel/missions/sentinel-1)

prepared and improved how we deploy algorithms and improve models in a ConvNN. Although this data is available in government facilities such as US Geological Survey, NASA, and ESA, none provide simultaneous access to petabytes of science-ready data, nor do they have capabilities that demonstrate the benefits of a robust ML-based system. (https://earthexplorer.usgs.gov/ and https://earthdata.nasa.gov/eosdis/daacs and http://www.esa.int/)

Recent advances in three technology areas make this massive data ingest possible and create the opportunity for innovation for SOI:

1) The explosion in commercial remote sensing capability in the form of small satellites and cubesats,
2) The advent of cloud-based storage, bandwidth, and computing with capabilities previously only available at national-level high performance computing facilities, and
3) The maturation of machine learning algorithms able to exploit data at scale and produce fast, accurate, and automated analysis of imagery with comparable accuracy to human analysts on limited sets of important applications using deep learning frameworks and ConvNN.

Also focusing on the development of these deep learning networks is the US Air Force's Space and Missile Command. As part of its Commercially Augmented Space Inter-networked Operations, aka CASINO (through the Defense Advanced Research Projects Agency, DARPA), it issued a satellite demonstration contract for a network of 20 satellites to be launched through the Department of Defense in 2021. The contract went to Ball Aerospace and Microsoft's cloud business for the computing infrastructure. The main operative question from executives leading the project is, "How do you handle large volumes of data from large distributed constellations [Fernholz, 2019]?"

The ConvNN begins to answer that question while it builds on these technologies to:

1) Harness the full volume and variety of signals and geospatial datasets to provide fused analyst-ready data, and

2) Leverage the power of interoperable software, operating on commodity elastic cloud storage, bandwidth, and data-adjacent processing to provide SOI at global scale, e.g., monitoring, analysis, and forecasting. This deep learning framework encompasses and draws information from other related land-imaging modes, such as synthetic aperture radar (SAR) with both amplitude and interferometry products, LIDAR, and hyperspectral and aerial imagery, all of which will aid in developing models for SDA.

In addition to distinguishing objects using high-dimensional spectral signatures, identifying sets of points that conform to an underlying dynamical model (such as Earth orbits) is a problem that occurs in a variety of domains [Kubica et al., 2006]. Linking isolated observations into trajectories is a conceptually simple problem, yet it can become computationally intractable with large numbers of observations in the presence of noise. The problem is compounded when we include multiple, disparate sensors that have different spatial and temporal sample rates, resolutions, and sensitivities. Tree-based algorithms for model-based spatial search are a prime example of how consideration of an underlying physical model can qualitatively improve the power of deep learning approaches. As Figure 10-10 shows, this approach holds the promise of breaking the classic Bayesian tracking model [Bertinetto et al, 2016] that typically requires explicit, almost a priori, understanding and modeling of sensor noise distributions, vehicle kinematics, etc.

10.2 Setting the Stage—Scenario for the Art of the Possible

10.2.1 Introduction

This section proposes a potential and novel framework for addressing the complex real space object identification need with advanced automated processes using deep learning. Extensive investments have been made in understanding and researching everything that is launched into space based on its physical, functional, measurable, operational, and kinematic characteristics, providing us with SOI. These investments focus on object detection, initial orbit establishment, steady-state tracking, and cross-section dimensions,

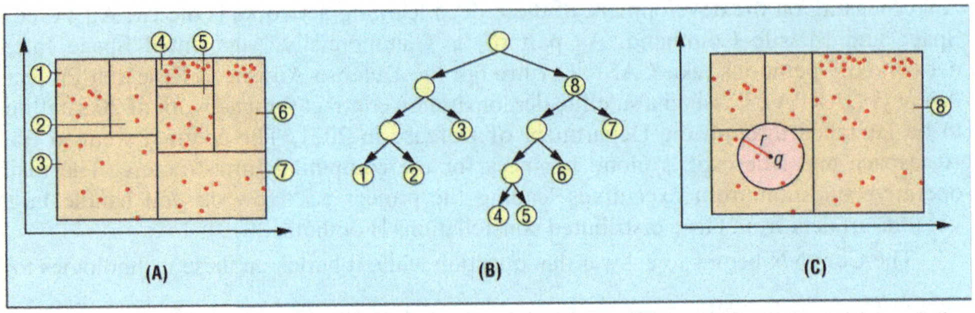

FIG. 10-10. Using a Tree Data-Structure to Speed up Spatial Search. The naive implementation of a query to find objects within radius *r* of point *q* would require a distance comparison with all *N* points in the system. Using the tree, all points in the cell labeled *8* can be eliminated with a single distance calculation, thus reducing the work from *N* comparisons to *log(N)*.

especially in the realm of radar data. This information is a means to keep custody of an object's track and identify anomalous changes.

Radar can also help perform first order recognition of an object; is it a satellite, rocket body, upper stage, or mission related debris? Use of radar polarization can assist with object motion determination (rotation, angular rates) and certain structural determinations, e.g., antennas and materials based upon the signal return. However, for all its appropriateness, processing of radar data may require more than one acquisition to deliver information. The problem is that active satellites are dynamic, with a continuing challenge being the rapid characterization of an object (the question being whether a satellite has become or will become a threat, such as a conjunction with another satellite, or as an interference element). On a pass-by-pass (acquisition-by-acquisition) basis, a satellite might make a maneuver such that its previous orbital kinematics (including orbital altitude and position) make it look significantly different based upon energy return, angle, etc.

Many sensors we use today to observe satellites produce data that is best described as "one-dimensional"—namely, data with relatively low content. This drives the need for multiple observations and multiple types of sensors, which might not be simultaneously located or available to perform an observation on the same orbital track. It can take several orbits to discern that an object is what we think it is, never mind the need to also figure out what it is doing or looks like it might be going to do. We need to collect data that is rich enough to support achieving "intelligence" about an object on a single orbital pass, with low false alarm rates and to do so fast enough to facilitate decision-making. We will look at this space intelligence challenge using a proposed 2D hyperspectral sensor for SOI coupled with a deep convolutional neural network to speed up the identification and "target surety" of the object. With the right data we can reformulate the problem definition of directly obtaining enough spatial and spectral pixels to generate abundant simulated data of several random shapes with varying noise levels. This will focus the network on learning rather than on the need for abundant training data. Such a proposed framework will be robust to noise and computationally fast. The results will highlight that a deep convNN is useful for SOI.

10.2.2 Proposed Method

Hyperspectral imaging (HSI) systems have the potential to produce unprecedented accuracy in the identification and functional classification of space objects. The tremendous amount of data associated with such systems poses a number of challenges in its analysis and interpretation, however. Even the appropriate dimensionality of the analysis is uncertain: the spectral, spatial, and temporal attributes of the data can be attacked individually, or in combination. Deep learning offers a variety of flexible approaches, and its use in remote sensing represents a concrete example of the current and future applications of HSI in the space domain.

Space object identification requires an operational, global, persistent, and scalable architecture, and an ability to ingest multiple terabytes or even petabytes of imagery per day with inline processing to build co-registered signal data archives that are analysis ready. The algorithmic inspection of that data runs automatically, and due to the pre-processing of each input, global scale computer vision and machine intelligence models can run in seconds or microseconds for data interrogation and exploitation. Currently, commercial clouds are used to run global scale machine learning applications over satellite, airborne, and terrestrial imagery, weather and vector data, and ground based objects. Descartes Labs, Inc. and other commercial companies such as IBM, Palantir, and Maxar have cloud-based platforms providing data to global commercial and defense clients. These core software stacks build upon many of the open source Geo projects, including Geospatial Data Abstraction Library (GDAL) and Mapserver.

Over 15 petabytes of public and private satellite imagery have been processed to date. By combining these open source tools with a FUSE-based virtual filesystem for cloud storage, a scalable compute platform has been developed that has demonstrated 100 terabytes / day (TB/day) of ingest performance. It generates global composites of 15-m Landsat 8, 20-m Sentinel 1 and 10-m Sentinel 2 composites requiring over 10,000 CPUs to process. (File System in Userspace (FUSE) is a software interface for Unix and Unix-like computer operating systems that lets non-privileged users create their own file systems without editing kernel code.) This platform is available through a public open source client library that can be used to query and access near real time imagery globally. The platform is built to scale as constellations are added (as well as being designed for potential new providers). The scalability and performance benchmarks in Table 10-1 are relative to current commercial based systems; Landsat 8 plus 40+ year archive, Sentinel 1, 2, Airbus, NASA, NOAA Weather, and other global datasets.

TABLE 10-1. Deep Learning System Scalability and Performance Benchmarks. These benchmarks apply to systems current as of 2020.

Current Archive	**15 Petabytes Raw, 5 Petabytes 14-bit JPEG2000 Encoded**
Rate of growth	1500 terabytes per month
Retrieval speed	Latency: ~100 milliseconds. Bandwidth: ~1 gigabyte/sec per node.
Ingest speed	12 TB / day (or higher for cloud-based imagery providers)

These data sets could yield "data products" relative to the space object fingerprint (Figure 1-10 in Chapter 1). For example, with ground systems configuration, it is possible to derive sufficient tracking measurables and kinematics on a single orbit, identifying the satellite based upon those "signatures." With enough observations as reference inputs and then using ML/AI, we can rapidly (on a single orbit) correlate on the physicals (including known external to internal components), functionals, and operational identifications. The 2D HSI will provide identification and the ML/AI will then provide rapid reference to a library, either associated specifically with that satellite or generally with known capabilities of other satellites. In addition, we can correlate the ownership and launch site (which is publicly available data). As data are collected on the antenna's position, we will be able to ascertain which ground stations are being used. These data products will yield on-demand space intelligence.

The key characteristic of this type of deep learning capability is that it can provide automated visual detection of similarities between spaceborne, terrestrial, and airborne imagery signals and apply machine learning to recognize different types of objects in space and across the globe. This is done through:

1) Tile source mapping with multiple overlapping grids, capturing features that would otherwise run across tiles,
2) Feature extraction for each image tile or signal through the convolutional neural network (ConvNN) for high dimensional vector space that includes shape, color, and texture in both visible and non-visible spectrum and through time,
3) Feature queries which establish the query features and features extracted from each image and signal input in a comparison data set, and
4) Matching results where the tiles with the smallest visual distance are matched to the query tile, as Figure 10-11 shows.

This capability is presently used in unsupervised change and anomaly detection mode, based on the observed spatial and multi-annual temporal statistics of every region of the world; the same type of methods would apply to SOI.

10.2.3 Potential Results

This automated capability allows for new feature alerts globally in near real time; ability to quickly build targeting decks; leverage imagery and signals to provide tip and cue; and establish temporal, spatial, and spectral searchable signatures for features throughout multimodal imagery and signal sources and interaction. Therefore, a reliable and interoperable capability will exist to not only help identify specific types of satellites in space but to provide the necessary intelligence (pattern-of-life) on what the satellite did, what it is presently doing, and what it might be preparing to do next. One cannot overemphasize the need to be able to independently accomplish this task. Even a small commercial operation such as Iridium, consisting of a 75-satellite constellation, will move satellites on an average of once a week [Foust, 2019]. The existence of mega-constellations (over 300 satellites per constellation) will increase the complexity exponentially and potentially make it an intractable and very expensive problem unless a new SOI capability

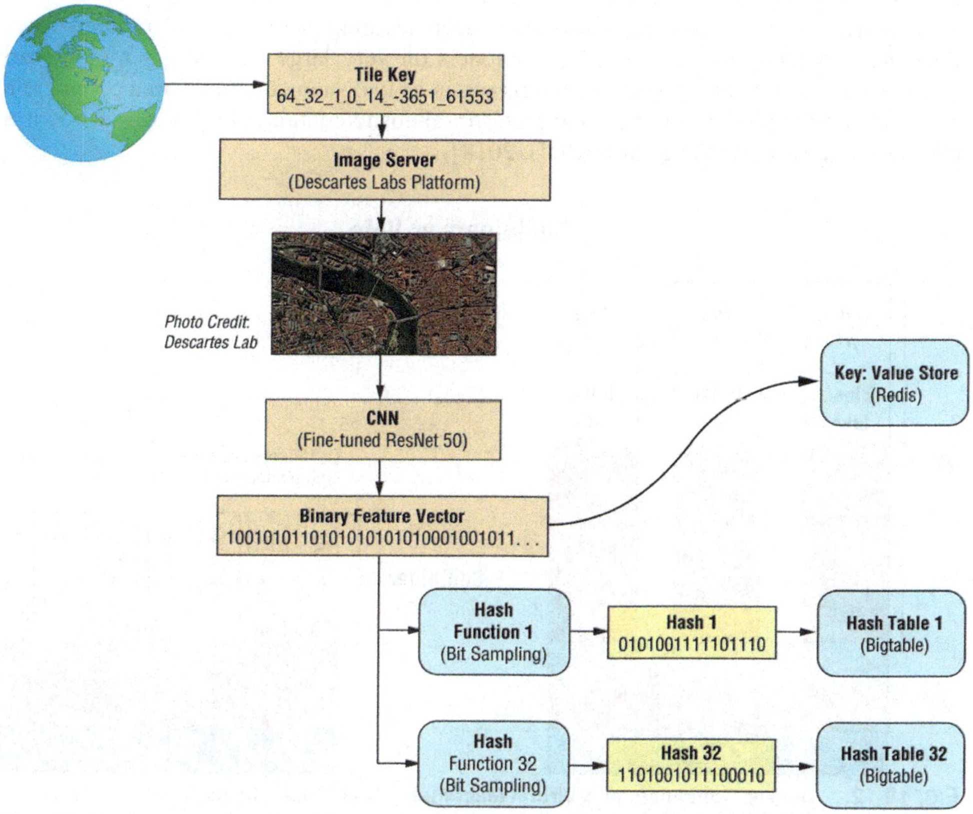

FIG. 10-11. **Automated Image Detection through a Convolutional Neural Network**. Earth is subdivided into a set of tiles (small overlapping images), and imagery is retrieved for each tile using the Descartes Labs, Inc. platform. Each image is passed through a convolutional neural network (ConvNN) to extract features. The resulting feature vector is binarized and stored in a simple key:value store, as well as a set of 32 hash tables (or map). This is a data structure that implements an associative array abstract type that can map keys to values and be useful for approximate nearest neighbor search.

is provided. And SpaceX filed paperwork to launch 30,000 additional Starlink internet satellites, so the complexity continues to increase, as does the critical importance of data sharing and management for the future of space situational awareness [Forbes, 2019].

10.2.4 Conclusion

The more data that is prepossessed in the system with high quality ground truth, the better the results of the search. The data and preprocessing provide that every pixel across every sensor is co-registered and pixel-to-pixel comparisons are consistent between sensors and time. This provides data consistency and only one re-registration required to match all the preprocessed data sets to other signals. Standard machine learning algorithms are extended by using the full spectral-spatial-temporal data hypercube, with peer-reviewed open source machine learning algorithms (deep learning networks and random forest classifiers) to train customized analysis models on very large amounts of multi-sensor imagery time-series data. These algorithms run in virtual supercomputers in a commercial cloud, exploiting modern open source platform-agnostic container technology (known as Docker, Figure 10-12 [Vaughan-Nichols, 2018]).

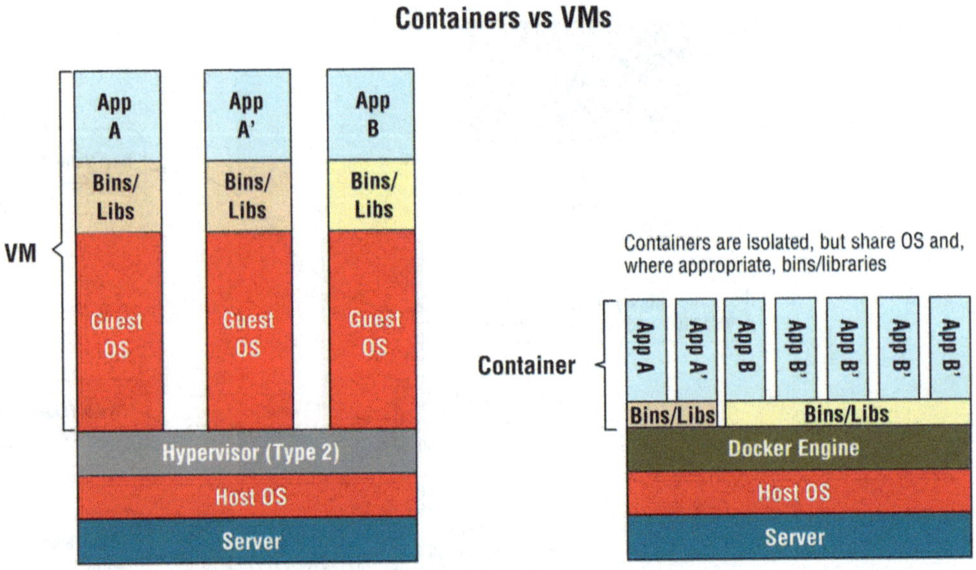

FIG. 10-12. Docker Containers and Virtual Machines (VMs). Docker helps provide portability and maintainability by packaging an application and its dependencies in a virtual container.

10.3 Context

10.3.1 Where This Fits into SDA and Why?

Militarily, space is a force multiplier in its ability to improve the effectiveness of conventional terrestrial military operations—on land, in the air, and at sea. The importance of these missions is well established. Satellites created the "national technical means" capability to project U.S. military power around the world and over great distances—something space-based capabilities are uniquely able to support [DoD, 2018]. Satellites augment vulnerable aerial reconnaissance flights (both manned and unmanned) in providing intelligence without intruding on territorial sovereignty; they provide early warnings of missile launches; GPS enables the exact location of aircraft, precision targeting, the synchronization of combat maneuvers, and clearer identification of blue force and red forces; satellite communications enable remote command and control (C2) operations anywhere in the world, including autonomous robotic aircraft, vehicles, and vessels.

Much of the traditional focus regarding SDA has been understandably on the physical vulnerability and space resiliency of our satellites; options for them to protect themselves from attacks are few. Attacks are possible from conventional anti-satellite (ASAT) weapons from a number of capable nations, to more sophisticated high-power microwave (HPM), electromagnetic pulse (EMP), or high energy laser (HEL) weapons (whether to blind or damage), to orbital satellite-on-satellite (including purposeful contaminates and debris) attacks. However, each of these attacks comes at a significant cost to the attacker, including the ability to tell whether the attack was successful in a timely manner and whether it was attributable. If the attack is attributable, then both political and counter-attack repercussions might ensue. Conversely, electronic warfare (EW) intrusions and interference are harder to detect and attribute, much more cost effective, and play well to our adversaries' strengths.

While the U.S. was busy using conventional means to fight lower intensity counter-insurgency and counter-terrorism conflicts since the early 2000s, the return of great power competition and near-peer adversaries have seen those state actors investing in increasingly sophisticated defense measures, particularly in EW [National Defense Strategy Commission, 2018], [Congressional Research Service, 2019], [Harper, 2019]. Compounding the issue is the proliferation of advanced technology around the world that erodes U.S. advantages and create new vulnerabilities in our ability to retain an overmatch position.

While most electronic warfare (EW) has been terrestrial (looking outward) or airborne (looking at range), less attention has focused on EW in and from space, though this threat has been known for some time [GAO, 2002], [Garino and Johnson, 2009]. Interference can be directed towards a satellite, from the ground or from another satellite, in an uplink mode, which affects all users in the satellite reception area. Conversely, downlink interference has a localized effect because it is directed at ground users, which might be a remote command and control node (SATCOM/TACSAT) [Boyd, 2017], or on-the-move ground forces units using satellite navigation (GPS) to determine their location. Of more concern is the concept of intrusion and spoofing to deceive a receiver by introducing a fake signal with erroneous information, which doesn't necessarily require breaking a signal's encryption [CSIS, 2019]. This is a particularly significant and real threat for ground stations because without them satellites would not be capable of supporting operations [Shelton, 2017]. The ramifications of these intrusions can be severe, from causing ships or aircraft to veer off-course to misinformation on weather conditions that might influence maneuvers to compromising of imagery from commercial sources used exclusively by the military. While "sniffers" and other radio frequency (RF) monitoring tools as mitigation are easier to employ on the ground as a means of passive intrusion discovery (though not always practical for small mobile units because of size and weight restrictions), systems in orbit do not have this "upgrade" luxury. Though EW can be challenging to attribute and distinguish from unintentional interference [DIA, 2019], [Manner, 2019], knowing that one is being targeted is essential before active counter-measures are appropriate. It is not enough to know that an intrusion or interference occurred—it must be properly sourced and attributed. If unwarranted counter-measures are engaged, a mobile military unit's position might be compromised or detrimental escalation of hostilities might occur, depending upon the action taken. We might even take mistaken action against an ally, causing an unwitting loss of capability.

Low-Earth orbit (LEO) is an ideal location for conducting intelligence, surveillance, and reconnaissance (ISR), environmental monitoring, and communications from small satellites [Erwin, 2019]. Because these objects are relatively close to Earth, they need less powerful transmitters for communications, achieve higher resolution imagery with smaller-sized apertures compared to objects in higher orbits, and are less costly to build and launch. LEOs have the disadvantage of only being in the view of a ground user or station for the short period when overhead. Therefore, continuous coverage requires a constellation of satellites spaced evenly around several orbital planes. Also, as 70 percent of Earth is covered with water and not within radio frequency (RF) range of a ground station (not limited to LEO satellites), there is a limited window to downlink and uplink data. Interference can slow down data transfers so much as to render time-sensitive data irrelevant. The problem is compounded by the explosion of cubesats and mega-constellations of small satellites, such as SpaceX's StarLink network and OneWeb, which propose over 30,000 additional objects in space (a five-fold increase in active satellites of most interest and potential for unintended consequences between satellites [Grush, 2019]).

In 2018 there were 13,237 satellites approved by the FCC under 47 CFR Part 25 [Zur, 2019]. According to U.S. Air Force Space Command (AFSC) Chief Scientist Joel Mozer, "The trend is…to smaller, distributed capabilities in orbit," wherein distributed constellations of small satellites will enable them to be replaced quickly as new technologies evolve, enabling our overmatch position to be maintained [Wilson, 2019]. Not only do all these objects need tracking, including the outdated and replaced ones that are not deorbited, but the complexity of the problem enables enemy satellites to more easily hide within the shadow of these commercial networks. In addition, it is not inconceivable that these commercial satellites could be physically hijacked by another foreign satellite and used as a surrogate weapon. Being able to quickly discern these threats will become paramount to successful SDA operations.

Along with the proliferation of new spacecraft impacting the ability to build successful SDA operations, the government is also focusing on the future of artificial intelligence (AI), where the potential exists for an AI-to-AI battlespace. Commercial entities are doing the same and are becoming data driven where quant (quantamentals) is the future. Quantamentals is the merger of two investment strategies combining quantitative and fundamental approaches where machine learning can be used on quantitative data as well as taking into account decades of historical inputs. This approach searches vast amounts of data and reveals similar conditions that could occur again. Commercial entities are investing large sums of money in the mechanization of their supply chains with the incorporation of vast data assets (corporate intellectual property and market domain knowledge) that they own from years of doing business. Theoretically, commercial organizations own the only durable form of advantage in the marketplace...their corporate data. The deep learning frameworks discussed here provide commercial organizations with an ability to "weaponize" their asset base with their own sensors to create new and unique data inputs in order to make better and faster business decisions.

Although this systematic approach to business does is not yet fully automated, we point out that the self-driving car must have data animation at a level where the vehicle is aware and reactive to its environment and continues to learn. The investment flowing into mechanizing the FUAI frameworks in the commercial space as they move to "self-driving enterprises" helps these systems evolve at an extremely advanced pace, as we see in Figure 10-13. The needs of SDA parallel the inputs of commercial organizations and the FUAI framework will continue to advance in its voracity and velocity to better support the SDA for spacecraft object identification [Smith, 2019].

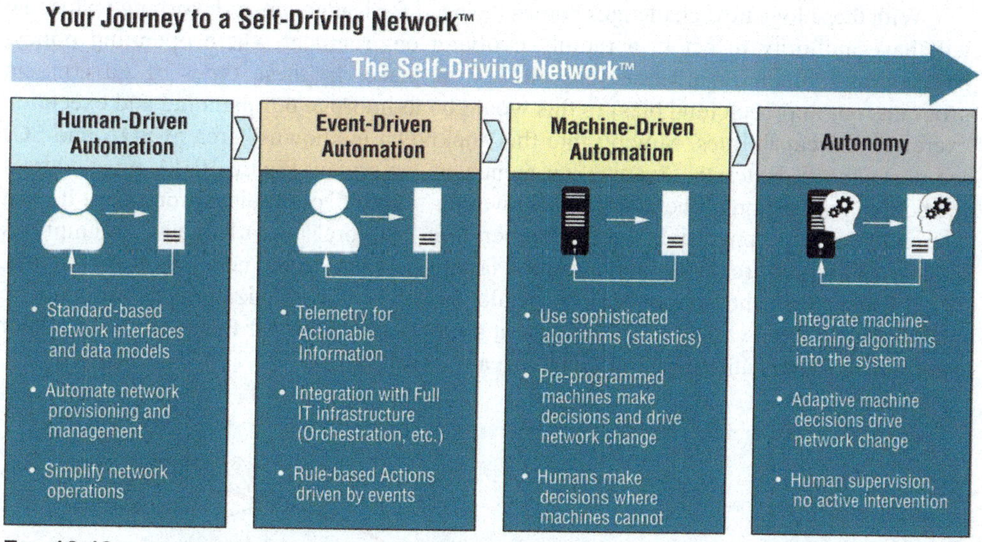

FIG. 10-13. **Self-Driving Network.** Hundreds of networks and platforms are being built daily to fulfill the needs of commercial industry data agnostic inputs. The learning obtained as these networks and platforms evolve continually improves the body of knowledge, quality, and capacity of the type of network needed for SDA space object identification FUAI architecture [Juniper, 2017].

Again, we can think of these deep learning frameworks as a way to "weaponize" the SOI asset base to manage a magnitude of inputs in order to make better and faster decisions. In other words, enabling smart to be scalable.

10.4 Process

How do we embrace, implement, and embody this type of system? Organizations typically have two types of problems: those which are predictable (the known knowns) and those which are new to them but have been addressed somewhere else so they can draw on the experiences of others (the known unknowns). The challenge of SDA and SOI, organizationally, is that it requires a new type of leadership and organization as we explore and embrace creativity as a process where failure is acceptable and rapid learning takes place in unstructured ways (the unknown unknowns.)

Communicating the volume of collected sensor data to the machine learning/artificial intelligence system for timely and low-latency processing on a global basis presents a logistical challenge. A combination of space-to-space systems and ground-to-space sensing systems is needed; no one sensor system will suffice either in capability or cost practicality. Even if affordable, an all space-to-space system is a primary (first-level) target for our adversaries. The use of ground systems creates a deterrent to attacking the space-based system, and radar contacts will remain a desired tip-and-cue to optical electromagnetic range sensors to characterize targets of interest.

With these logistical challenges comes the need for leadership and working teams that will have authority to act in a rapidly evolving environment where operating policy, budgets, and funding guidelines are already in place. In these types of government programs our approach (and bias) to this would be to improve our planning and execution exercises and capabilities. Moving into this "unknown unknowns" area of SDA and SOI requires a multi-dexterous organization, which, as we review Figure 10-14, must address the bottom left section of the graph (where we can "exploit" our typical processes). It must also embrace the upper right section (where we "explore" and challenge assumptions quickly and where creativity (and accepted failure) is a part of the new processes [Geurts, 2019]. The upper right section requires a leadership body and organization that are adaptive and bring together a variety of very different sectors of the U.S. Air Force, as well as other government sectors and the commercial private sector.

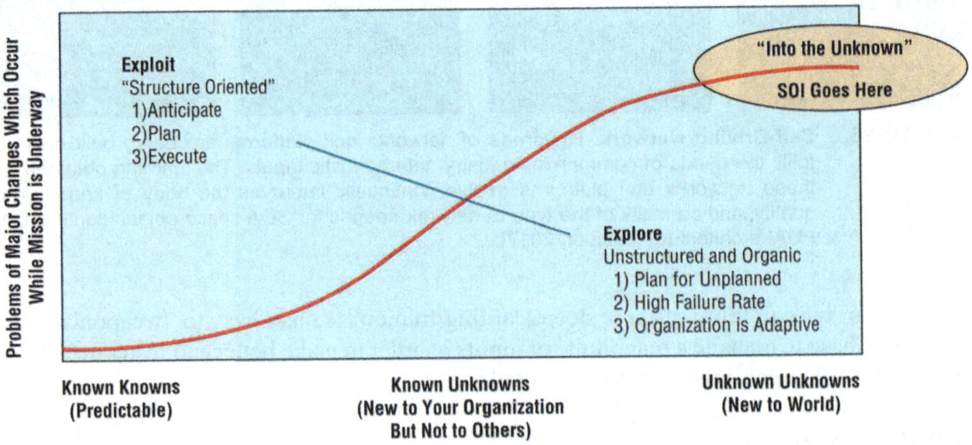

FIG. 10-14. Leadership Requirements for SOI Organizations. Leadership in the government must embrace some public sector entrepreneurship skills in order to incorporate the learning and body of knowledge created in the commercial sector sensor and machine learning domains. If this new leadership model is not undertaken by the government, we will continually fall behind the public sector's global advance in space [Leonard, 2019].

A network of developers and users from diverse sensor domains and disciplines must be embraced and invited into this system. This network must be agile in orientation with the ability to assemble and disassemble quickly (always checking humility at the door) in order to allow all ideas to be heard, tested, and killed easily. This network must be able to address and support the questions and challenges that will arise and revolve around the components of SOI which are physical, functional, measurable, operational, and kinematic for the actual object identification.

These processes focus on a new form of research and development that draws from a wide variety of talent, allowing the research to be alive and constantly changing. This approach requires leadership and an organization that learns, changes, and improves in near real-time. What we are suggesting is that while never forgetting mission focus, the organization needed to support the SOI system should not focus on specific products for delivery to individual groups or users but should be platform focused allowing multiple users and inputs to the platform all at the same time. This means the government will need to be able to operate outside of their current problem/product focused approach in a secure and classified environment with very specific government tradecraft for SOI. Instead they must move to a platform environment where many "network" participants work on a problem/solution at one time who are specialists from a variety of academic and applied science commercial backgrounds (whether they are government employees, subject matter experts, civil servants, etc). This suggestion is in keeping with the changes that have been rapidly implemented by the government for essential versus non-essential government employees. Due to the coronavirus pandemic—COVID 19—the government changes allows for thousands of non-essential employees to work from home from their own computers utilizing a public network and a variety cloud platforms for communication and work submittals.

This platform is scalable, where many can come and go to experiment boldly and rapidly; it does not cater to an old way of thinking or typical product/developer users. The platform learns from the multiple users and provides an environment to solve problems where

1) Conditions exist to allow these users to innovate based on the constant machine learning inputs, and
2) Users are not afraid to fail or ask others to help them reinforce, change, and improve their processes. Figure 10-15 highlights how the move to a solutions architecture or platform provides an environment where we can quickly explore possibilities concerning the new SOI inputs and still embrace efficiencies.

This type of platform approach to solutions architecture (once the sensor data are obtained and available) provides a continuous management oversight dashboard, network user access, and improvement tools with a set of embedded processes. This network becomes the de facto "expert modelers" who work in the unknown unknowns and build unconventional intervention methods to help drive and build protections for SDA. This is similar to the design dilemma faced in civil aviation when they simulate managing risks. They must develop ways to learn before an event occurs and this risk management drives them to use an expert model as well [Sparrow, 2019]. But the reason for this solutions architecture platform is to enable the organization to be problem-centric.

This approach is not entirely new to the U.S. Air Force and is being employed by the AFWERX program, "which is a catalyst for agile Air Force engagement across industry, academia, and non-traditional contributors to create transformative opportunities and foster an Air Force culture of innovation. The ultimate aim is to solve problems and enhance the effectiveness of the Air Force through a small business innovation research (SBIR)

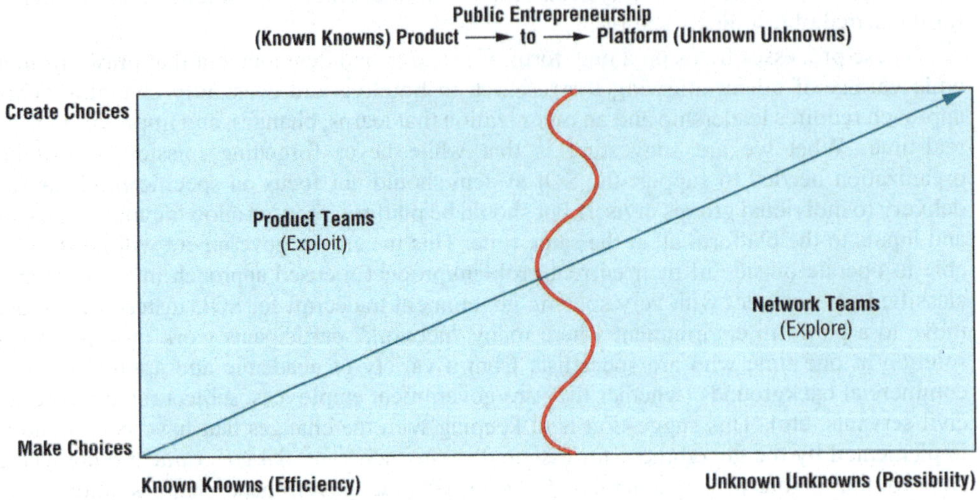

FIG. 10-15. Public Sector Entrepreneurship Moving to Platform-Based Teams for Problem-Centric Choices in Exploring Unknown Unknowns. This approach requires flexibility in terms of the government's ability to embrace teams of experts (regardless of their industry sector) who come together for a short time to solve immediate and high value problems; work on a joint network; provide their input; and then depart that network [Leonard et al., 2019].

program" [Lauver, 2019].This same type of business process could be expanded and used for the SDA SOI.

10.5 Functional Needs

The functional needs for the SDA SOI to fuse, use, analyze, and inform (FUAI) will require the usual DoD operational or administrative capabilities for support. However, what will be unusual is the amount of time that must first be spent on recognizing, understanding, and developing the public value offering of the SDA SOI. Creating the definition of what public value is within the SDA for SOI will in turn drive and guide operational capacity and help mobilize and build the legitimacy and support arguments and facts needed for the authorizing environment. The way in which personnel are acquired and used for such a sophisticated system will require an entirely different approach for SDA success. Subject matter experts will come from all walks of the space industry and commercial industries on an as-needed basis, which will require a different way of thinking about security and risk. This area must be addressed with the authorizing environment and with the public value well-defined in order to be operationally successful.

The public value point on the strategic triangle in Figure 10-16 challenges public managers to lay out for themselves, their organizations, and their overseers (authorizing environment) the ultimate values their enterprises seek to produce and to embody their operations [Moore, 2013]. Meeting this requirement will entail identifying the value that

SDA intends to provide, how it will provide the value, and the costs and budget associated with it. In terms of budgeting for these capabilities, a new acquisition model is necessary for deep learning and ConvNN networks in defense settings. Today's government acquisition and procurement methods whereby derived data is purchased on a per-area basis typically results in poor outcomes for both the supplier and buyer. There are several reasons for this, but the core issue is that SDA requires global, persistent surveillance inputs constantly running on a deep learning framework that will require constant access to the original data in order to support a variety of scenarios on the ConvNN.

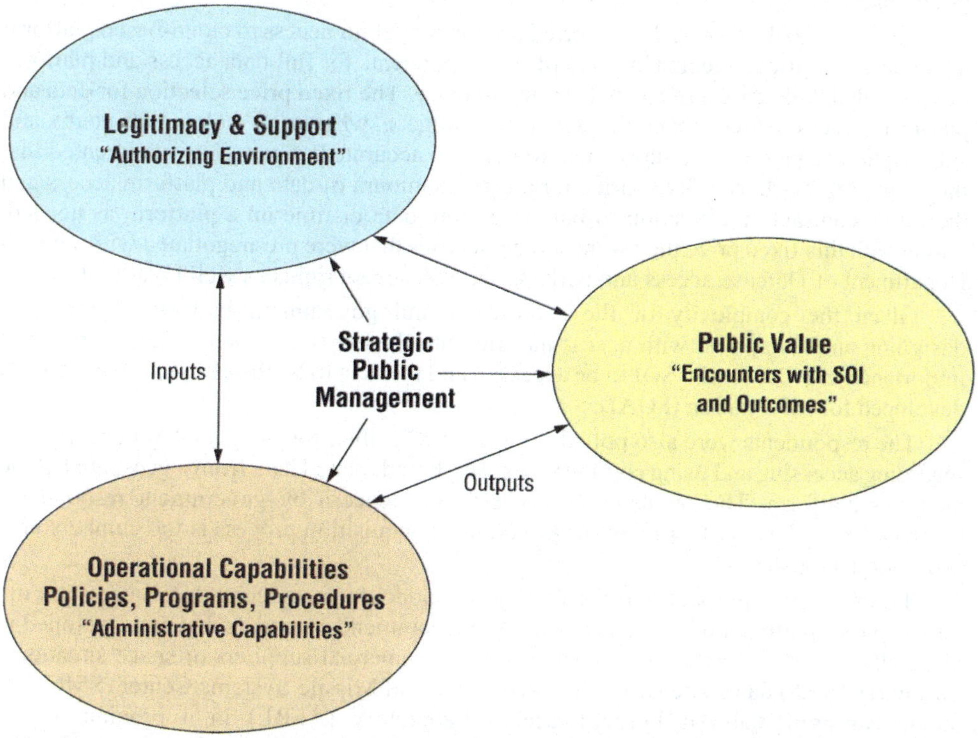

FIG. 10-16. Strategic Triangle. The strategic triangle illustrates the general categories that must be re-engineered in the government in order to focus on the types of functional needs required to enable the success of the SDA space object identification FUAI platform and its problem-centric focus [Moore, 2013].

The SDA needs identified in this book require multiple layers of global, commercial, ground-based data as well as the SOI data inputs from hyperspectral sensors. The new budget scenario must create a structure for the government that features decreasing costs over time of data access, ingest, processing, and storage along with ever-increasing data amounts and quality outcomes that require processes to run repeatedly. The budget structures must allow for this constant updating as well as for what will surely become constantly changing SDA scenarios as the government requests specific solutions to the key questions impacting the machine learning framework. Ground-based systems are less expensive but more are required when compared to space-to-space (or space-to-ground) systems.

The easiest and most useful approach to access all this data through machine learning platforms will be through service level agreements with the vendors, allowing for access across the government enterprise on a platform delivery tool with a fixed price model for accessing data by the network of users. Approximately 200 Department of Defense and Federal civil government and intelligence community analysts and managers were polled the summer of 2019 during several government events focused on commercial geospatial intelligence.*

As Figure 10-17 shows, the preferred pricing model for access to cloud-based platform data was fixed price, selected by 51% of the respondents for full data access and platform access with data license agreements already in place. The fixed price selection for data and platform access allows them to plan their budgets, whereas a volume discount and subscription model did not allow them to build an accurate budget. They commented that they would rather have a fixed price for a certain amount of data and platform access and then do a contract modification to purchase more data or time on a platform as needed. Along with this fixed price they wanted data licenses that were pre-negotiated with Title 10 Department of Defense access and perhaps Title 50 access rights as well [Wall, 2011].

Given the complexity of the acquisition and government budgeting processes, designing an SOI system with access that fits into the existing structure is of paramount importance in order for the SOI to be accessed and the data to be fused, used, analyzed and developed for information (FUAI).

The respondents were also polled to find out what their top accessibility concern was, regarding accessing and using commercial space-based related data from a geospatial cloud analytics platform. The number one accessibility concern by government respondents (>40%) across the board was that the government acquisition process is too cumbersome, as Figure 10-18 shows.

This platform approach and the fixed price model for new technology access are not new to the Department of Defense (DoD). We recommend processes that are developed in close alliance or as a network extension of the commercial suppliers of space situational awareness (SSA) data. The U.S. Air Force Space and Missile Systems Center (SMC) has teamed up with the Air Force Research Laboratory (AFRL) in a program called "Commercially Augmented Mission Operations" (CAMO). This is a response to the fact that commercial-based sensor growth is outpacing the DoD.

Another program, the Unified Data Library (UDL), also a joint SMC-AFRL project, is creating a commercial data repository. Although the basic premise of the SOI platform is that the data are readily available for a network of users, the UDL could be accessed as a repository when network users want to try new data sets with their algorithms and models as they use the current cloud-enabled platform available SOI data. This would make an even greater amount of data available to a much larger and broader pool of military and civilian government users. Perhaps as this book is published the CAMO will have made

---

* Global Marketing Insights

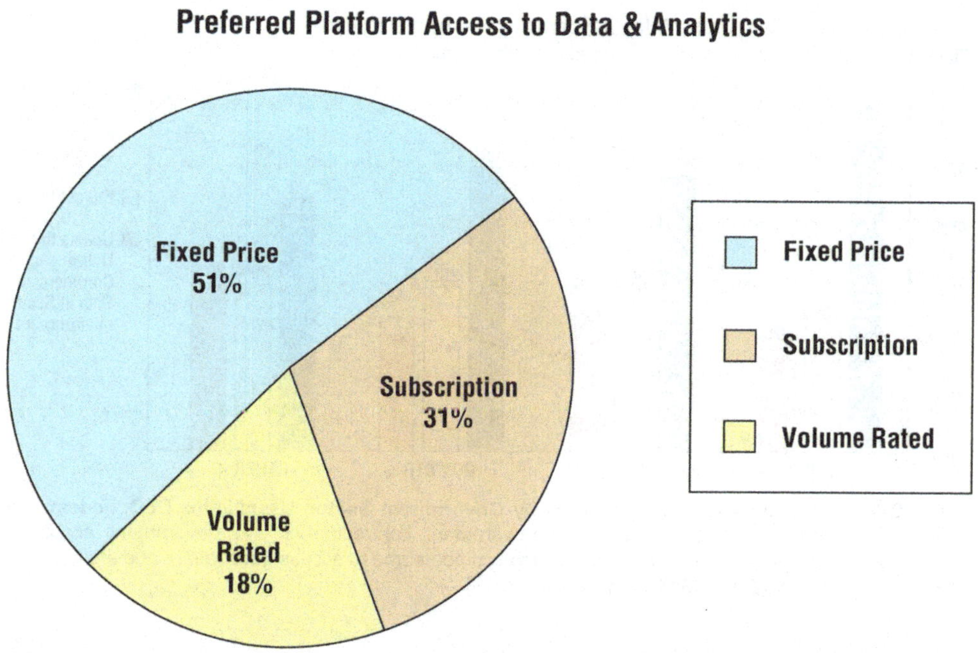

Fig. 10-17. Preferred Platform Access to Data and Analytics. Preferred pricing model selected by the DoD, Federal civil, intelligence community and government contractor respondents focused on fixed pricing in order to access geospatial cloud analytics platform and data [Global Marketing Insights, 2019].

significant progress as a contracting avenue for military buyers to acquire data from vendors with minimum red tape. Recently a Space News article focused on the CAMO and an industry source stated, "Commercial can take 98% of the burden off (current DOD efforts) [Erwin, 2019]." The Air Force could rely on SSA companies to monitor the entire sky all the time and to track events like conjunctions, re-entries, and breakups. The government could then focus its sensors on specific areas of concern or threats. The shift from military to commercial is difficult. [Erwin, 2019] However, over time the military has successfully purchased commercial telecommunications bandwidth, so a private sector commercial access model does exist.

10.6 Challenges and Gaps: Sensor and Systems

Today's reality of hyperspectral sensor (HSI) vs. what we would need to provide as inputs to FUAI is a key area of focus. Spectral information will support more information about satellite materials and capabilities including

- Antennas that might otherwise be sub-pixel from a purely spatial perspective
- Solar panel make-up

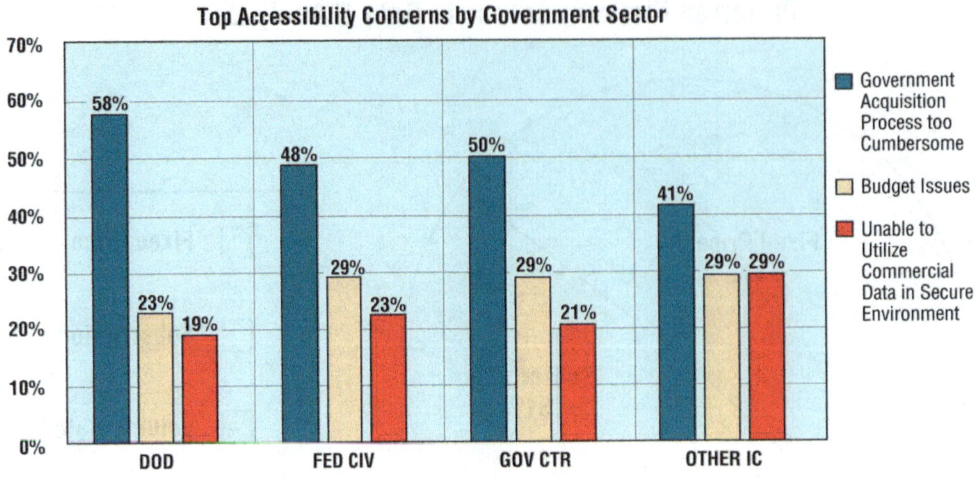

FIG. 10-18. Top Accessibility Concerns by Government Sector. Users in the DoD, Federal civil, intelligence community, and government contractors in the government acquisition process hold that the process is too cumbersome to access geospatial cloud analytics and data. [Global Marketing Insights, 2019].

- Shielding
- Optical capability (from coatings)
- Orbital maneuver propulsion methods (from chemical plume discharges)
- Other elements such as radar absorbing material to confound radar's ability to determine a satellite's true purpose) that, while not completely spatially resolved, are sufficient for spectral positive identification of the object of interest

The limitation of traditional hyperspectral sensors lies in either their push-broom or whisk-broom acquisition mode based upon a single or a few line detectors. As the line detector moves over the target (by motion of its platform), it collects one line at a time (for a unit of time) and then in post-processing the collected lines are stitched together, forming a composite image. This compares to a traditional camera concept that instantaneously collects an entire area. The challenge is multiplied when both bodies (the Earth and the satellite) are moving and one is trying to "hold the line" on the object of interest. The traditional method is a very fine and expensive pointing and tracking subsystem, whether on the ground or in space (via reaction wheels). Often the result is a distortion of the object of interest because it is moving (Figure 10-19). While there are post-processing techniques to correct these artifacts, traditional shape-based algorithms are naturally highly influenced by these supposed "solutions," especially when there is close similarity between objects.

An alternative hyperspectral design is a two-dimension (2D) framing hyperspectral imager, which can capture an area in much the same way as a conventional camera, eliminating the need to stitch lines together to form an image and simultaneously preserving spatial integrity. As a 2D HSI, the integration time can be controlled, supporting

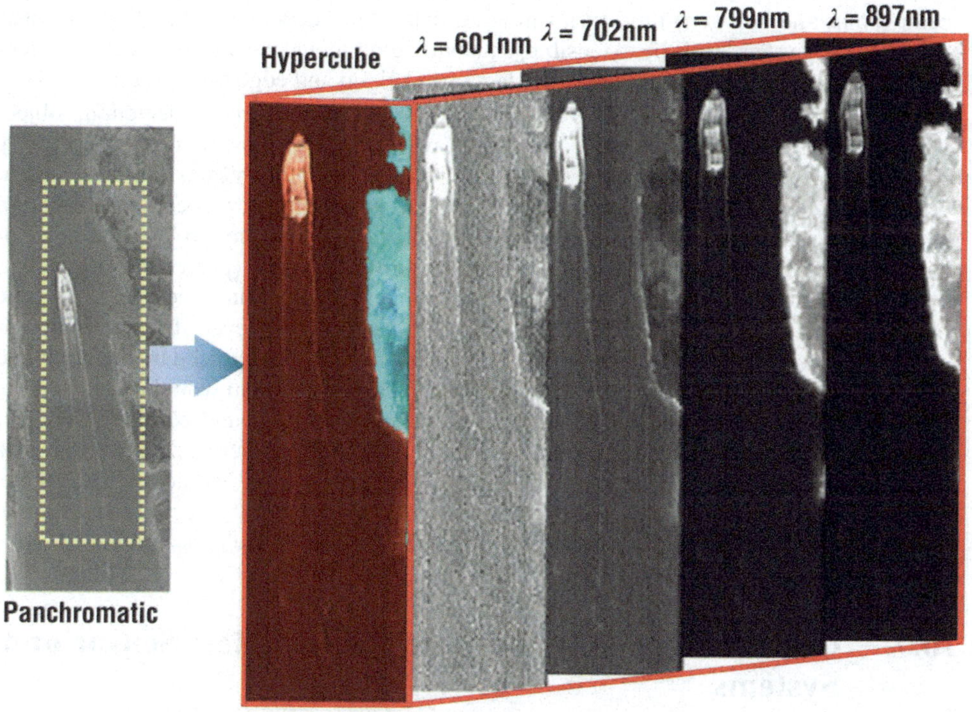

FIG. 10-19. Traditional 1-D Scanning Raw HSI Collection. We can see the spatial warping effect on the ship in the hypercube compared to a companion panchromatic image, the former due to the movement of the ship and the simultaneous movement of the satellite. The HSI value is that the hypercube produces a unique signature of the ship that we can use to reacquire, identify, and track it anywhere, making it very difficult to hide its movements. [Courtesy of Kestrel Corporation]

the ability to image very dim (or camouflaged) objects. In addition, the 2D element lets us use a less sophisticated pointing and tracking system, improving the affordability of the system (whether in space or on the ground). We discuss the 2D HSI in detail later. In addition, if the 2D HSI solution is ground-based, adaptive optics (AO), historically used by the astronomical community to look vertically (zenith), can increase a single sensor's total field of regard, at angles well off of zenith [Kestrel Corporation, 2013], [Military & Aerospace Electronics, 2019]. By employing AO, we can compensate for atmospheric turbulence and other optical distortions to enable the right energy to be placed in the appropriate detector well, thus reducing the "spectral smear" and increasing spectral fidelity, which will increase SOI processing efficacy.

Typical challenges today are in the areas of object detection, object identification, change detection, state estimation, pattern of life, and anomaly detection. The ability to create a layer of analytics that analysts share and can see connections (precedent, dependent, overlap, etc.) image data and signal analytics system provides for automated near real-time ingest. This maintains pixel provenance (meta-data) as well as non-image data ingest of vector (point and polygon) time series training data (e.g., locations of buildings or large vehicles). The system must be extensible, standardize on metadata, and provide consistent geo-registration, spectral and signal consistency, and cloud and edge portability.

The dynamic changing nature of satellites complicates object detection, object identification, change detection, state estimation, pattern of life, anomaly detection, etc. By providing a higher level of spectral resolution information, probabilities are increased at breaking the conventional fixed, static descriptions of a satellite. New spectral information goes beyond spatial information and simple illumination differences to create a layer of analytics that can see connections and new signatures over time (precedent, dependent, overlap, etc.) in the image data and support a system capable of automated near real-time correlations. Such a system would be able to recognize a satellite even if it was in a non-cooperative posture mode in the same way that a person can be recognized by their gait and mannerisms, an unparalleled capability. The mechanics of the system would maintain pixel provenance (meta-data) at both the spectral and spatial resolution, and correlate to higher levels of non-image data such as vectors (point and polygon) time series data to assist with understanding on-object component changes. The ideal system would be extensible, standardize on Metadata, provide consistent geo-registration, spectral and signal consistency, and be able to accept both space-to-space, as well as ground-to-space acquisitions of data.

10.7 Gap Analysis Solutions and Possibilities: Sensor and Systems

Terrestrial data sources and on-ground inputs are commercially available and growing daily in terms of sensor inputs. These additional inputs will be reviewed in order to determine additional characteristics of global ground station networks (stationary and mobile, Figure 10-20) and support networks such as launch facilities, for further identification parameters of the SOI.

While detection, identification, and tracking of targets in any environment is challenging, it is especially so for objects at high angular speed and at extreme range, such as those in space. When these objects are also relatively small and correspondingly dim, such as at the horizon and at low angles from the observation position, the target may be hidden in the clutter and background. This is further complicated when the objects do not want to be observed and are supported by enemy or a threat camouflage, concealment, and deception (CCD) techniques and electronic warfare (EW) capabilities. Detecting threats first, while remaining undetected ourselves so as to not give away our own knowledge advantage, buys us more reaction time to best counter those threats.

Given that infinite zoom and motion compensation are not practical, an object is going to present itself as a collection of mixed pixels. Present passive remote sensing approaches for SOI have focused on increasing spatial resolution for a given aperture size to reduce the portion of mixed pixels. The general approach to achieve *truer pixels* is by decreasing the detector well size (pixel pitch), yielding greater focal plane array (FPA) pixel density. However, optimum pixel dimensions are limited by diffraction effects from the aperture and

FIG. 10-20. Kongsberg Satellite Station (KSAT). This figure provides a view of multiple types of operational ground receiving antennas as seen from the KSAT conference rooms and offices in Tromso, Norway. [Courtesy of Global Marketing Insights]

are, in turn, wavelength-dependent. Also, while building a matching "diffraction-limited" detector is possible, it often comes at the cost of light-gathering sensitivity as the pixels get smaller or denser. The result is a drop in image quality because the pixels don't have sufficient photonic area (independent of optimized dark current, pixel hybridization, pixel delineation, and unit cell readout capacity that define sensitivity), resulting in contrast that's too low to provide *effectual* resolution for traditional shape-based identification approaches.

Granted, it is possible to build a larger aperture for a given high-resolution detector, but that comes with increased costs associated with manufacturing, deployment (whether on the ground or in space), supporting infrastructure, and operations. It is also possible to integrate longer to collect more photons, but that comes at a cost in terms of increased system acquisition, pointing, and tracking (APT) complexity, collection speed (number of objects observed for a specific unit of time), and trades with regards to what is optimal in one part of the spectrum is saturation in another. While FPA development advances will continue with meta- and nanotechnology-based materials such as quantum dots, nanotubes, layered carbon graphene structures, and others, scaling these to the cost-affordable densities and defect levels already reached by complementary metal oxide semiconductor (CMOS) applications continues to remain an unanswered question [NRC, 2010].

In reality, we cannot solve the mixed pixel problem by simply increasing the spatial resolution This is especially true for moving objects that are constantly changing their incidence angle relative to the observer wherein skew and keystone of pixels misrepresents the sampling area. In practice, with higher resolution the proportion of mixed pixels can actually increase because the finer detail resolves new features that were not recorded previously. Furthermore, at higher resolutions, the within-class variation increases as local differences become more dominant. This can inject unwanted noise (versus increased signal) into a solution, which can disadvantage an automated process, lowering the unmixing/ID probability rates. Conversely, another reason that an increase in spatial resolution is limited in practice is because of the redundancy in the resulting 3D spatial-spectral cube of collected data. Redundancy can occur because higher spatial resolution usually comes at the expense of spectral resolution. While the measured spatial resolution might be physically higher (smaller area per pixel), the mixed pixel signals used to separate those very pixels are lower. This is due to the reduction in energy that was traded in support of a smaller instantaneous field of view (IFOV), producing the higher spatial resolution, which is balanced by widening the individual spectral bandwidths. This in turn correspondingly reduces the number of total bands available for classifying a pixel. As a result, when pixels start to look statistically the same, consolidation occurs and identification rates can actually become lower.

Of course there is always a trade-off between *spatial* and *spectral* resolution (Equation 10-1). The digital representation of an image is defined by a linear function of the detected radiance, where dA and $d\Omega$ combine to be the spatial resolution, and $d\lambda$ is the spectral resolution. With a finite amount of photons available per element of time, there must be a practical compromise between the size of a detector and the width of the spectral bands (as part of the total spectrum captured), in order to keep the radiometric precision and desired signal-to-noise ratio (SNR) of the imager high.

$$dc = \text{gain} \times \iiint \text{radiance} \times \; dA \times \; d\Omega \times \; d\lambda + \text{bias} \qquad (10\text{-}1)$$

dA = differential area of a pixel

$d\Omega$ = differential solid angle through the optics onto the object of interest

$d\lambda$ = spectral resolution

It is the combination of spectral resolution, spectral sampling interval, and number of spectral bands (along with the spectral range sampled) that allows us to discriminate more components in an object. The many bands, each with different spectral sensitivities, can increase our understanding of the spatial (i.e., subpixel) content of a scene despite spatial resolution limitations. In general, employing high *spectral* resolution provides a 4X to 8X improvement in discrimination and ID over spatial resolution alone. This assessment is based upon the well-known ability of spectral unmixing algorithms. It is not uncommon to identify four to eight materials with high confidence from single pixels [Newland, 1999]. Of course, for this approach to succeed, the spectral sensor must collect the *right* spectral data. In addition, it is helpful to have a limited number of end-members that represent the possible materials that are likely present or a means to constrain the number of possible end-members available for identification [Martin and Plaza, 2012]. We have traditionally done this by knowing there are some pure pixels in the scene (e.g., a solar panel) as starting points or by having an extensive library of likely materials. The good news is that most space objects use a limited number of external materials, keeping the size of the library tractable. With this information and approach, a previously unknown object can become typed.

Consequently, a collected set of sufficiently robust narrow band spectra adds a needed spectral dimension and is particularly effective in military find, fix, track, target, engage, and assess (F2T2EA) applications. This is due to the enhanced ability to still discriminate materials and detect and identify objects that might otherwise be viewed by traditional automatic target recognition (ATR) software as containing insufficient spatial resolution. With enough spectral bands, we can define a spectral-based *tag* (in support of tagging, tracking, and locating (TTL) capabilities) that is uniquely associated with the target object and thereby overcome the limitations of spatial resolution. This spectral-based TTL approach is virtually impossible for traditional systems because their spatial-based approach is affected by changes in a target's aspect ratio as it moves. The traditional systems' focus on ever-higher spatial resolution has compromised their spectral fidelity and inhibits their ability to easily reacquire a target by spectral signature if it is momentarily lost or other similar shaped objects are within the field of view (e.g., following a particular white Toyota pickup truck in heavy traffic with other white Toyota pickup trucks, while they also pass momentarily out of sight due to an underpass). An object with a spectral tag can be quickly identified, even if the spatial resolution is less than originally imaged or the orientation is different, which could happen at long slant ranges from the observer or if the object has significantly changed orbital altitude. A spectral tag for a space object would also be particularly useful in providing dynamic tracking motion control as part of the system's feedback loop. Also, with an optical tag in three dimensions, we can quickly discern the orientation of the object, providing increased intelligence about it.

***How and to what degree a 2D hyperspectral imaging (HSI) approach is relevant*:** Despite the advantages of hyperspectral imaging systems for a number of widely varying missions [Hege et al., 2004], [Briottet et al., 2006], [Barnes et al., 2016], conventional HSI systems use a line scanning schema, which collects and assembles the composite lines into an image (which takes some time). A classic example here is the MightySat II.1 payload, launched 19 July 2000, developed by Kestrel Corporation, which was the first operationally successful true hyperspectral imager in space [Yarbrough et al., 2002] (Figure 10-21). Therefore, a scanning sensor (much like a radar sweep) can "step over" and lose a target, especially a moving one, during the collection. Likewise, traditional scanning systems cannot halt to "lock-on" to a target. Once a target is lost it is exceeding difficult to reacquire. This is quite problematic when multiple simultaneous targets are involved and discrimination and prioritization of targets is required. Simply speeding up the scan rates to cover the field-of-view (FOV) has yet to be successful, in part because of the fewer photons that become available. The use of scanning further impacts the issue when the target is moving fast through a particular FOV, resulting in less time to acquire a robust signal. This is why traditional hyperspectral systems are not appropriate for many applications today, despite their other benefits.

Conversely, a *true* 2D HSI design combines the advantages of a high spectral resolution imaging spectrometer (full spectral data in each frame of data) with a novel 2D simultaneous spatial sampling that enables the system to collect *a 3D hypercube* in each standard image frame. This has come about by redesign of the spatial and spectral elements in a unique and optimal way to instantaneously cover a 2D area (a single spatially oriented

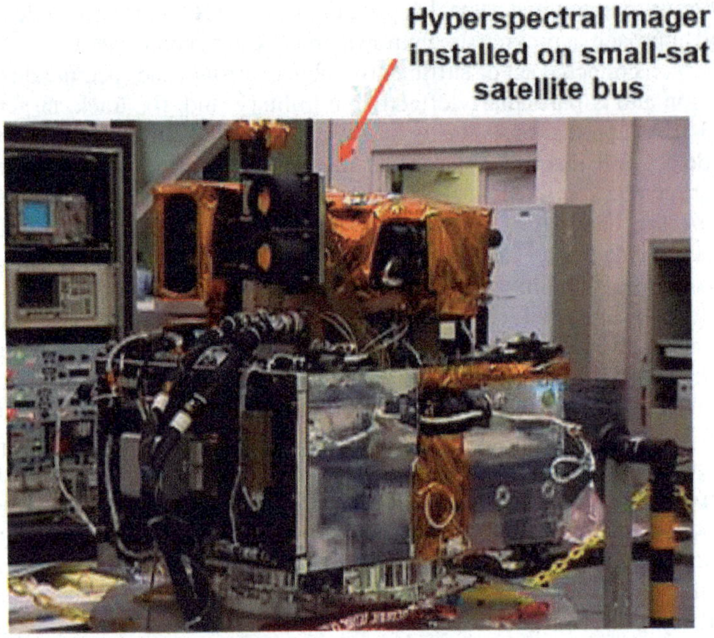

FIG. 10-21. MightySat II.1. A view of the hyperspectral imager installed on the small-sat satellite bus. MightySat II.1 demonstrated that hyperspectral LEO observations were practical and valuable. Even though the spatial resolution was relatively coarse, the system could positively identify and separate by their spectral signatures objects that were 6X smaller. [Courtesy of Kestrel Corporation]

x/y frame) while producing a full frame of hyperspectral data, essentially achieving a hyperspectral "camera". The architecture is flexible across the full spatial and spectral resolutions, from ultraviolet (UV) to the long-wave infrared (LWIR). The 2D HSI approach can be custom-fitted to the wavelengths, aperture size, mission CONOPS and platform, providing maximum optimization to the mission at hand.

There are several particular benefits to the SOI mission. One benefit from a "camera" design is that it forgoes the need for the ultra-demanding pointing and stability required by traditional scanning systems to keep a moving target within a frame without losing it. As an area imager, the 2D HSI camera design can stare at a location and provide persistent surveillance of multiple simultaneous targets in the FOV. The camera approach also compensates for view angle occlusions as the object moves, scene illumination, and speed of a moving target. Another benefit of a hyperspectral camera (currently available at some capability in the commercial marketspace, and under development for military purposes) is its inherent ability to integrate over time (equivalent to an adjustable exposure). This allows it to collect essential hyperspectral data even for very dim targets or those with low light illumination, therein increasing the signal. This is an unprecedented capability for an HSI system, since it permits us to extract relatively weak target signals from the background while retaining the full richness of the signal, even in ultralow light (Figure 10-22). The frame technique also provides co-registration elements for subsequent frames. As the object changes relative to the observer, the 2D HSI approach might also produce beneficial fringe patterns. It might be possible to correlate the fringe patterns and spectral information with machine learning to improve the realized resolution of the target [Feng et al., 2019]

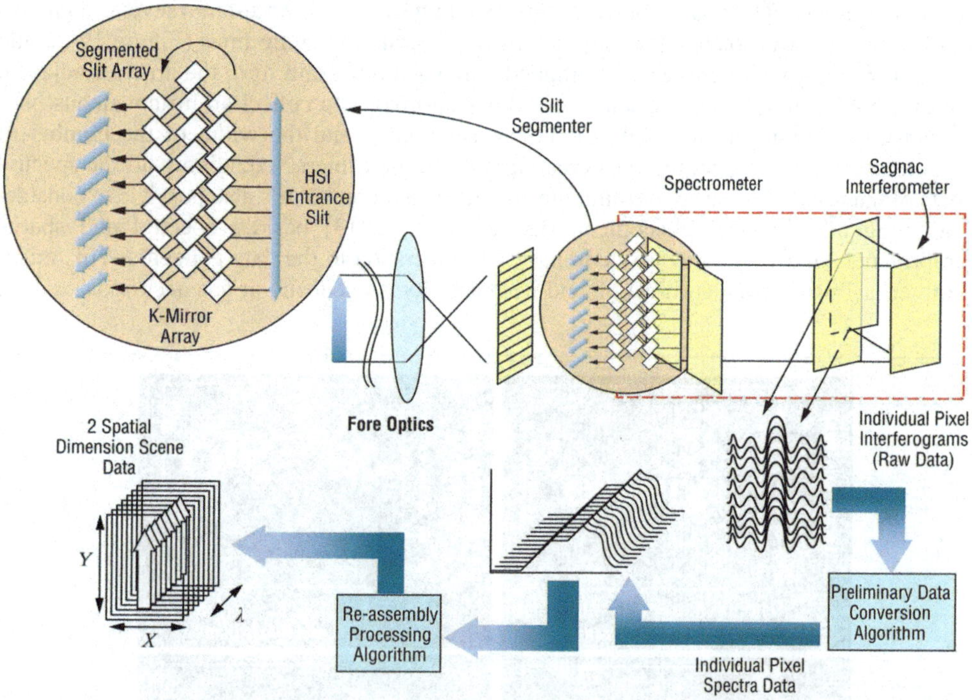

FIG. 10-22. True Two-dimensional (2D) Hyperspectral Imager (HSI) Block Diagram. This diagram shows how a 2D area is collected and simultaneously converted into a hyperspectral image without scanning. The result is a "hyperspectral camera" with variable integration time that is able to image exceedingly dim targets and extract sufficient spectral and spatial information to drive rapid object identification. [Courtesy of Kestrel Corporation]

Lastly, as alluded to above with regard to the moving cargo ship, a 2D HSI camera addresses the past issues of extended spectral smear (blur), especially with moving objects. Previous attempts at 2D designs stepped through the spectrum over time, causing the blur. The time element was exaggerated when motion was involved, when either a target or the platform holding the sensor was moving. This blur reduced the spectral quality of the data, complicated automated processing of signatures, and reduced confidence levels. Contrariwise, a 2D HSI would have the entire spectrum collected simultaneously for a particular area, delivering the highest quality spectral data possible. This characteristic increases automation possibilities and heightens confidence levels—critical to supporting decision-making when time is short.

How it would work: People have attempted several different ways to develop a 2D HSI; most of these approaches compromise some portion of the spectral data collected by reducing the number of bands collected (which also significantly reduces cost) or they are not operationally useful for the SOI mission at hand (either temporal collection or spatial assignment is not fully harmonious) [Coffey, 2015]. A *true* 2D HSI is possible when the design breaks an FOV image area and corresponding hyperspectral slit into several segments and simultaneously rotates them into an array of parallel imaging lines (Figure 10-22 and 10-23). These rotated segments are aligned with each other and form the array of adjacent lines in the resultant image space. This allows an imager to collect an instantaneous two-dimensional image of the height of the slit segments—and the width of the number of segments—without scanning, thus creating a framing "camera" containing all the spectral data of interest and instantly forming the classic 3D hypercube. It's also possible to consider capturing other optical phenomena [Evans et al., 2013] besides spectral and spatial information, and simultaneously adding a 4th dimension to the data collected. This might further facilitate both identification and what activity is occurring or has occurred.

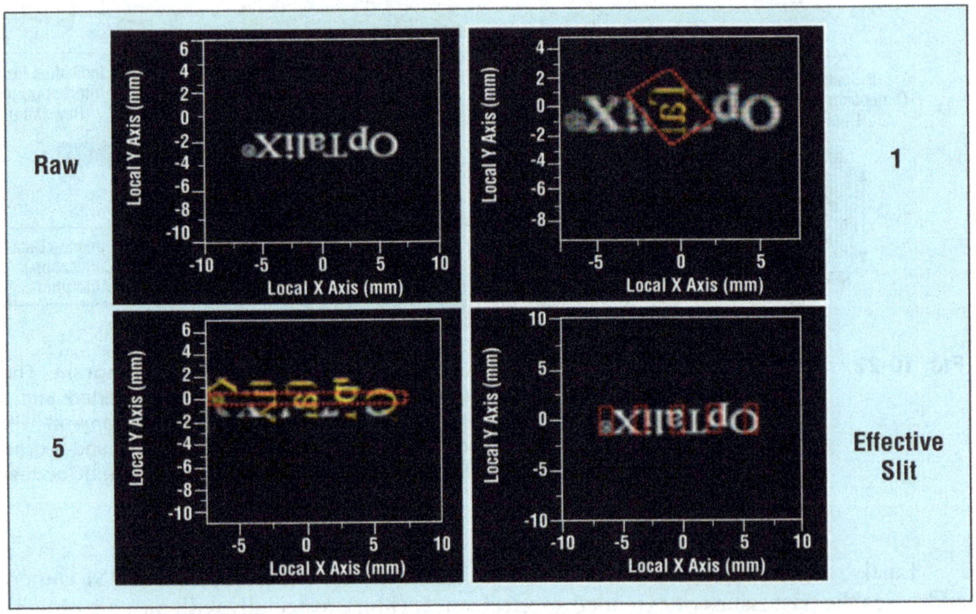

FIG. 10-23. Field of View Quadrants. Upper left-quadrant - original raw image; upper right quadrant - rotational of 5/5 of the FOV; lower left-quadrant - rotation of 5/5 of the FOV; lower right-quadrant - reconstruction of the entire FOV. [Courtesy of Kestrel Corporation]

Using the example of a Fourier transform hyperspectral imager with an n × m element detector array, with n elements along the slit and a centered interferogram, there are n × m/2 hyperspectral data points collected in each frame of data. The n spatial points may be remapped as desired with no loss of spectral resolution. If lower spectral resolution is acceptable, it is possible to increase the spatial resolution beyond n points – this is a straightforward trade-off of spectral resolution for improved spatial resolution. Alternatively, it is also possible to design a combination approach by stitching multiple focal plane arrays (FPAs) together to increase both the FOV for detection and the inherent 2D spatial density in support of tracking and targeting. The coverage of each instrument will be driven by detector response, optical factors (e.g., coatings, transmission, etc.), and designed cut-off wavelengths and resolution. This approach would enable creation of a wide FOV 2D HSI, one that is operationally useful, affordable, and would allow the military and defense community to fully exploit the capabilities that hyperspectral information can deliver [Yuen et al., 2006], [Weatherbee at al., 2012], [Curtiss, 2015], [Hanham, 2017].

Present 2D HSI development work has focused on the challenge of uniquely identifying and tracking individual high-speed moving objects at 600+ km (from either airborne or space-borne assets) that might present as sub-pixel to few pixels (i.e., a low aspect ratio) in spatial resolution when using a relatively small aperture [Kestrel Corporation, 2014].

This work has resulted in a high-speed design with upwards of 512 high-bit spectral bands covering more than one octave of energy—a rich signal for applicable HSI target recognition algorithms, even in the presence of background noise and clutter and purposeful adversary countermeasures, surpassing today's EO/IR and 1-D HSI approaches. With ML algorithms and data fusion even small, fast moving satellites can be properly sourced and attributed in near real-time with limited observation time. In turn, a properly configured system could observe more objects per unit of time, while driving down the cost of each observation. This approach would enable a network of small observation systems to provide affordable SOI around the world.

Identifying the SOI differences can be directly supported through global on-ground training data. For example, characterizing solar photovoltaic panels (solar PV) would help to identify and understand the characterization of SOI panels. This work is currently underway by deploying a solar panel photovoltaic model over the world (175 concurrent GPU-enabled tasks) using a variety of current commercial sensors such as Landsat 8, Sentinel 2 and vectorized results with the Airbus SPOT data model. The results of the ground-based outcome of the global solar PV will aid in testing against SOI solar arrays to identify structural (and therefore spacecraft) differences.

10.8 Summary

While the resolutions of Landsat 8 and Sentinel 2 satellites are coarse by comparison to what is necessary for SOI, the data collected and the processing approaches are analogous to what can be done from ML and the 2D HSI. Just as the 2D HSI does not rely primarily upon spatial resolution to determine details about a space target of interest, Landsat 8 and Sentinel 2 are Earth-observing satellites consisting of multispectral imagers that derive identification and classification details about areas of interest by performing mixed-pixel and sub-pixel analysis.

Such processing is ideal for machine learning/deep learning algorithms to analyze and make judgments based upon localized heuristics that can influence decision-making in adjoining pixels. In addition, the spectral-based machine-learning approach can be "warm started" by performing relevant ground-truth-based 2D HSI characterization of known structural components, such as solar photovoltaic (PV) panels, communication arrays and dishes, heat exchangers, telescopes, standard bus shapes and sizes, etc., at varying resolutions and angles of regard (mimicking different observation slant ranges and angles).

References

Adamczyk, E. 21 August 2019. *Space Command, National Reconnaissance Office to Form Joint Command.* Space War. At *http://www.spacewar.com/reports/Space_Command_National_Reconnaissance_Office_to_form_joint_command_999.html*

Barnes, J.L., et al. 17 February 2016. A Shot in the Dark – *Shedding Light on Exoatmospheric Situational Awareness with Alternate Sensor Utilization.* Air & Space Power Journal, Vol. 30., No. 1.

Berthier et al. 2 March 2016. *Prediction of transits of Solar system objects in Kepler/K2 images: an extension of the Virtual Observatory service SkyBoT.* Royal Astronomical Society.

Bertinetto, L. et al. 3 November 2016. *Fully-convolutional Siamese Networks for Object Tracking.* European Conference on Computer Vision. Springer.

Boyd, A. November 2017. *Satellite and Ground Communication Systems: Space and Electronic Warfare Threats to the United States Army.* The Institute of Land Warfare. No. 115.

Briottet, X. et al. May 2006. *Military Applications of Hyperspectral Imagery. Targets and Backgrounds XII: Characterization and Representation.* Edited by Watkins, W. R. and C. Dieter. Proceedings of the SPIE, Volume 6239, id. 62390B.

Businesswire. 27 March 2019. https://www.businesswire.com/news/home/20190327005660/en/Satellites-Built-Launched-2027-10-Year-Forecasts-Including

Center for Strategic and International Studies (CSIS). April 2019. *Space Threat Assessment 2019.* Aerospace Security Project (ASP).

Coats, D. (Director of National Intelligence). 29 January 2019. *Statement for the Record: Worldwide Threat Assessment of the U.S. Intelligence Community.* Select Senate Committee on Intelligence.

Coffey, V. C. October 2015. *Hyperspectral Imaging for Safety and Security.* Optics and Photonics News.

Congressional Research Service. 24 May 2019. *Defense Primer: Electronic Warfare. IF11118.*

Curtiss, B. 1 July 2015. *Hyperspectral Sensing for Defense and Security Applications.* Tech Briefs. At https://www.techbriefs.com/component/content/article/tb/supplements/ptb/features/articles/22463

Defense Intelligence Agency (DIA). February 2019. *Challenges to Security in Space.*

Department of Defense (DoD), Joint Chiefs of Staff. 10 April 2018. *Space Operations.* Joint Publication 3-14.

Department of Defense (DoD). 2018. *Summary of the National Defense Strategy of the United States of America: Sharpening the Military's Competitive Edge.*

Erwin, S. 18 June 2019. *Top General Says Large Constellations of Satellites in Low Orbit Could Address Key Needs.* Space News.

Erwin, S. 8 September 2019. Military Sees Value in Commercial Data, But Needs to Figure Out How to Buy It. At https://spacenews.com/military-sees-value-in-commercial-data-but-needs-to-figure-out-how-to-buy-it/

Evans, H. et al. 2013. *The Phenomenology of Intelligence-focused Remote Sensing.* Riverside Research.

Feng, Shijie et al. Mar/April 2019. *Fringe pattern analysis using deep learning.* Advanced Photonics, Vol. 1, No. 2.

Fernholz, T. 17 September 2019. *The Pentagon taps Microsoft and Ball to pilot new military satellites.* Quartz.

Foust, J. 7 October 2019. *How to Better Manage Space Traffic.* SpaceNews.

Garino, B. and J. Gibson. 28 October 2009. *Space System Threats.* United States Air Force, Air University, AU-18_Chap21.

Gershgorn, D. 11 April 2017. https://qz.com/954530/five-years-ago-ai-was-struggling-to-identify-cats-now-its-trying-to-tackle-5000-species/ Quartz.

Geurts, J. F. 22 July 2019. Guest lecture, Senior Managers in Government Program, Harvard Kennedy School of Government, Cambridge, MA.

Global Marketing Insights, Inc. July 2019. *Pricing Model Data Collection. Geospatial Cloud Analytics; 195 government respondents.* ESRI User Conference Defense and Intelligence Summit. San Antonio, TX.

Grush, L. 3 September 2019. *A bug in SpaceX's communication system kept the company in the dark about potential satellite collision.* The Verge. At https://www.theverge.com/2019/9/3/20847243/spacex-starlink-satellite-european-space-agency-aeolus-conjunction-space-debris

Hanham, M. and B. Lim. July 2017. Hyperspectral analysis set to expand in coming decade. IHS Jane's Military & Security Assessments Intelligence Centre.

Harper, J. 23 July 2019. *Electronic Warfare Spending on the Rise.* National Defense Industrial Association.

Hege, E. K.et al. 7 January 2004. *Hyperspectral Imaging for Astronomy and Space Surveillance.* SPIE Conference. Imaging Spectrometry IX, 5159.

Juniper. 24 October 2017. *Brutal Automation is the Only Way to Succeed.* At https://forums.juniper.net/t5/Industry-Solutions-and-Trends/Brutal-Automation-is-the-Only-Way-to-Succeed/ba-p/314706

Kestrel Corporation. 2013. SBIR Topic AF121-072 -Dual-mode Wavefront Sensor for Space Surveillance. Air Force Contract: FA9453-12-M-0353.

Kestrel Corporation. 2014. MDA Contract: HQ0147-14-C-7014: (Ph IIe work on-going under MDA's Acquisition, Tracking and Pointing (ATP) Technologies Initiative, Focus Area II: Airborne hyper-spectral sensors for ballistic and airborne target applications).

Koksal, I. 2 November 2019. *SpaceX Planning To Launch 30,000 Additional Starlink Global Internet Satellites.* Forbes.

Kongsberg Satellite Services (KSAT). 2020. https://www.ksat.no/globalassets/ksat/files/ksat_a-company-presentation web.pdf

Kubica, J. et al. 2006. *Variable kd-tree algorithms for spatial pattern search and discovery.* In Advances in Neural Information Processing Systems (pp. 691-698).

Lambakis, S. September 2017. *Foreign Space Capabilities: Implications for U.S. National Security.* National Institute for Public Policy.

Lauver, S. 30 July 2019. At https://www.youtube.com/watch?v=J8qtyxkeU7o

Leonard, H. et al. 16 July 2018. SOFWERX: Innovation at U.S. Special Operations Command, Case Study, Harvard University, Cambridge, MA

Leonard, H. 22 July 2019. Classroom Diagram - Senior Managers in Government Program, Harvard Kennedy School of Government. Cambridge, MA.

Manner, J. 18 January 2019. *Spectrum Interference in the New World of Space and 5G.* At https://www.satellitetoday.com/innovation/2019/01/18/spectrum-interference-in-the-new-world-of-space-and-5g/

Martín, G. and A. Plaza. April 2012. *Spatial-Spectral Preprocessing Prior to Endmember Identification and Unmixing of Remotely Sensed Hyperspectral Data.* IEEE Journal Of Selected Topics In Applied Earth Observations And Remote Sensing, Vol. 5, No. 2.

Mittal, A. 19 May 2017. https://elearningindustry.com/machine-learning-process-and-scenarios eLearning Industry.

Moore, M. H. 2013. *Recognizing Public Value, 111, 121-124.* Senior Managers in Government Program, Harvard Kennedy School of Government, Cambridge, MA.

National Air and Space Intelligence Center (NASIC). 15 January 2019. *Competing in Space. F NV711 0002.*

National Defense Strategy Commission. 13 November 2018. *Providing for the Common Defense – The Assessment and Recommendations of the National Defense Strategy Commission.*

National Research Council (NRC). 2010. *Seeing Photons: Progress and Limits of Visible and Infrared Sensor Arrays*. Washington, DC: The National Academies Press. At https://doi.org/10.17226/12896

Newland, Daniel. 1999. *Evaluation of Stepwise Spectral Unmixing with HYDICE Data.* Thesis, Rochester Institute of Technology, Center for Imaging Science.

Shelton, W. 29 March 2017. *Threats to Space Assets and Implications for Homeland Security.* Testimony, House Armed Services Subcommittee on Strategic Forces and House Homeland Security Subcommittee on Emergency Preparedness, Response and Communications.

Smith, Adam. 1 May 2019. "Co-Founder, Head of Business Development." Descartes Labs.

Sparrow, M. K. 1 August 2019. *Managing Risk Control Functions of Government: Operational Risk-Management in Government - Mapping the Terrain Course*. Senior Managers in Government Program, Harvard Kennedy School of Government, Cambridge, MA.

Spoorthi et al. 2018. *PhaseNet: A Deep Convolutional Neural Network for Two-Dimensional Phase Unwrapping*. IEEE.

TechTarget. 2020. https://searchenterpriseai.techtarget.com/definition/convolutional-neural-network

U.S. Air Force (USAF). 19 February 2019. *Air Force Approaching Industry for Adaptive Optics Technologies in Ground-Based Space Situational Awareness*. Military & Aerospace Electronics. At https://www.militaryaerospace.com/computers/article/16722168/air-force-approaching-industry-for-adaptive-optics-technologies-in-groundbased-space-situational-awareness

U.S. General Accounting Office (GAO). August 2002. *Critical Infrastructure Protection: Commercial Satellite Security Should Be More Fully Addressed.* Report to the Ranking Minority Member, Permanent Subcommittee on Investigations, Committee on Governmental Affairs, U.S. Senate. Report: GAO-02-781

Vaughan-Nichols, S. 21 March 2018. *What is Docker and why is it so darn popular?* ZDNet.

Wall, A. E. 2011. *Demystifying the Title 10-Title 50 Debate: Distinguishing Military Operations, Intelligence Activities & Covert Action.* Harvard National Security Journal, Volume 3.

Weatherbee, O. et al. 2012. *Advanced Hyperspectral Imaging Solutions for Near Real-Time Target Detection.* Proceedings Volume 8542, Electro-Optical Remote Sensing, Photonic Technologies, and Applications VI; 854223.

Wikipedia. 2018. https://en.wikipedia.org/wiki/Moore%27s_law

Wilson, J.R. June 2019. *Technology and Politics of the New U.S. Space Force.* Military & Aerospace Electronics, Vol. 30. No. 6.

Yarbrough, S. et al. 17 January 2002. *MightySat II.1 hyperspectral imager: summary of on-orbit performance.* Proc. SPIE 4480, Imaging Spectrometry VII. At https://doi.org/10.1117/12.453339

Yuen, P. W. T. and G. Bishop. October 2006. *Hyperspectral Multiple Approach Fusion for the Long-Range Detection of Low Observable Objects: MUF2.* Proc. SPIE, vol. 6396, pp. 63960C-1-63960C-12.

Zur, C. 29 July 2019. Small satellites are Big Business. Space News.

Made in United States
Orlando, FL
04 November 2026

90205878R00227